Climate Modes of the Phanerozoic

Climate Modes of the Phanerozoic

The history of the Earth's climate over the past 600 million years

LAWRENCE A. FRAKES
JANE E. FRANCIS
JOZEF I. SYKTUS

Department of Geology and Geophysics
University of Adelaide, South Australia

CAMBRIDGE
UNIVERSITY PRESS

CAMBRIDGE UNIVERSITY PRESS
Cambridge, New York, Melbourne, Madrid, Cape Town, Singapore, São Paulo

Cambridge University Press
The Edinburgh Building, Cambridge CB2 2RU, UK

Published in the United States of America by Cambridge University Press, New York

www.cambridge.org
Information on this title: www.cambridge.org/9780521366274

First published 1992
Reprinted 1994
This digitally printed first paperback version 2005

A catalogue record for this publication is available from the British Library

Library of Congress Cataloguing in Publication data
Frakes, Lawrence A.
Climate Modes of the Phanerozoic : the history of the earth's
climate over the past 600 million years / Lawrence A. Frakes, Jane
E. Francis, Jozef I. Syktus.
p. cm.
Includes bibliographical references and index.
ISBN 0-521-36627-5
1. Paleoclimatology. 2. Paleontology. 3. Historical geology.
I. Francis, Jane E. II. Syktus, Jozef I. III. Title.
QC884.F69 1993
551.6–dc20 92-4727 CIP

ISBN-13 978-0-521-36627-4 hardback
ISBN-10 0-521-36627-5 hardback

ISBN-13 978-0-521-02194-4 paperback
ISBN-10 0-521-02194-4 paperback

Contents

Preface *page ix*

1 *Introduction* 1

2 *The Warm Mode: early Cambrian to late Ordovician* 7
 Age and distribution of the latest Precambrian glaciation 7
 Lithological indicators 9
 Other indicators 10
 Evidence of climates from palaeontology 12
 Other factors possibly relating to climate 13
 Chronology of the Warm Mode 14

3 *The Cool Mode: late Ordovician to early Silurian* 15
 Distribution and age of the glacials 15
 Palaeolatitude of the glaciations 16
 Other sedimentary indicators 20
 Palaeontological indicators 23
 Chronology of the Cool Mode 24
 Other factors bearing on climate 25

4 *The Warm Mode: late Silurian to early Carboniferous* 27
 Trends from oxygen isotopes 27
 Distribution of lithological indicators 28
 Late Devonian and early Carboniferous glaciation 33
 Palaeontological indicators 34
 Chronology of the Warm Mode 34
 Other factors bearing on climate 35

5 *The Cool Mode: early Carboniferous to late Permian* 37
 Distribution and age of the glacials 37
 Palaeolatitudes of the glacials 41

Contents

	Other climatic indicators	44
	Rhythmic sedimentation	45
	Other sedimentary indicators	47
	Palaeontological indicators	50
	Chronology of the Cool Mode	51
	Other factors bearing on climate	54
6	*The Warm Mode: late Permian to middle Jurassic*	56
	Oxygen isotopes	56
	Biogenic reefs and carbonate abundance and distribution	58
	Variations in humidity	58
	Evidence of climates from palaeontology	60
	Other factors possibly related to climate	61
	Chronology of the Warm Mode	64
7	*The Cool Mode: middle Jurassic to early Cretaceous*	65
	Geological setting	66
	Ocean temperatures	68
	Ocean faunal realms	69
	The carbon record	72
	Evidence of ice	74
	Climate significance of fossil floras	77
	Clay minerals and climate change	79
	Seasonal climates	80
	Chronology of the Cool Mode	82
8	*The Warm Mode: late Cretaceous to early Tertiary*	83
	Sea-level fluctuations and circulation	84
	Ocean temperatures	87
	Organic-rich sediments	90
	Land temperatures	92
	Chronology of the Warm Mode	98
9	*The Cenozoic Cool Mode: early Eocene to late Miocene*	99
	The onset of polar glaciation	101
	Ocean records	104
	The carbon record	105
	Ocean faunas	107
	Tectonic events	108
	Continental climates	109
	Chronology of the Cool Mode	113

Contents

10 *The late Cenozoic Cool Mode: late Miocene to Holocene* 115

 Summary of marine climates from oxygen isotopes 115
 Late Neogene climates 117
 Quaternary climatic events 138
 Late Quaternary climatic history 155

11 *Causes and chronology of climate change* 189

 The search for cyclical climate change 189
 The development of Cool Modes 191
 The onset of Warm Modes 195
 The carbon record in Warm and Cool Modes 201
 Climate forcing 208
 Conclusions 218

 Bibliography 220
 Index 271

Contents

The Land, the Law, the Language and the People

the time there is despotes

Organic philosophy

the Christian emphasis

total development of persons

both resorted nature

nature of law

Preface

This book covers the history of climate for the last 600 million years. Part of the book is a summary of palaeoclimate information for different slices of geologic time. We believe it is necessary to update and attempt to synthesize the great body of research on palaeoclimates generated over the last ten years or so. But there was a further purpose in the collection of these basic data and that was to establish a framework which we could use to compare and contrast similar climate states of the past, and from there to recognize some causes of climate change. We have divided climate history into Warm Modes and Cool Modes, in a way not unlike Fischer's (1982) 'Greenhouse' and 'Icehouse' states, but our Modes are of shorter duration and contain some controversial elements – we have questioned the theory that the Mesozoic climates were uniformally warm and ice free and instead propose a Cool Mode in the middle Mesozoic. We have also included a chapter on the climates throughout the Quaternary, something which is often missing or abbreviated in texts on geologic climates, not surprisingly considering the huge volume and the increased scale of detailed information available for such a relatively short time.

It is assumed that the reader will already be familiar with Earth history and have some understanding of the basic principles and techniques of palaeoclimatology, such as the measurement and interpretation of oxygen and carbon isotopes, and the significance and weaknesses of evidence from sedimentary rocks and fossils etc. Descriptions of palaeoclimatological methods and accounts of the historical development of ideas, both important aspects of palaeoclimatology, are abundantly available elsewhere and are here kept to a minimum for the purpose of economy; we apologize in advance for thus not including all such pertinent information.

It has become apparent in recent years that the carbon cycle is one of the major influences in climate change, both in the near future with the likely greenhouse warming, and during the Quaternary ice ages. Was the carbon cycle so dominant in the geologic past? This has been the question we have had in mind when compiling this palaeoclimate information. In the final chapter we bring together information about climate change from previous

chapters and discuss the influence of the carbon cycle and the geological factors which may have influenced it. We do not suggest that we have found the answer to past climate change, but have highlighted some events and possible consequences that will hopefully inspire further analysis and exploration.

The climate system is now and was in the past affected by, and itself affects, a huge diversity of factors, including the carbon cycle, the nature of the oceans and their circulation patterns, distribution and types of vegetation, mantle composition and circulation, and extraterrestrial forces, to name only a few. Therefore, the sphere of literature that has to be examined to draw together clues about ancient climates is, not surprisingly, immense. To print every reference which has some use for palaeoclimate analysis and which we have consulted would take up a huge volume in itself. We have therefore tried to be more economical with our bibliography and have preferred to refer to many excellent review papers, or compiled volumes about palaeoclimates or relevant themes. We recommend that the source references in these papers be consulted for more thorough treatment of some aspects of climate systems or geological histories that we have been able to refer to only briefly.

One of the major problems that we have encountered when trying to compile and compare palaeoclimate information is the discrepancy between different time scales. Examine, for example, Fig. 7.3 in this book. We have illustrated the time scale used by Haq, Hardenbol and Vail (1987) for their sea-level curves alongside the DNAG (Decade of North American Geology) time scale (Palmer, 1983), matched by the absolute ages. There is a large discrepancy between the durations and positions of the Stages – the Cretaceous Berriasian of DNAG correlates with the Jurassic Kimmeridgian of Haq *et al.* Drawing parallels or correlations between different parameters will always be difficult and include some degree of uncertainty until more consistency between time scales is achieved.

We chose to use the time scale of DNAG (Palmer, 1983) since it was the most compatible with the majority of scales used in publications. If necessary we have mentioned in the figure captions how we matched information to the time scale (by matching absolute dates or Stage boundaries). To make full use of geological data for palaeoclimate work it is essential to state in each publication which time scale is being used when mentioning Stages/ Ages – our task was made more difficult by the fact that many papers do not provide this information.

There are many people who have inspired, helped, discussed and argued with us during the preparation of this book. We express our gratitude to all of them. In particular, we would like to thank the reviewers: Judy Parrish, John Crowell and Thomas Crowley. The enigma of Cretaceous ice, which was the stimulus for the new climate modes, was discussed in the field in the

deserts of central Australia with Neville Alley, Graham Kreig, Nick Lemon and Tan van Doan, and in Arctic Canada with Ashton Embry. Colleagues at the University of Adelaide provided a testing ground for our ideas, so our thanks are extended to Brian McGowran, Malcolm Wallace, Vic Gostin, George Williams, and to those involved in the practical preparation, including Sophia Tsemitsedis, Sherry Proferes, Rick Barrett and Fran Parker. Financial assistance was provided by the Australian Research Council and the University of Adelaide.

Lawrence Frakes
Jane Francis*
Jozef Syktus.

* Present address: Department of Earth Sciences,
University of Leeds,
Leeds LS2 9JT, UK.

1

Introduction

There are many reasons for investigating the climatic history of the Earth. First and foremost is the need to know how the climate of our planet has evolved over the last 600 million years (m.y.). Only by understanding past climate states of the Earth can we discern the driving mechanisms for global climate change, set boundary conditions for numerical modelling and learn how to predict future climates. Improved predictability of climate in the short-term future can be included among these advantages to be gained in the applied sense, as can using palaeoclimate information to predict the distribution of economically significant commodities such as petroleum, phosphorite, bauxite and the accumulation of other sedimentary minerals related to climatically controlled redox changes.

This book first describes, in a concise way, the salient features of Earth climates over the last 600 m.y. or so. With this background information in hand, our purpose is to recognize and then compare similar climatic states in history (e.g. the warm intervals in the early Palaeozoic and in the late Mesozoic), so as to determine whether major episodes display any significant similarities. In cases where strict parallelism seems to exist, whether for relatively warm or cool, wet or dry, or seasonally comparable global conditions, the description of climate for any one interval perhaps can be broadened by inferences from conditions in another interval, for which circumstances may be better known. Thus, we adopt an approach of comparative palaeoclimatology, in the hope that pooled information will provide new insights into the evolution of climate on Earth.

Palaeoclimatologists appear to be inveterate seekers after causes of change and the authors of this book are no exception. Our effort is, after summarizing modern hypotheses of change, to test the ability of several hypotheses, such as the carbon cycle, orbital influences and tectonic events, to explain the results of our comparative studies.

The Earth's global mean surface temperature depends on its orbital parameters, the luminosity of the Sun, and the planet's distance from the Sun. It also depends on the planetary albedo (surface and cloud reflectivity) and on the composition and dynamics of the atmosphere and hydrosphere.

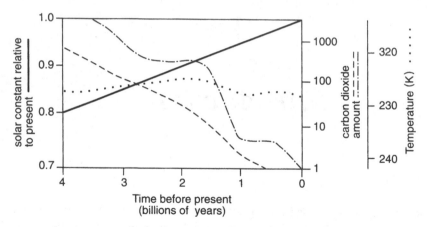

Fig. 1.1. *Variation of surface temperature, solar constant and atmospheric* CO_2 *over geologic time. Temperature curve (dotted line) refers to average surface temperature at 35° latitude. Carbon dioxide curves from Kasting (1987, top curve) and Hart (1978, lower curve). Adapted from Kuhn et al. (1989).*

In the Earth's atmosphere the dominant greenhouse gas is CO_2. The CO_2 content of the atmosphere has changed with time in response to changes in the rates and patterns of tectonic activity which cause sections of the Earth's sedimentary reservoir to be heated to the point where sediment-bound carbon is converted into CO_2 gas, which is released back to the surface with volcanic fluids. Over geologic time these changes led to variations in the Earth's climate.

Over geologic time, there is a tendency towards a decrease in CO_2 concentration in the atmosphere, owing to cumulative burial of carbon in sediments. This trend of decreasing CO_2 (and hence cooling) is counterbalanced by increasing solar luminosity over geologic time (Fig. 1.1). Tectonic activity is thus linked to a non-organic part of the carbon cycle. The organic part of the carbon cycle is also linked to the climate through reduction of the CO_2 level in the atmosphere by vegetation and phytoplankton.

The Earth's climate is also sensitive to planetary orbital parameters: the obliquity, eccentricity and precessional cycles. These parameters undergo cyclic change as the result of the influence of the other planets in the solar system on the Earth's orbit. It is likely that changes in the seasonal and latitudinal distribution of solar energy reaching the Earth as a result of orbital cycles cause the Earth's polar ice caps to wax and wane. The orbital climate oscillations are superimposed on tectonic-scale climate trends.

Redistribution of incoming solar energy within the Earth's climate system is a function of several climate parameters, such as surface albedo, land–sea distribution, cloudiness etc. The carbon cycle may be the most important factor responsible for the long-term evolution of climate in the Phanerozoic.

However, regional and seasonal changes in climate related to internal energy distribution and storage as a result of, for example, opening and closing of ocean gateways, changes in land–sea distribution, ocean heat transport, etc., have also caused significant changes in climate during the Phanerozoic.

The Earth's climate has been far from static over the last 600 m.y. It has passed from phases of extreme cold and glaciation to periods of global warmth and aridity. This is demonstrated by physical and biological evidence from sedimentary sequences at individual sites, or from comparisons of contemporaneous suites at locations around the globe. Thus, at a site in western Australia, warm conditions indicated by Devonian reefs were followed by cold climates in the Carboniferous and Permian as indicated by glacial deposits. Of course, whether such a change is testimony to a global climate change depends on whether the site changed its position relative to the Earth's thermal gradients along lines of latitude.

Through the use of reliable palaeogeographic maps based on rock palaeomagnetism, which were developed in the 1960s and 1970s, this factor of continental motion can now be dealt with for the Phanerozoic. The problem of contemporaneity of geological sites and of the reliability of their palaeoclimate information increases with age of the deposits, owing to decreasing resolution of age data. However, in many instances resolution is sufficient to permit establishment of approximate synchroneity, thus allowing definition of regional, continent-wide or global climate events, and comparison of climate from one interval to another. Climate change on a global scale is clearly demonstrated through the Phanerozoic.

Our efforts in this book are directed towards the last 600 m.y. of Earth history and therefore only a brief outline of even earlier climates is given here. Since evidence of cold climates is lacking for most of Precambrian time, it is commonly accepted that warm climates prevailed. This notion is supported by the widespread occurrence of shelf carbonate sediments, particularly in Proterozoic sequences. A major cold episode is indicated by tillites and related glacial features in Huronian rocks of North America and their approximate correlates in southern Africa and western Australia. These are dated at about 2300 Ma and are followed by carbonate rocks suggesting renewed warmth. The late Precambrian glacial episodes took place between about 860 and 600 Ma and glacial debris from this interval has been described from all continents, except Antarctica.

In the majority of previous palaeoclimatological studies it has been common to study the climate of a particular geological Period, starting at the beginning and finishing at the end of the Period, with little mention of the climates before and after that particular interval. Although this may be the most logical way to divide the history of the Earth, it leads to the potential for missing important and influential processes and events that preconditioned

Table 1.1. *Climate Modes of the Phanerozoic*

	Mode	Approximate age range (Ma)
Early Eocene to present	cool	55–0
Early Cretaceous to early Eocene	warm	105–55
Late Jurassic to early Cretaceous	cool	183–105
Latest Permian to middle Jurassic	warm	253–183
Early Carboniferous to late Permian	cool	333–253
Early Silurian to early Carboniferous	warm	421–333
Late Ordovician to early Silurian	cool	458–421
Earliest Cambrian to late Ordovician	warm	560–458
Latest Precambrian to earliest Cambrian	cool	615–560

Note: Time scale from Palmer, 1983; Fig. 1.2.

climates. We have therefore chosen to ignore Period boundaries and divide the climate history of the Earth into climate modes, or intervals of time during which similar climates prevailed. In particular, we have divided the palaeoclimate history into Cool Modes and Warm Modes (Table 1.1).

Our Cool Modes are defined as times of global refrigeration during which ice was present on Earth. This includes the range from phases of intense glaciation when the polar regions were covered by large permanent ice caps to intervals during which high-latitude regions were only seasonally cold but were, however, sufficiently cold for the formation of ice during winter. The record of ancient glaciations, represented by such well-known features as tillites, striated pavements and grooved clasts (see Frakes, 1979), is not difficult to confirm, whereas the record of seasonal ice is much more controversial and indistinct. The presence of ice-rafted dropstones in fine-grained mudrocks is the best evidence for seasonal ice. Where these are found in rocks representing glacial times the activity of ice is not questioned. However, when reported in rocks that show no other obvious signs of glacial activity, interpretation is more difficult.

The reports of ice-rafted deposits from late Jurassic to early Cretaceous high-latitude regions form the basis for our recognition of this interval as a Cool Mode, quite at odds with the generally accepted view that this was one of the warmest times in Earth history. However, evidence for cool polar climates at these times is now accumulating from previous reports, our own observations in the field, from geochemical data and climate modelling (see Chapter 7). We probably have to admit to being somewhat biased in deliberately searching for signs of cold climate, but we feel that over the past years the notion that the Mesozoic was so warm has tended to be accepted without question and has influenced the interpretation of climate data towards global warmth. Whether we are correct about these Mesozoic cool climates, or have been blinkered ourselves, will remain to be seen as more

DECADE OF NORTH AMERICAN GEOLOGY
GEOLOGIC TIME SCALE

Fig. 1.2. The Decade of North American Geology 1983 Time Scale (Palmer, 1983).

palaeoclimate data are collected. At least we hope our interpretation will stimulate further searching for more evidence for palaeoclimates in the field, the laboratory and through computer modelling.

Our Warm Modes are defined as times when climates were globally warm, as indicated by the abundance of evaporites, geochemical data, faunal distributions, etc., and with little or no polar ice. Because our Modes span rather long intervals of approximately 150 m.y., they do include brief intervals of contrasting climates. For example, our late Silurian to early Carboniferous Warm Mode includes the very brief glaciation that occurred in South America. This event was so localized that we felt it did not justify erecting another Cool Mode or lengthening a later one. But, as discussed in the final chapter, the very factors that enforced the Warm Mode may have operated to prevent the development of this small glaciation into a more global event. Likewise, during the late Cretaceous to early Tertiary Warm Mode we have found signs of cooler (though perhaps not deeply freezing) climates, and in the late Jurassic to early Cretaceous Cool Mode there are signs of relatively warm intervals. As more climate data are collected in future, it may be shown that our divisions into climate modes are far too simple or completely wrong. But for now they provide a framework that has enabled us to handle the vast amount of palaeoclimate data being accumulated.

Because we have also chosen blocks of time that span across Period boundaries we have had difficulties in collating information from sources that have restricted their discussion to within the boundaries. Thus, one part of a Mode may have been reported in great detail while the other part has scant information. This was particularly apparent when trying to compile summary diagrams – this explains why some of our figures appear to have data or curves that span only a portion of the Mode.

2

The Warm Mode: early Cambrian to late Ordovician

The first Warm Mode began at the termination of the late Precambrian glaciation, probably in the earliest Cambrian; definite Cambrian glacial deposits are not known. Over the approximate 100 m.y. of the early Palaeozoic Warm Mode (\sim560 to \sim458 Ma) continents were mostly dispersed over the low-latitude zones and glacial deposits are lacking. This Warm Mode eventually terminated with the initiation of glaciers in North Africa in the late Ordovician (Caradocian). Major problems exist in trying to explain, first, the occurrence of ?late Cambrian to Precambrian glaciation at low latitudes, and second, the termination of the previous glaciation and the beginning of this early Palaeozoic Warm Mode.

Age and distribution of the latest Precambrian glaciation

Past research has established only a poor framework for dating and distribution of late Precambrian glacial deposits, and as a result the synchroneity versus diachronism of these deposits is not resolved. When the geochronological dates are assembled (Fig. 2.1) it is apparent that glaciation extended over at least 230 m.y. (\sim800 to \sim570 Ma). These data, taken from Hambrey and Harland (1981), are of variable quality and several, included on the diagram, are based on stratigraphic estimates. Moreover, age data are limited to Africa, Asia, Australia and Europe.

The shape of the histogram in Fig. 2.1 is not obviously polymodal but it is not suitable to consider the results as a normal distribution of data about a mean value because, when considered in detail, continents have individual dating patterns. Africa for example, contributes all pre-800 Ma dates and, together with Asia, all dates younger than about 610 Ma.

Latest Precambrian glacial deposits that are well dated within narrow ranges include the Tiddiline Tilloid of Morocco (615–580 Ma; Leblanc, 1981), the Ibeleat Group, Mauritania (630–595 Ma; Clauer and Deynoux, 1987), the Louquan Formation of central China (620–600 Ma), and the

Fig. 2.1. *Age ranges and frequency of age determinations of glacial deposits in the late Proterozoic. Dates are mainly from Hambrey and Harland (1981).*

Churochnaga Tillite of the north-eastern Urals (650–630 Ma; Chumakov, 1981). These dates give a range of ~650 to ~580 Ma for glaciation on three separate continents, an interval narrow enough to suggest synchronism of their respective glaciations. Whether glacial deposits of North and South America, in similar stratigraphic positions, correspond to this event cannot yet be determined.

An attempt at global reconstruction for the late Precambrian (675 Ma; Morel and Irving, 1978) yields a Pangaean continental mass straddling the equator but reaching to middle and high latitudes in a sector which includes China and northern Gondwana (western Australia, East Antarctica, India, north-east Africa and the Arabian peninsula). While high-latitude positioning might thus explain the glaciation in the Chinese localities, the west African glacials were apparently situated near the equator. The tillites of the Ural mountains lie in mid-latitudes on the reconstructions of Bond, Nickerson and Kominz (1984) and Piper (1987). However, it is well established that the pole occupied a position near North Africa both in earlier times and in the Cambro-Ordovician (e.g. Morel and Irving, 1978, figs. 7 and 16). An interval of rapid polar wander/drift is implied, which could account for any glaciation in West Africa. Yet, such an explanation of glaciation arising from high-latitude positioning does not encompass the whole of the late Precambrian glacial mode. This is further complicated by the fact that palaeomagnetic results from older glacial beds show low latitudes of formation (Embleton and Williams, 1986; Sumner, Kirschvink and Runnegar, 1987). This issue is beyond the scope of the present work.

The above discussion relates solely to the latest Precambrian tillites, which constitute convincing evidence of glaciation immediately before the early Palaeozoic Warm Mode. However, there is some indication that glaciation continued on into the Cambrian. Hambrey and Harland (1981, p. 947) note the occurrence of possible glacial deposits in the Cambrian of West Africa and of problematic deposits of less certain age in North and South America, Asia and Europe. The possibility of Cambrian glaciation, together with other factors, has influenced us to place the beginning of the Warm Mode in the early Cambrian (\sim 560 Ma).

Lithological indicators
Evaporites

According to earlier work (e.g. Meyerhoff, 1970; Zharkov, 1981), the Cambrian period was a time of major halogenic accumulation in Earth history. Specifically, the early Cambrian was the most significant interval (Fig. 2.2), during which almost 39% of all Palaeozoic evaporites were deposited. Despite this abundance, early Cambrian evaporites are not widespread on the globe. The bulk of early Cambrian strata accumulated in Siberia and the southern Asian region between Saudi Arabia and India, but Cambrian deposits also occur in central Australia, Morocco and North and South America. Very low levels of evaporite formation in Asia, Australia, Europe and North America characterize the remainder of the Warm Mode.

On the reconstructions of Smith, Hurley and Briden (1981) and Ziegler *et al.* (1979), early Cambrian evaporites of Asia and South America are seen to lie in low southern latitudes ($< 25°$) and the southern Asian occurrences appear to occupy west coast positions in Gondwana at less than 30° latitude, situations ideal for generation of desert conditions. The early Cambrian evaporites of Morocco are anomalous in that they occur at palaeolatitudes of around 50° and in an east coast situation where humid climates might be expected. Evaporite deposition expanded from somewhat lower latitudes in the late Proterozoic to latitudes of about 25° in the Warm Mode and then retreated again in the latest Ordovician. Although these trends are weak and therefore doubtful, a real difference is apparent between the late Proterozoic and the Ordovician. Also, the abundant early Cambrian evaporites of Asia were deposited at slightly higher latitudes than were earlier and later evaporites.

Carbonates

Subaqueously deposited carbonates on the continents accumulated at relatively low and progressively decreasing rates through the Warm Mode, with the Ordovician rate of carbonate deposition being about half that of the Phanerozoic average (Ronov, 1980; Hay, 1985). In the Cambrian, reef

structures were restricted to North America, Australia and Antarctica but were more widespread in the Ordovician, including on the above continents (except Antarctica) plus Europe and the Asian blocks. Webby (1984) concludes that these distributions are best explained by several climatic fluctuations. Non-skeletal aragonite was not common in the early Palaeozoic and the distribution/abundance of early Palaeozoic ooliths has not yet been quantified.

Fig. 2.2 shows the latitudinal distribution of carbonates in which the Warm Mode is characterized by widespread carbonates to relatively high latitudes. The Asian blocks show this best (from $\sim 30°$ in the Vendian to $\sim 45°$ in the early and middle Ordovician). In the Cambrian, archaeocyathid reefs and other prominant carbonate build-ups extended to $\sim 20°$. New reef types, made of corals and other taxa that originated in the middle Ordovician, may have attained 40° latitude in Baltica.

The above features are supported by a compilation of the distribution of major rock types (Ziegler *et al.*, 1981a). A distinctive feature of the early Palaeozoic as seen in the latter study is that carbonates of the Cambrian are abundant to about 50° palaeolatitude, in contrast to most other times in the Palaeozoic.

Other indicators

Calculations from oxygen isotope compositions of late Cambrian chert–phosphate pairs suggest low-latitude surface sea-water temperatures of 50–60 °C (Karhu and Epstein, 1986). Data from Ordovician carbonate cements (cited in Lindström, 1984 and Popp, Anderson and Sandberg, 1986) also suggest warmth. It is unknown whether such extremely light isotopic values might be explained by low salinity effects; we view the problem with caution in the absence of supportive data.

Regarding the great mass of early Cambrian evaporites, it must be remembered that evaporite distributions cannot strictly be used to determine palaeotemperatures because evaporation takes place wherever the gradient in air temperature exceeds that in the water column. This may occur at any temperature above 0°C, although effectiveness of the evaporation process increases with temperature. Thus, evaporitic regimes were not necessarily located in warm zones in the early Cambrian. Rather, it appears that appropriate combinations of rainfall, topographic and wind conditions for arid climates were well developed along the western margins of the Asian blocks at this time.

The above relates to the fact that few calcretes and other indicators of humid climates are represented in the early Palaeozoic record. A few middle and late Cambrian lateritic profiles are known from Laurentia (Chafetz, 1980; Van Houten, 1985), and bauxites are limited to a few occurrences in

Fig. 2.2. Variations in climate indicators during the early Palaeozoic Warm Mode: Column 1 shows the latitudinal extent of glacial deposits; column 2 volume of evaporites; column 3 latitudinal extent of evaporites; column 4 latitudinal extent of sedimentary carbonate rocks. In the last two columns data from Laurentia are represented by the heavy lines, from Asia by the thin solid lines, and Gondwana by the broken lines. Time scale from DNAG (Palmer, 1983).

the Vendian to early Cambrian of Asia, (central Siberia; Bardossy, 1979). These would have all formed at palaeolatitudes of less than 20°, suggesting that low-latitude zones during this Warm Mode were of mixed character, comprising both arid and humid climates. It is likely that humid climates, as well as extremely dry conditions leading to the formation of evaporites, were regional rather than zonal in nature.

11

Cyclicity appears to have been a prominant characteristic of Cambrian and Ordovician shelf-carbonate sequences of North America (Aitken, 1966). Chow and James (1987), studying these 'Grand Cycles' in the northern Appalachians, attribute their formation to variations in the rate of sea-level rise during the middle to late Cambrian. They suggested a eustatic cause on the basis of three distinct and correlative Grand Cycles across North America: one in the late middle Cambrian and two others in the late Cambrian. Palmer (1981) suggests 12 North American cycles through the Cambrian. Eustatic curves produced by Vail, Mitchum and Thompson (1977) and Hallam (1984b) are not sufficiently detailed to confirm either chronology, although they are in agreement with irregular black shale cycles recognized by Leggett *et al.* (1981).

Evidence of climates from palaeontology

Inferences from palaeontology about climates of the early Palaeozoic are weakened by the need to consider the rapid spread and diversification of the early shelled organisms (Sepkoski, 1979). Here, more than at other times in Earth history, there was an abundance of unoccupied environmental niches and a minimum of impediments to diversification. Also, the rates of evolution among fossil groups were at their most variable. As a result, studies of early Palaeozoic diversity as a guide to climate are less useful than for other times.

An example of diversity variation is seen in the data for echinoderms (Sprinkle, 1981). Initially, the numbers of genera and species were low (early Cambrian) and there was a general increase in diversity in the middle Cambrian. Perhaps surprisingly, the late Cambrian saw a diversity decrease, well in advance of the appearance of organisms which were predatory to echinoderms. It is uncertain whether the late Cambrian diversity decrease resulted from climate change or some other environmental change.

The climates inferred for the Cambrian are dominantly tropical to subtropical, with Europe being interpreted as temperate by Ziegler *et al.* (1979). The Ordovician represents a time of major change in the organic composition of the oceans. Featured in this change are (1) the rise and spreading of the brachiopods, corals and other groups, and (2) an unparalleled increase in total diversity. Both displacement of taxa and widespread environmental changes contributed to the diversity change (Sepkoski, 1981). The nature of environmental change is not easily deciphered from the preserved faunas, although distinct shallow-marine faunal provinces are distinguishable (Palmer, 1972; 1979; Jell, 1974; Ziegler *et al.*,1979). The preglacial Ordovician is best characterized by warm and cool temperate realms according to Berry (1979), the latter including Baltica. Spjeldnaes (1981) opts for global warmth to explain the middle and late Cambrian faunas,

followed by a progressive but irregular Ordovician cooling through the Llandeilian to the beginning of the following Cool Mode glaciation.

Other factors possibly relating to climate

There seems little doubt that the Cambrian was a time of moderate continental volcanicity and major marine transgression (Vail *et al.*, 1977; Ronov, 1980; Hallam, 1984b). It has been suggested that a post-glacial eustatic rise was initiated within the early Cambrian (Harland and Rudwick, 1964; Matthews and Cowie, 1979). The chronology of this widespread event is poorly established but it might be bracketed between 590 and 560 Ma and hence indicative of the end of an early Cambrian glaciation.

Fluctuations of sea level have been postulated for other times in the Warm Mode (see Fig. 11.1 below) and our suggestion is that these reflect tectono-eustatic events. Hallam recognized a regression at the end of the early Cambrian, quickly followed by another transgression. The succeeding fall of sea level is placed by Hallam at the end of the Cambrian but Vail *et al.* reckon this major culmination, comparable in scale to that of the late Cretaceous, occurred in the early Ordovician. McKerrow (1979) favours a culmination, after several oscillations, in the late Ordovician. Oolitic ironstones were scarce in the Cambrian and abundant in the early to middle Ordovician (Van Houten, 1985), the latter condition suggesting a correlation with the sea-level high stand.

Organic-rich shales have been used to define times of transgression and assumed eustatic high stands. In Europe these are concentrated in the middle and late Cambrian and in the later half of the Ordovician (Thickpenny and Leggett, 1987), while shales in part of the early Cambrian and the Tremadocian to late Llandeilian were characterized by oxidized conditions and a paucity of organic matter (Leggett *et al.*, 1981). Increased productivity resulting from oceanic upwelling can explain about half of these dark shale occurrences, which tend to be concentrated along west coasts of continents, at least in the late Cambrian (Parrish, 1982; Parrish, Ziegler and Humphreville, 1983).

The late Proterozoic and the early Cambrian were times of enormous phosphorite accumulation. A steady decrease in accumulation occurred until the late Ordovician (Cook and McElhinny, 1979). According to Cook and Shergold (1986) lower Cambrian phosphorites are prominant in the Asian blocks, Europe and Australia. On the early Cambrian reconstruction of Smith *et al.* (1981) most of these fall near the equator and might be attributed to low-latitude coastal upwelling, although the Australian ones attain $\sim 35°$ latitude. The often quoted relationship between phosphorite genesis/upwelling and cold global climate therefore would appear not to hold for the early Palaeozoic Warm Mode. Other evidence suggests that

upwelling was not common; the record of sedimentary chert is sparse in the Cambro-Ordovician (Hein and Parrish, 1987).

Isotopically, oceans during this Warm Mode were distinct in that they were very light in carbon; $\delta^{13}C$ values in carbonates were unique in the early Phanerozoic in being consistently negative (range from about -1 to 0). During the succeeding Cool Mode there was an increase to positive values. Fluctuations in the carbon isotope record during this Warm Mode included an increasing trend through the Cambrian and a sharp decrease to the Phanerozoic minimum near the beginning of the Ordovician. Thereafter, $\delta^{13}C$ increased regularly throughout the Ordovician (Holser, 1984). The latter broad trend is supported by detailed studies by Popp *et al.* (1986). Sulphur isotopes display the expected reversed trends.

Chronology of the Warm Mode

At some undetermined time in the early Cambrian, the late Proterozoic to Cambrian glacial phase ended and the Earth began to warm. The first discernible evidence for this is in the latitudinal expansion of carbonate sedimentation, but since expansion was gradual, it is inferred that the process was slow and did not reach a culmination until the middle Ordovician. At first there was a major increase in aridity, judging from the abundance of low- to mid-latitude evaporites, but this terminated abruptly at the end of the early Cambrian and thereafter the globe was dominated by humid climates. The Warm Mode ended in the middle Ordovician with the development of ice sheets in northern Africa.

3

The Cool Mode: late Ordovician to early Silurian

Ranked after the late Palaeozoic and the late Cenozoic, the Ordovician–Silurian is the most extensive and intensive Cool Mode of Phanerozoic time. Glacial effects were felt predominantly in Africa, and in displaced terranes subsequently derived from there, and were also felt in South America. A major ice sheet developed in North and central Africa. The age of glacial deposits covers probably 35 m.y., as opposed to a minimum of 65 m.y. for the late Palaeozoic. Despite this widespread evidence of glaciation, it appears that the Ordovician–Silurian glaciation was limited to high-latitude land masses in the southern hemisphere; cooling effects are discerned with difficulty elsewhere.

Distribution and age of the glacials

Ordovician–Silurian glacial deposits, unlike those of the late Palaeozoic, are found in Gondwana and in regions commonly considered as parts of the Laurentia and Baltica blocks. The latter areas probably were attached to Gondwana in the early Palaeozoic and have since been rifted away to new sites in North America and Europe. The evidence of glaciation in these displaced continental fragments provided a strong incentive for revising continental reconstructions for the early Palaeozoic.

Glaciation at this time was centred on North Africa, the best documented deposits being those of the central Sahara region (Beuf *et al.*, 1971). Tillites and associated glacial features occur over a wide area from Algeria to Libya and Mali. A total of four separate tillites and striated pavements showing generally northward flow have been recognized. Deposits of about the same age occur to the west and south-west in Morocco, Mauretania and Sierra Leone; these include both terrestrial tillites and glacial-marine strata. A boundary separating marine and non-marine facies runs roughly parallel to the coastline in northern and north-western Africa (Frakes, 1979, fig. 5.1), with glacial marine facies lying to the north and west. An additional thin

tillite has been reported from the northern Arabian peninsula (Young, 1981).

In South Africa, Ordovician–Silurian glacial deposits are not typical tillites (Rust, 1975) and these deposits, while nearly coincident in age, lie at the opposite end of the continent from those in North Africa. In Europe several basins apparently received glacial-marine sediments (in Scotland, Thuringia in eastern Germany, Normandy and Spain). Subglacial geomorphic features have been identified in Spain but early Palaeozoic terrestrial tillites are notably lacking in Europe. These facts pose difficulties in interpretations of Ordovician–Silurian palaeogeography.

Other occurrences of glacial deposits are located in Canada (Nova Scotia and Newfoundland), the former with both tillites and glacial-marine deposits and the latter only glacial marine. In South America, the Amazon basin displays both tillites and glacial-marine strata and in the region from southern Peru to northern Argentina, the Cancaniri Formation may be totally of marine glacial origin (Caputo and Crowell, 1985). In the same paper these authors describe the Vila Maria Formation of the Parana Basin as the glacial-marine facies of the Iapo Formation, long suspected as of glacial origin.

The range of ages given for early Palaeozoic glacial strata is summarized in Fig. 3.1, and the histogram reveals that the occurrences were mostly concentrated in the late Ordovician (Caradocian–Ashgillian). A substantial number carry an early Silurian assignment (Llandoverian–Wenlockian) and two poorly dated occurrences might extend back to the middle Ordovician. The most reasonable interpretation is that glaciation began in the Caradocian, peaked at about the Ashgillian and terminated somewhat gradually thereafter until the Wenlockian. Glacial deposits nearest in age to the Ordovician–Silurian ones are the questionable Cambrian tillites in North Africa, and certain late Devonian deposits in Brazil. Crowell (1983) ascribes all Palaeozoic glaciations to the passage of the pole over Africa and South America through the Palaeozoic, while Hambrey (1985) explains the South American glaciations by development of mountain glaciers in South America under a similar scenario.

Palaeolatitude of the glaciations

On palaeomagnetic evidence the Ordovician–Silurian glaciation appears to have been a high-latitude glaciation centred on the South Pole. However, there is a problem relating to the placement of what are now eastern fringe areas of North America and southern/central Europe. Prior to their recognition as displaced terranes, the occurrences of glacial and ice-rafted strata in Nova Scotia, Newfoundland and the Baltic were considered as part of North America and Europe in the early Palaeozoic and appeared to have formed at

Fig. 3.1. Frequency of age ranges by locality (histogram) and palaeolatitudinal distribution by age of glacial deposits during the early Palaeozoic Cool Mode. Dates are from Hambrey and Harland (1981), palaeolatitudes from reconstructions in Smith et al. (1981).

Fig. 3.2. *Variations in climate indicators during the early Palaeozoic Cool Mode. Column 1 represents the latitudinal distribution of glacial deposits (including ice-rafted clasts); column 2 the volume of evaporite deposits; column 3 the latitudinal extent of evaporites; and column 4 the extent of sedimentary carbonate rocks. In the last column Laurentia is represented by the heavy solid line, Asia by the light solid line and Gondwana by the broken line. Time scale from DNAG (Palmer, 1983).*

lower than about 35°S latitude (Hambrey,1985, Fig.2). The mid-Ordovician map (Llandeilian–Caradocian) of Scotese, Van der Voo and Ross (1981), on the other hand, positions the block referred to as Baltica at 60–85°S latitude. Baltica, representing much of Europe, lies adjacent to North Africa on this reconstruction. Plotting the European early Palaeozoic glacials on this map means that they all lie between ~ 55° and 70° latitude and close to the north African glacials, a result more easily explained. As pointed out by Hallam (1984b), adopting such a reconstruction causes other problems: late Ordovician carbonate reefs of Sweden would then have formed at 60° latitude or greater. This is perhaps an indication that Baltica, as presently defined, is in fact a composite block with a complex history.

For the construction of Fig. 3.2, palaeolatitudes were used according to Scotese *et al.*'s (1981) position of Baltica. The Canadian early Palaeozoic glacials were displaced from the North African region along the lines suggested by Van der Voo (1988) (Fig. 3.3). Subsequent to the late Ordovician, these are assumed to have travelled independently to Laurentia from positions alongside North Africa.

Fig. 3.3. Palaeogeography of Laurentia and part of Gondwana in the middle to late Ordovician, showing original positions of displaced terranes now located in North America and Europe. From Van der Voo (1988).

Whether or not these assumptions about the position of Laurentia prove to be justified, the consistency in the distribution of early Palaeozoic glacials is further perturbed by the South African glacials, which lie near 30° latitude whichever reconstruction is used. The most likely explanation is that the glacials of South Africa are slightly younger than North African ones, and that they therefore lie further along the apparent polar path, as suggested by Crowell (1983).

Most early Palaeozoic glaciations thus appear to have occurred in response to polar motion, with the earliest events centred at high latitudes in North Africa but with ice-rafted deposits originating either from ice shelves or marine periglacial effects in nearby regions that are now parts of Europe and Canada. Rapid motion of the pole towards southern Africa and South America in the Silurian can explain the occurrences in South Africa, the Parana basin and the Andean region in the manner suggested by Crowell (1983). However, the initial glaciation in the late Ordovician in North Africa was both the most intensive, with probably four main advances of the ice, and the most extensive, covering some $6-8 \times 10^6$ km² (Hambrey, 1985). Other glaciations along the apparent polar path were mostly short lived and

restricted geographically; even those in the Andean region are now known to cover only some 300000 km². For these and other reasons discussed below, we interpret the early Palaeozoic as a discrete Cool Mode of late Ordovician to early Silurian age, and separated from the late Palaeozoic Cool Mode by a Warm Mode (mid-Silurian to early Carboniferous).

Other sedimentary indicators

Evaporites

Guides to precipitation/evaporation during the early Palaeozoic Cool Mode (Caradocian–Wenlockian) are not abundant owing to general scarcity of evaporites and the lack of land plants. On the other hand, interpretation is made easier because early Palaeozoic marine faunas have been studied extensively and the period has been broken down stratigraphically into time units having relatively short durations (on average, about 10 m.y.). Recent compilations of Palaeozoic evaporite abundance by age (Ronov, 1980; Zharkov, 1981) allow comparisons to be made through and beyond the Cool Mode (Fig. 3.2). Overall, the Ordovician–Silurian is deficient in evaporites, with only about 2.1% by volume of total Palaeozoic evaporites, whereas the interval constitutes about 17% of Palaeozoic time. The preceding late Cambrian contained moderate amounts of evaporites, and the rate of accumulation increased again strongly in the mid-Devonian. There is only one time of significant evaporite deposition during this Cool Mode: during the Ashgillian (latest Ordovician), at the height of the glaciation. The Pridolian represents the culmination of a late Silurian increase in aridity following the Cool Mode. There was insignificant evaporite accumulation in both the Tremadocian and the early Silurian.

Plotting of evaporite data (from Ronov, Khain and Seslavinski, 1984) on palaeogeographic reconstructions of Scotese *et al.* (1979) reveals that the glacial interval was characterized by sharp contractions of evaporites towards the equatorial zone; agreement between distribution and abundance data would seem to eliminate sea-level change as a primary cause of evaporite formation at this time. The time of greatest evaporite contraction was the late Caradocian to early Llandoverian and it is best seen in Asian data (about 20° of latitude contraction). Partially coincident with this, the Ashgillian increase in volume of evaporites apparently took place largely in Laurentia, but the latitudinal spread of evaporite deposition decreased again in the early Silurian. Despite this, Asia (Siberia) was the site of the most widespread evaporite deposition throughout this Cool Mode, owing to the presence there of large intracratonic basins and to an extensive west coast configuration conducive to arid conditions.

The approximate coincidence in time of glaciation and large-scale evaporite deposition fits with the idea that a cold Earth is a dry Earth (Frakes,

1979). Another example is the late Miocene, with expansion of Antarctic ice followed by evaporite formation in the Messinian of the Mediterranean. Precise dating of these Ordovician deposits might establish whether the analogy is valid; there appears to be a difference in that the peak of Cool Mode glaciation was in the late Ashgillian and a strict analogy would require that evaporite deposition was concentrated immediately afterward (i.e. in the earliest Silurian).

Carbonates

The abundance and accumulation rate figures for the Phanerozoic (Ronov, 1980; Hay, 1985) reveal that carbonate sediments constituted ~34% of sediments deposited on the continents in the Ordovician but that the gross accumulation rate for carbonates was low relative to that of the Cambrian and Devonian. In the Silurian, both the proportion of carbonates (~21%) and the accumulation rate were even lower. Rates for both the Ordovician and Silurian are less than half the Phanerozoic average. Unfortunately, figures are not available for breakdown into epochs but data on the global distribution of carbonates suggest the Ashgillian to Llandoverian interval featured the lowest proportions and the lowest rates of carbonate accumulation during the Cool Mode.

Carbonate sediments occurred at latitudes of less than ~35° throughout the Ordovician–Silurian, except for extending almost to 50° in the pre-Caradocian and the post-Wenlockian in Asia (blocks designated as Siberia, Kazakhstania and China by Ziegler *et al.* (1979) on maps for the Llandeilo-Caradocian and the Wenlockian). Thus, the glacial interval was accompanied by latitudinally restricted carbonate sedimentation on the continents (Fig. 3.2). Gondwana, however, appears to have experienced a slight expansion of carbonates (from ~30° to ~35° S latitude) in the Caradocian–Ashgillian, probably owing to rotation of the Australian sector into more favourable latitudes.

Three additional features of carbonate rocks are important in determining palaeoclimate: (1) oxygen isotope composition, (2) the presence of reef structures, and (3) development of ooliths. With regard to isotopes, preliminary results on early Ordovocian carbonates of Sweden (Baltica) give $\delta^{18}O$ values between −5.5‰ and −8‰ (Lindström, 1984), well below those of most comparable modern environments. Popp *et al.* (1986) point out that there is scant isotopic information from the Ordovician–Silurian (Pridolian only); these data are in keeping with generally light $\delta^{18}O$ values in the early Palaeozoic, suggesting temperatures of 29°C to more than 40°C. However, some recent work on low-latitude Ordovician brachiopods yielded $\delta^{18}O$ values as heavy as −2‰, approximately 20°C (Middleton *et al.*, 1988). Karhu and Epstein (1986) determined equatorial temperatures of ~22°C for the Ordovician, and Geigtey (1985) suggested an average decrease in tempera-

ture of $\sim 10°C$ through the Ordovician from isotopic data. As with all Palaeozoic isotopic determinations, those cited above are constrained by uncertainties about the isotopic composition of the ocean.

It has been suggested that warm deep water of high salinity and enriched in $\delta^{18}O$ was characteristic of the Ordovician oceans (Railsback *et al.*, 1990). Such a sequestration of ^{18}O at depth would explain the very light isotopic values suggested by most analyses of shelf-dwelling brachiopods and would be more in keeping with cool ocean surface waters thought to accompany late Ordovician glaciation. The coincidence in the Caradocian of an apparent inverse gradient in temperature with the occurrence of abundant organic sediments suggests high organic productivity in a stratified ocean as envisaged by Brass, Southam and Peterson (1982).

The distribution of reefs, since they possibly may not reflect biotic temperature tolerances similar to those of reef-builders today, is best considered together with the distribution of ooliths and/or non-skeletal pelletal limestones (bahamites), which require at least subtropical temperatures to form. Although early Palaeozoic ooliths are not abundant (Wilkinson, Owen and Carroll, 1985), reef–oolith associations are known from the early Ordovician of eastern North America and Argentina, from the late Ordovician of the Baltic (Webby, 1984), and the Wenlockian of central China, Wales and North America (see Seslavinskiy, 1978), all of which were formed at latitudes of 30° or less. Considered alone, the various types of reefs and build-ups, each probably having different ecological restraints dependent on their biological composition, nonetheless all plot at even lower latitudes. Possible exceptions are the early Silurian reefs of the Siberian block which may attain latitudes as high as 45°. Otherwise there appears to have been a restriction of reef growth to 15° latitude or less in the early Ordovician and the Ashgillian.

Indicators of humidity

Since coals are lacking, other indicators must be used to designate early Palaeozoic humid climates. Bauxites are of little assistance, being only scantily known from mid-Ordovician rocks of the equatorial zone. Detrital kaolinite, representing a provenance of deep weathering and hence high humidity, occurs below Ordovician tillites in North Africa (Fairbridge, 1970), but only in abundance in central Asia (Seslavinskiy, 1978).

Oolitic ironstones and sedimentary manganese deposits have been considered as representative of deep weathering resulting from humid climates (Seslavinskiy, 1978; Van Houten, 1985). Certainly, dissolution and transport of the highly mobile iron and manganese elements would be aided by increased precipitation. The latitudinal distribution of Ordovician ironstones on the map of Perroud, Van der Voo and Bonhommet (1984) shows a disjunct character, as follows: several localities in Baltica, Laurentia and

Siberia between 0° and ~25° latitude; and Gondwana and Amorica localities between ~45° and ~85° latitude. A Silurian plot (Ludlovian) shows a similar result: low-latitude occurrences in the northern continents (to ~35° N latitude) and high-latitude ones in Gondwana to ~70° S. Manganese deposits, on the other hand, appear to be concentrated in the low latitudes, generally less than 35°. One can speculate that these distributions define two belts of humid climate, with the low latitude one (to ~35°) having conditions also suitable for manganese accumulation.

The low-latitude ironstone belt at least seems to have been in existence through much of Ordovician and Silurian time. Abundance data (Van Houten, 1985) suggest that in the Ordovician, ironstone formation was at a major peak (second only to the Jurassic during the Phanerozoic). Although there was a significant drop during the late Ordovician to early Silurian glacial mode, ironstones subsequently resumed forming at a high rate. Evaporites and ironstones together suggest global high humidity during post-glacial times and increased aridity during glacial periods of the Ordovician–Silurian interval. This is in keeping with observations on sequences in other Cool Modes.

Palaeontological indicators

The rapid evolution and diversification in both shelly and graptolite faunas has permitted detailed analyses of early Palaeozoic fossils, including their palaeoclimatic significance (Holland, 1971; Boucot and Johnson, 1973; Boucot, 1975). In fact, Spjeldnaes (1961) predicted the occurrence of glaciation on the basis of faunal characteristics well before the discovery of Ordovician–Silurian glacial deposits in Africa. A general picture of palaeobiogeographical provinces or organic events based on global sequences is as follows:

1 Tremadocian–Llandeilian well-defined provinces
2 Caradocian fluctuations of province boundaries
3 Ashgillian–Llandoverian widespread extinctions
4 Wenlockian increasing diversity
5 late Silurian increasing cosmopolitanism

Steps 1 to 5 can be interpreted in terms of climate, with step 1 representing more or less uniformly warm climates similar to those of the late Cambrian. Step 2 records the beginnings and step 3 the middle and late stages of the glacial phase. Step 4 can be seen as the first stages of warming, which continues on into step 5. These interpretations are similar to those of Spjeldnaes (1961, 1981) who, like others, cautions against an overemphasis on climate as a cause of changes in provincialism and diversity. He suggested trangressions and regressions as mechanisms to explain some variation, particularly in cosmopolitanism versus endemism.

Chronology of the Cool Mode

The early Palaeozoic Cool Mode developed after uniformly warm climates of the late Cambrian and the early/middle Ordovician. It seems likely that the first glaciers were formed in Caradocian time and with intensification an ice sheet, possibly with marine ice shelves, developed in polar northern Africa in the Ashgillian. North African ice began to wane in the Llandoverian and had vanished by the end of the Wenlockian, but meanwhile other smaller centres grew in southern Africa and in southern Brazil as this segment of Gondwana moved over the pole (Llandoverian). A reversal in motion of the pole led to growth of small centres in the Andean belt and probably somewhat later in northern Brazil. All glaciers apparently had melted by the end of the Wenlockian, including the final phases in South America and in the Amorican fragments. The latter eventually drifted across the Iapetus Ocean to become joined to Europe and north eastern North America. Again, polar motion and the drift of these localities into lower latitudes were responsible for termination of this phase.

An enigma of Palaeozoic history is the apparent lack of glaciation and any ice-rafting during the interval from the late early Silurian until the late Devonian (an interval of some 47 m.y.). Late Silurian to middle Devonian glacial deposits would be expected in South America and Africa, given the postulated apparent polar path across the two continents. However, they have not been detected in extensive late Silurian and Devonian sequences in eastern South America nor in north western and southern Africa. It is possible to explain this as due to a lack of a moisture source in those areas, all of which were in interior positions within the Gondwana supercontinent. Without precipitation, there would have been no opportunity for glaciers to grow.

Temperatures in low latitudes during the Cool Mode may have been about 22°C, judging from isotopic evidence (Popp *et al.*, 1986). This supports the contention of Spjeldnaes (1981) that Ordovician subequatorial climates were more similar to modern ones (~ 24°C) than to those of the Cambrian and Devonian, which featured temperatures of ~ 40°C and ~ 45°C respectively, according to Karhu and Epstein (1986).

The first strong indication of cooling from carbonate data is the Caradocian contraction of carbonate deposition from latitudes of ~ 45° to near 30° in the Asian blocks. The subtropical carbonate belt remained restricted to these latitudes until post-glacial (Ludlovian) time in Asia but only until Wenlockian time in Laurentia. Also, reefs were more restricted at the height of the glaciation (Ashgillian), to latitudes of less than ~ 15° (compared with ~ 30° at present).

The precipitation/evaporation history of the Cool Mode in the Ashgillian, as revealed by evaporites and various weathering profiles, is one of widespread arid regimes in Laurentia contemporaneous with narrow arid belts

elsewhere (less than 20° latitude). The succeeding Llandoverian witnessed an enormous decrease in evaporite deposition in Laurentia and, indeed, in global evaporite abundance. However, there was also a marked increase to ~45° in the width of the arid zone in the Asian blocks. Some of these variations were probably caused by changes in the positions of continents relative to ocean currents and wind systems, but overall the evidence suggests that the Llandoverian was a time of decreased global aridity corresponding to the later stages of the glaciation.

Ordovician humid climates are delineated by oolitic ironstones within two belts, the first located between the equator and ~25° latitude, and the second beyond ~45°. Both zones shifted poleward during the Silurian to as high as ~35° in the first case and to latitudes higher than 70° in the second. This suggests that the post-Llandoverian increase in aridity as indicated by evaporite distributions was effective mainly in middle to high latitudes (Asia, Gondwana).

On these lines of evidence it appears that, relative to the present, the pre-glacial Ordovician was somewhat wetter, the glacial mode was drier and the post-glacial Silurian was similar to the present. In terms of temperature, high-latitude climates may have been comparable to the present only during the Ashgillian–Wenlockian. At other times conditions may have been warmer.

Other factors bearing on climate

Tectonic activity in the northern continents was intense but occurred at different times during the early Palaeozoic, with the Taconic orogeny of eastern North America being prominent. This event probably began in the early/middle Ordovician transition and its effects included mountain building and rapid siliciclastic sedimentation. The Caledonian deformation of northern Europe took place somewhat later, and was probably centred in the early Devonian (Perroud *et al.*, 1984). On a global basis, associated volcanism was high in the Ordovician but very low in the Silurian (Ronov, 1980).

Sea level was at a Phanerozoic high in the Ordovician; the highest level is placed in the late Ordovician (Caradocian) by Hallam (1984b) (based in part on detailed global studies by Fortey, 1984), but put in the early Ordovician by Vail *et al.* (1977). Both studies recognized a sea-level high in the Caradocian, significant sea-level fall near the Ordovician–Silurian boundary (~50–100 m; see Brenchley and Newall, 1980) and a large fall in the late Silurian. Details are also seen in the transgression–regression curve of Spjeldnaes (1961), who suggested the largest transgression was between the middle Llandeilian and the late Caradocian. Additional significant regressions (early Llandoverian etc.) were noted by Spjeldnaes but are given less importance in other studies. Three short pulses of change (1–2 m.y.) are

attributed by McKerrow (1979) to changes in ice volume in the Llandeilian to early Llandoverian interval.

Dark organic-rich shales are a prominent feature of early Palaeozoic sedimentation; Ronov (1980) estimates the quantity of carbon in these Ordovician shales as second only to that of the early Carboniferous for the Palaeozoic. The Silurian, by this inventory, is much less endowed and contains approximately the same quantity as the early Permian. Such dark sediments, implying high productivity or low oxygen at the deposition site, have long been recognized as an integral part of the early Palaeozoic marine realm and are a key component of polytaxic or greenhouse modes as defined by Fischer and Arthur (1977). Thickpenny and Leggett (1987) have delineated the concentration of abundant dark shales in European sequences as being in the intervals middle and late Cambrian, the Caradocian, and the Llandoverian to Wenlockian. Additional intervals of concentration occur on smaller scales in response to local geography and hydrology. Conditions opposite to the greenhouse mode include oxygenation of deep basin floors. Abundant evidence for these has been found in (1) late Cambrian to middle Ordovician; (2) Ashgillian; and (3) mid-Wenlockian to Pridolian times (Berry and Wilde, 1978; Leggett *et al.*, 1981).

Correlation of organic shales with climates is not satisfactory on a fine scale, although many times of extensive organic accumulation appear to coincide with warm Phanerozoic climates (as determined by other means). Rather, the correlation seems to be with marine transgressions (Irving, North and Couillard, 1974; Arthur and Schlanger, 1979; Summerhayes, 1981). This is apparent in the early Palaeozoic Cool Mode; both Hallam and Vail show a Caradocian sea-level high stand coincident with extensive black shale deposition in Europe. There is also an apparent association of organic shales with marine cherts and phosphorites, probably through the upwelling mechanism (Parrish, 1982).

A rough measure of marine chert abundance (Hein and Parrish, 1987) suggests moderate amounts in the Ordovician as compared with a Phanerozoic low in the Silurian, and in parallel with the early Palaeozoic trends in volcanism. Similarly, phosphorites were deposited abundantly in the Ordovician but are almost totally lacking in the Silurian (Cook and McElhinny, 1979).

An interesting distribution of $\delta^{13}C$ is seen in early Palaeozoic carbonate rocks. Holser (1984) showed a sharp decrease with negative values in the pre-Ashgillian Ordovician, followed by a steep increase culminating in the Llandoverian and a sharp decrease in the Wenlockian. The Ashgillian values are in accord with the high rates of burial of organic carbon suggested for the Ordovician by Ronov (1980). A last point has to do with the apparent scarcity of aragonitic mineralogy in early Palaeozoic sediments (Sandberg, 1985), which suggests a change in the oceanic Ca/Mg ratio.

4

The Warm Mode: late Silurian to early Carboniferous

The period from the end of the early Silurian until the beginning of the Namurian (early Carboniferous) was characterized by globally warm climates. It might seem that the late Devonian and Visean glacial intervals were exceptions to this but glaciation at these times was apparently limited only to high-latitude regions in South America. A general warming trend took place throughout the Warm Mode and can best be seen in a progressive expansion of the evaporite and carbonate sedimentation belts. Carbonates and evaporites persisted in mid-latitudes until late in the Namurian, after the succeeding late Palaeozoic Cool Mode had begun.

Trends from oxygen isotopes

The tendency for oxygen isotope ratios to be progressively lighter back in time through the Palaeozoic has been confirmed by Popp *et al.*, (1986) and by Hudson and Anderson (1989). This work, on brachiopod shells that show evidence of being chemically unaltered, also showed similarities to $\delta^{18}O$ determined on diagenetic cements and other components of carbonate sediments. Karhu and Epstein (1986) estimated the isotopic composition of Palaeozoic oceans at $-1‰$ PDB and calculated temperatures for subequatorial North America over the interval of the Warm Mode to be in the range 36–64°C. Previous measurements on unaltered Warm Mode carbonate fossils gave values between $\sim21°C$ and $\sim45°C$. Unaltered brachiopods studied by Popp *et al.* included only one Warm Mode sample, but other samples gave a range from about $-12‰$ to $-2‰$. Assuming no polar ice and an oceanic composition of $-1‰$, this range yields temperatures estimated at between about 65°C and 20°C. The unaltered sample ($-4‰$ to $-3.5‰$, Givetian, low-latitude North America) suggests a temperature of about 27–29°C, not significantly higher than present equatorial temperatures.

Estimation of palaeotemperatures from oxygen isotopes remains an unsatisfactory technique for the Palaeozoic, owing to several factors. While post-

depositional alteration may be detectable and the isotopic composition of sea water may be theoretically constrained, the extent of regional dilution of $\delta^{18}O$ due to run-off and rainfall, and of concentration due to evaporation, cannot always be anticipated. Further, in fossils this old it is not possible to establish to what extent organisms fractionated isotopes. Finally, the relative contribution of land ice in determining the oceanic composition can seldom be discerned.

Distribution of lithological indicators

Information is now available on the global distribution of major rock types of significance in palaeoclimate analysis for the Palaeozoic. Zharkov's (1981) summary on the geographic and stratigraphic distribution of evaporites and a compendium of global lithologies including carbonate rocks (Ronov *et al.*, 1984) are valuable. We have attempted to determine latitudinal variations of evaporites and carbonates for the Silurian–Carboniferous interval on each of several palaeogeographical reconstructions now available (Smith *et al.*, 1981; Ziegler *et al.*, 1981a; Livermore, Smith and Briden, 1985; Scotese, Barrett and Van der Voo, 1985). While similarities in the maps allow general comments on rock distribution, lack of agreement on continental positioning leads to marked dissimilarities for particular times. However, a compilation by Scotese and McKerrow (1990) takes account of the views of many workers and these have provided the basis for our analysis. The mid-Silurian and mid-Devonian maps are reproduced here (Figs 4.1 and 4.2).

Biogeography can assist in selecting the most reasonable reconstruction and this approach has been used by Boucot and Gray (1983) and Boucot (1985), in presenting sketch maps for the early Palaeozoic. In the case of the early Silurian, the map of Scotese *et al.* (1985) takes account of most biogeographic limitations (isolation and mixing of faunas, migration routes) and their late Silurian reconstruction is in apparent agreement with the sketch map of Boucot (1985). For the early Devonian, maps from both Scotese *et al.* and Livermore *et al.* show fair agreement with the Boucot and Gray sketch map and with each other, at least with regard to latitudes of continents. Southern Asia is the major exception, with a discrepancy of some 40° of latitude. Model A for the late Devonian (Scotese *et al.*, 1985) agrees well with the Livermore *et al.* (1985) Eifelian–Tournaisian map. It should be kept in mind that each reconstruction has a certain bias and a particular measurement error, which in all cases is unstated. Fig. 4.3 is constructed by use of the above maps.

The salient feature of the compilation (Fig. 4.3) is a trend through time towards higher latitude limits for carbonates and, to some extent, for evaporites. From the carbonate curve it appears this change was initiated in the middle Devonian, followed by a significant widening of the carbonate

Fig. 4.1. Global palaeogeography in the mid-Silurian (Wenlockian). From Scotese and McKerrow (1990).

Fig. 4.2. Global palaeogeography in the mid-Devonian (Givetian). From Scotese and McKerrow (1990).

Fig. 4.3. *Distribution of palaeoclimatic indicators in the late Silurian to early Carboniferous Warm Mode. The first column gives the latitudinal distribution of glacial and ice-rafted deposits. Column 2 shows the volume of evaporites (Zharkov, 1981); column 3 the latitudinal spread of evaporites; column 4 the spread of carbonate sedimentary rocks. In columns 3 and 4 ENA denotes Europe and North America, G denotes Gondwana and S denotes Siberia. Time scale from Palmer (1983).*

zone in the early Carboniferous. This was accompanied by a doubling in the global carbonate accumulation rate (from an average of about 0.3×10^{21} g/m.y. for the early Palaeozoic, to about 0.6×10^{21} g/m.y. for the Middle Devonian). Mid-latitude carbonates and high accumulation rates persisted through the Namurian when conditions reverted to being similar to the early Palaeozoic. The middle Devonian also saw the beginning of a long-term increase in carbonate $\delta^{13}C$ and a corresponding decrease in sulphate $\delta^{34}S$ (Holser, 1984).

The widening trend of evaporite distribution is gradual and more irregular than for carbonates but there is a more obvious shift from lower to higher latitude limits at about middle Devonian time. A large peak in evaporite abundance also occurs in the middle Devonian (Zharkov, 1981), related

mainly to massive deposits in the west Canadian Basin, Siberia and the Tindouf Basin, North Africa. Evaporites retain a wide zone of accumulation until Namurian time.

Carbonates

The distribution of carbonate sediments during the Warm Mode shows an extraordinary expansion, suggesting a progressive warming. This interpretation is in keeping with the common view that carbonate deposition is indicative of warm shallow seas. Certainly, modern carbonates are largely restricted to latitudes of less than about 30°, although limited accumulation extends to beyond 40° in both hemispheres in the form of the so-called temperate carbonates. This concept of temperature control has been challenged, however, by Ziegler *et al* (1984), who postulate the penetration of seawater by light, which diminishes strongly above latitude 40°, as the dominant control on carbonate distribution. Their conclusion arises from evidence that for the Mesozoic and Cenozoic at least, there was minimal shifting of the low-latitude carbonate belt with documented climate changes. However, the method used by Zeigler *et al.* was based on broad time intervals and, given the rapidity of climate change in the Mesozoic–Cenozoic, this approach may have, in effect, averaged any changes in opposite directions. The Miocene, for example, featured both warm and cold times; the average would therefore lie somewhere between the extremes.

In the Silurian–Carboniferous Warm Mode there was a large latitudinal shift in the mean poleward limit of carbonate distribution, from about 35° in the early Silurian to at least 50° by the early Carboniferous (Fig. 4.3). A consistent series of shifts is best seen in the curves for the Asian blocks and for Gondwana, and the magnitude of these suggested changes would probably exceed the measurement error for the palaeomagnetic data. Widespread middle Devonian carbonates also plot in high latitudes ($\sim 60°$, Kamchatka area) on the palaeogeographic reconstruction of Bambach, Scotese and Zeigler (1980).

Thick carbonate sequences, together with ooliths and bioherms, were taken by Heckel and Witzke (1979) as indicative of relatively low-latitude deposition. Their carbonate sequence distributions, replotted on the late Devonian map, show good correspondence with information on carbonate occurrences from the maps of Ronov *et al* (1984). Ooliths show a distribution to latitudes possibly as high as 50° in the Asian blocks, well outside their present limitation to about 30° latitude or less, with those in Gondwana reaching to $\sim 45°$ and Euramerican ones to $\sim 25°$. It is difficult to evaluate oolith distributions since the data of Heckel and Witzke are not specific as to age other than being Devonian. (Non-skeletal carbonates of the Warm Mode

are generally of the low-Mg calcite type; a transition to aragonitic types takes place in the early Carboniferous (Sandberg, 1985), perhaps in response to the initial climate cooling.)

Devonian bioherms and other organic build-ups, as shown by Heckel and Witzke (1979), have a similar distribution to thick carbonates and ooliths when replotted. Asia again shows the highest latitudinal spread, with carbonate build-ups reaching to about 50° latitude; Gondwana reefs attained about 35° and Euramerican ones about 30°. As pointed out by Heckel and Witzke, some of these structures may have been analogous to modern high-latitude build-ups of deep water origin and their climatic significance therefore is uncertain. Early Silurian reefs and carbonate sequences extended to near 50° latitude in Siberia (Ziegler *et al.*, 1977).

The complex assembly history of the Asian continent prompts a cautious approach to interpreting carbonate distributions there during the Silurian–Carboniferous Warm Mode. However, the parallelism of the Asian carbonate curve with that derived for Gondwana constitutes a case for accepting the Asian distributions as shown in Fig. 4.1. A global warming accordingly is supported for the late Silurian to early Carboniferous interval. The effects of warming seemingly included a broad expansion of carbonate sedimentation into middle latitudes except in Euramerica, where carbonates extended to the limits of the continent at $\sim 30°$–$40°$ latitude.

Evaporite and coal deposition

Only a few coal deposits were formed in the late Devonian and it is thus not possible to map the evolution of the evaporite–coal boundary through the Warm Mode. Frasnian and Famennian coals were used to define an equator-ial humid zone by Boucot and Gray (1983). This zone extends to about 25°N latitude on the late Devonian map, including localities in the Canadian Arctic and the northern Urals, but there is no suggestion of a comparable southern humid zone. These coals appear to be the forerunners of Euramerican Carboniferous coals which also formed in low latitudes ($< \sim 30°$) and which had largely disappeared by Permian time.

Because of its low-latitude position, Euramerica accumulated evaporites to latitudes no higher than $\sim 30°$ during the Warm Mode, in contrast to their much wider distribution elsewhere. Devonian redbeds are abundant in Euramerica (Old Red Sandstone, Catskill Formation etc.; Glennie, 1987). For low-latitude Euramerica, the picture is of dominantly arid climates, with coal-forming environments only along its northern fringe; in other words, evidence from Euramerica suggests a Warm Mode with low-latitude cli-mates similar to (or warmer than) modern ones. Moreover, northern Eura-merica, with its extended east–west coastline in low latitudes, occupied a classic monsoonal configuration, and coal formation can therefore be attri-buted to strong seasonal precipitation near the northern coast. On the other

hand, evaporites occur in large marine embayments of cratonic basins (West Canadian, Hudson and Moose River basins and on the Russian platform) well removed from the monsoon zone. These embayments were the recipients of some of the largest accumulations of evaporites known in the Phanerozoic, during the Eifelian.

The data on evaporite distribution (Zharkov, 1981), derived from late Silurian and late Devonian maps, can be used as indications of climate variability, together with their relative stratigraphic abundances, and in the absence of coals (Fig. 4.3). Gondwana shows considerable variation over time. Here, evaporites do not extend past ~30° latitude in the Silurian, extend to 20–40° (e.g. Tindouf Basin, North Africa) through the Devonian and to ~50° in the Namurian. Similar mid-latitude limits characterize evaporites on the Asian blocks, but evaporites in North America and Europe ranged only as far as ~30° (Devonian). Evaporites of both Gondwana and Asia also tended to occupy positions within cratons, being distant from moist maritime climates. Overall, the information available indicates that broad zones of mostly slow and intermittent evaporite formation extended to 40–50° latitude in the Asian blocks and Gondwana. Low-latitudes exhibited a marked expansion of the evaporite belt in the middle Devonian and this was also reflected in Gondwana and Asia.

Late Devonian and early Carboniferous glaciation

Despite the generally warm conditions indicated for late Silurian through early Carboniferous time, a glaciated region is well documented in northern Brazil (Acre, Solimoes, Amazonas and Parnaiba Basins; Caputo, 1985). Caputo also refers to other possible late Devonian glacial deposits in West Africa, which apparently require further study. The age of the Brazilian tillites is given as middle Famennian, which would seem to indicate a rather short glacial episode (less than 3 m.y.), and their distribution is at palaeolatitudes greater than about 70°.

Tournaisean and Visean glaciation also took place in Brazil and in the Andean ranges (northern Argentina and Bolivia; Caputo and Crowell, 1985). In the Andes the glaciation may have spanned some 25 m.y. or may have been much shorter; Brazilian early Carboniferous refrigeration would appear to be entirely of Visean age.

Palaeogeographic reconstructions of Gondwana are rather contentious for the Devonian owing to a paucity of well-determined palaeopoles. It is within the limitations of the data that the apparent polar path followed a large westward excursion from central Africa into the eastern Pacific in the Silurian–Devonian. As suggested by Crowell (1903), the later, northern, arm of this loop may have crossed northern South America in the Devonian thus generating the Brazilian glaciation. Presumably, the briefness of the episode

would be attributed to rapid relative polar motion, as the pole travelled on to southern Africa within the span of the Carboniferous. Tournaisian and Visean glaciers of South America might have formed between 70° and 40° latitude in such a scenario.

Palaeontological indicators

The warming which followed the early Palaeozoic Cool Mode is seen in both lithological and palaeontological data. Increasing cosmopolitanism among the graptolites beginning in the early Silurian has been attributed to progressive warming (Berry, 1973). By the early Devonian, however, marine faunas had diversified significantly and there was a large reversion to cosmopolitanism by late Devonian time (Boucot and Gray, 1983). Early Devonian biogeography is dominated by the inferred cool-water Malvinokaffric realm located in the Antarctic, south African and South American segments of Gondwana.

Copper (1977) noted the lack of tropical faunal elements throughout the Devonian in north-eastern South America, which is in keeping with the development of glaciers there in the late Devonian. Globally cool climates continued through to the end of the Famennian, according to Copper (1986), owing to the diversion of cold currents around subtropical parts of newly formed Pangaea. However, the timing of continental collisions and interpretation of the thermal state of the oceans in the Devonian remain contentious issues (Boucot and Gray, 1983; Stanley, 1984; Caputo, 1985; Copper, 1986, 1987; Raymond, 1987). There is general agreement that early Carboniferous faunas were cosmopolitan in nature, implying low temperature gradients (Boucot and Gray, 1983; Raymond, Kelly and Lutken, 1987;) and this is supported by studies on land plants (Raymond, 1985). The latter study suggests that the end of the mid-Palaeozoic Warm Mode was in the earliest part of Namurian A time.

Chronology of the Warm Mode

It would appear from evidence from carbonate distribution and from fossils that, following the late Ordovician to early Silurian Cool Mode, there was a slow and gradual global warming, which persisted until the Famennian and possibly continued unabated until Namurian A time. Under the latter interpretation, the late Devonian and Visean glaciations are considered as regional events affecting only South America, in the same way as early Silurian glaciations developed on that continent during the early Palaeozoic Cool Mode. All these glacial events reflect proximity to the apparent polar path and were short-lived owing to rapid relative motion of the pole. However, late Devonian glaciation and ice-rafting extended to about 60°

latitude and the early Carboniferous possibly to about 40°. There is thus some suggestion of strong late Devonian cooling even in the subtropics. More information is required before the significance of late Devonian and early Carboniferous glaciations can be fully assessed.

Concurrently with early to middle Palaeozoic warming, there was a significant though irregular increase in aridity as indicated by distribution by latitude of evaporites. Indications of humid climates (coals, bauxites) are not abundant over this interval of time; bauxites and laterites first become abundant in the late Devonian of the Soviet Union (Bardossy, 1979; Van Houten, 1985) and reflect continued humid climates in low to middle latitudes of the early Carboniferous globe.

Other factors bearing on climate

Efforts have shown that continents were assembled to form Pangaea in the mid-Palaeozoic, as deduced from palaeomagnetism and palaeobiogeography. Such a supercontinent arrangement should have caused extreme continentality. What can be said, however, is that the scarcity of indications of extreme continental climates argues against the assembly of Pangaea in the middle Palaeozoic. Deformation events and volcanic activity over this interval show that tectonic activity was concentrated in the late Silurian (Caledonian orogeny) and the early to middle Devonian (Arcadian orogeny) and may have contributed to global warming. While Caledonian events may have affected global climates through CO_2 release in volcanism, the late Devonian orogenies in South America and Australia, occurring just before glaciation began in South America, may have contributed to that episode by raising the mean elevation of land masses (Veevers, 1990).

Continental volcanism shows a spectacular growth during the Warm Mode. According to the compilation of Ronov (1980), continental volcanogenic rocks attained their greatest Phanerozoic accumulation in the late Devonian; amounts increase progressively from rather low quantities in the Silurian. Lower Carboniferous volcanogenics are only slightly less abundant than in the upper Devonian. The relationship to global climate via CO_2 liberated to the atmosphere cannot be doubted. Enhancement of the Silurian–Carboniferous Warm Mode was greatly assisted by widespread volcanism.

Steady rises in relative sea level (Vail *et al.*, 1977; Hallam, 1984b; Johnson, Klapper and Sandberg, 1985) closely reflect Warm Mode warming in the Devonian. The curves by Hallam and Johnson *et al.* show a sharp fall in the Famennian, in keeping with the glacial-eustatic regression discussed by Caputo (1985) but any immediate post glacial transgression is problematic. Additional data suggest major regressions in Euramerica at the end of the Pragian and in the Givetian (Johnson *et al.*, 1985) and both Vail *et al.* (1977)

and Hallam (1984b) document a relative fall at the end of the Silurian. These regressive events are perhaps related to continental docking, or to intervals of slowed sea-floor spreading; there are no corresponding climatic events known. The end of the Warm Mode is clearly indicated by falling relative levels in the Namurian.

Organic matter buried in sediments accumulated in relatively small quantities in the Silurian and early Devonian (Ronov, 1980), but increased in parallel with volcanism through the middle and late Devonian, reaching a peak with coal deposition in the Visean. True anoxic events were perhaps lacking in the early Palaeozoic Warm Mode, except possibly at the Frasnian–Famennian boundary, where a peak in $\delta^{13}C$ of about 1.5‰ occurred and organic matter and black shales accumulated abundantly in shelf settings. These accumulations and the associated marine extinctions have been attributed to oceanic overturning during cooling by Wilde and Berry (1984), but Geldsetzer *et al.* (1987) prefer a bolide impact as an explanation (see Wilde and Berry, 1988). The isotope peak correlates well with a relative high stand of sea level, and again, there is rough correlation with abundances of marine chert and phosphorite (both are low in the Silurian and moderately high in the Devonian and Carboniferous) (Cook and McElhinny, 1979; Hein and Parrish, 1987). Values of $\delta^{13}C$ show an increase (~ 2‰) during the Warm Mode (Popp *et al.*, 1986). Fluctuations occur in the Givetian through Famennian and these are roughly matched by large inverse shifts in $\delta^{34}S$ (Holser, 1984).

5

The Cool Mode: early Carboniferous to late Permian

A great deal is known about the late Palaeozoic glacial phase owing to intensive study since its recognition in the early part of this century. We now know that glaciations of this time were centred in polar to subpolar latitudes, that they were developed in parts of Laurasia as well as Gondwana, that their inception was regionally related to both collisional and extensional orogenic activity, that ice sheets developed in the centres of continents, and that glaciers underwent waxing and waning in response to changes in their palaeolatitudinal position. There are many other aspects of the late Palaeozoic Cool Mode about which we are relatively ignorant. What was the full duration of glacier development? What were the climates of the non-glaciated parts of the globe? Why did the glaciation cease?

Distribution and age of the glacials

Late Palaeozoic glacial deposits occur widely in Gondwana and also in displaced terranes which are now part of Asia (Fig. 5.1). These have been summarized most recently by publications in Hambrey and Harland (1981), Frakes (1979) and Crowell (1983). Africa is the continent with the most widespread deposits from this glaciation; glacial deposits are known from South Africa to the Arabian peninsula and from the east (Madagascar) to the west (Namibia). The southern exposures consist of the Dwyka Tillite, which occurs in the Karroo basin across the full length of the Cape Fold Belt and eastward into coastal Natal. An ice sheet occupied the northern part of South Africa and adjacent countries to the north (Zimbabwe, Botswana, Zambia) and radiated lobes to the south-east, and south-west (Kaokoveld lobe).

A smaller and perhaps isolated centre was located near the eastern margin of the Congo (Zaïre) basin and fed ice both westward and south-eastward into then-adjacent Madagascar. This or, more likely, another ice centre contributed debris to an undated Palaeozoic sequence now exposed in

Fig. 5.1. *Frequency of age ranges by locality (histogram) and palaeolatitudinal distribution by age of glacial deposits during the late Palaeozoic Cool Mode. Dates are from Hambrey and Harland (1981) and palaeolatitudes from reconstructions by Smith et al. (1981).*

Tigre, Ethiopia and others in the Arabian peninsula (Yemen Arab Republic, Saudi Arabia, Oman; McClure, 1980; Braackman *et al.*, 1982; Kruck and Thiele, 1983). All the African–Arabian glacial sequences are poorly dated but fall into the 'Permo–Carboniferous' age range (Fig. 5.1), except for a possible late Devonian occurrence in northern Niger (Hambrey and Kluyver, 1981).

In South America early Carboniferous (Visean) glacials were deposited in the Andean trend in Argentina and Bolivia and on the craton in Brazil (Amazonas and Parnaiba Basins) (Caputo and Crowell, 1985). These early glacials were followed by Namurian and younger glaciations farther to the south.

As part of Gondwana, South America received debris from the Kaokoveld lobe into the Parana Basin of Brazil, as well as from other lobes from another centre situated on the Uruguayan Shield. South America, however, does not seem to have developed a major cratonic ice sheet, as did Africa. A small ice cap did occupy the Asuncion Arch and other masses were scattered along highlands developed from collisional tectonics in the Andean trend (Argentina, Bolivia). Late Devonian glacial deposition took place even earlier in the Amazonas, Solimoes and Parnaiba Basins (Caputo and Crowell, 1985).

The oldest glacials known from Australia are Namurian in age. As in South America, these were derived from alpine/piedmont glaciers formed on highlands within a compressive regime (the Lachlan Fold Belt). Subsequently, major ice sheets formed in the south-east, in part fed by a transcontinental ice sheet flowing from adjacent Antarctica. Other ice centres were situated in the Australian interior and adjacent to west coast basins. Tillites, representing direct deposition from melting ice masses, are dated as no younger than Sakmarian, but ice-rafted deposits continued on until the Kazanian.

Glacial deposits of the Indian subcontinent occur as thin diamictites within rift basins and glacial marine layers in the Himalayan ranges (Salt Range, Pakistan; Nepal; Bhutan). Again, most age determinations are only to the extent of 'Permo-Carboniferous' although Carboniferous ones have been recognized in the Himalayas. A major problem concerns the relationship between large ice masses, including an ice shelf in the Salt Range region, and those required to explain the glacial deposits of the Arabian peninsula and of southern Thailand. The latter (Tantiwanit, Raksakauwong and Montajit, 1983; Altermann, 1986) are questionable glacial-marine deposits formed in a displaced terrane which possibly originated from the north-western Australian margin (Ridd, 1971).

Late Palaeozoic glacial deposits in Antarctica were laid down in three basins partially exposed along the Transantarctic Mountains and in West Antarctica (Ellsworth Mountains). Except for the latter occurrences and those in the Pensacola Mountains, which are subaqueous in origin, the Antarctic glacials appear to be largely non-marine. Ice probably occupied

highlands between the basins and radiated outward onto Africa (Natal) and southern Australia. All these deposits are poorly dated, generally being designated 'Permo-Carboniferous'.

Whereas all the preceding relates to deposits formed in Gondwana, deposits related to ice are also known from non-Gondwanan land masses. These comprise middle Carboniferous and late Permian glacial-marine strata from the Siberian and Kolyma blocks, now located in north-east Asia. Unlike the Gondwana occurrences, there is no association with tillites in these widespread Asian examples (Ustritsky and Yavshits, 1971; Epshtyn, 1972; Pavlov, 1979). In the case of the Carboniferous ones (late Namurian to Stephanian) their age ranges overlap with the Gondwanan glacials. However, the Permian glacial-marine deposits correspond in age almost precisely with the final phase glacial-marine deposits of Australia (Artinskian to Kazanian).

There are two main problems in dating late Palaeozoic glacial deposits. First is the problem of determining precisely when the episode began and ended. The second problem is stratigraphic and relates to correlations between widely separated palaeobiological provinces.

We have indicated the beginning of the Cool Mode at the beginning of the Namurian epoch (~ 333 Ma) because this is the age of the oldest known glacials which continue into the Permian. The Namurian age is accepted even though there was an earlier local Visean glacial event in northern Brazil (~ 352–333 Ma), though this may not have long predated the Namurian glaciation of Argentina. It is also possible that even older Tournaisian glacial deposits occurred in South America in the time gap between the late Devonian and the Visean (Caputo and Crowell, 1985).

The age of termination of the late Palaeozoic glaciation is not known for certain either, because there is a fairly long time gap between the youngest tillites (Sakmarian ~ 268 Ma) and the youngest ice-rafted deposits of Australia and the Soviet Union (Kazanian ~ 253 Ma). If the latter deposits resulted from rafting by icebergs, the presence of glaciers further onto the continent are implied; alternatively they may have originated from shore or river ice formed only in cold seasons and not necessarily related to full-scale glaciation. The absence of tillites of Artinskian–Kazanian age suggests that glaciers probably did not exist at this time but again, since any evidence of glaciation may have been destroyed by the highly erosional post-glacial environments or buried in basin centres or beneath modern ice in Antarctica, there may be alternative interpretations. It is also important to note that the occurrence of high-latitude ice-rafting at sea level in both hemispheres signifies climates cold enough to support glaciers in the polar zones until well into the late Permian.

Biostratigraphic problems abound in assigning ages to Gondwanan glacial deposits, largely because of the difficulty of correlating high-latitude south-

ern hemisphere faunas and floras with those of the low-latitude European type sections. Initially, this resulted in most glacials being designated as Permo-Carboniferous in age. Recent attempts to zone Gondwanan fossils (Roberts *et al.*, 1976; Kemp *et al.*, 1977) have succeeded in establishing consistent relative ages which are tied, where possible, to internationally recognized time units.

The distribution of known ages for Namurian and younger glacial deposits (Fig. 5.1; data from Hambrey and Harland, 1981) has a bimodal nature, with peaks in the Westphalian and the Sakmarian. In fact, the intervening Stephanian epoch is represented only as part of long-ranging deposits, that is, those poorly bracketed in terms of age. This is a striking change from ages inferred several years ago, when Stephanian–Sakmarian was given as the range for most Gondwanan glacials (e.g. Frakes, 1979). Part of this change is due to the refinement by stratigraphers of the Stephanian, which has now been shortened to ~ 10 m.y. duration, thus expanding the Asselian epoch of the Permian. Another major change, as a result of new work in South America, is the reassignment of some glacial strata that were thought to be Visean in age to Namurian and Westphalian ages.

Whether the early (Carboniferous) dates, represented by the first peak on Fig. 5.1 and dominantly from South America and Australia, in fact correspond to a separate glacial phase that was distinct from the more widespread Permian ones (second peak) will require thorough reviews of palynological and faunal zonations on a global scale. However, the glacial sequence in Tasmania provides strong support for such an interpretation. In Tasmania the youngest tillites are dated as late Carboniferous (Clarke and Farmer, 1976; 1982; Truswell, 1978). The succeeding Lower Parmeener Supergroup rocks (late Carboniferous to Asselian) were deposited during a major regression. This regression can perhaps be related to rejuvenated glaciation after the Stephanian warming phase. Ice-rafting was the only ice-related process in Tasmania at ~ 80° palaeolatitude during the Asselian–Kazanian interval, suggesting that the Permian glacial phase either was less intense than that of the Carboniferous or, if comparable, was focused elsewhere in Gondwana.

An interesting feature of the age–distribution histogram (Fig. 5.1) in both major glacial intervals is the apparent gradual expansion of ice over 20 m.y. or more in both intervals, followed by comparatively rapid declines. It is not possible to determine whether this characteristic is present in individual deposits, owing to a lack of precise stratigraphic control.

Palaeolatitudes of the glacials

The palaeogeographical distribution of late Palaeozoic glacial deposits can be derived by plotting their locations on global reconstructions based on

41

palaeomagnetism. For this purpose three sets of maps are available (Scotese *et al.*, 1979; Smith *et al.*, 1981; Scotese and McKerrow, 1990). Results are similar whichever set is used, despite the fact that Smith *et al.* provide Namurian, Sakmarian and Tatarian maps, while those of Scotese *et al.* are for the late Westphalian and the Kazanian. An advantage of the Smith *et al.* maps is that the value of alpha (radius of the circle which includes 95% of the measurements) is provided for each map. These values are for Gondwana 8.3° and for Laurasia 6.1°. Scotese *et al.* made an effort to include displaced terranes on the basis of their individual rock palaeomagnetism.

From the Smith *et al.* (1981) maps, Antarctic late Palaeozoic glacials fall between ~80° and 90° palaeolatitude while those from other Gondwanan segments lie between ~80° and 35°. The result is not greatly different using other maps, nor does a significant difference arise from plotting all occurrences first on the Carboniferous and second on the Permian maps. Dating and distribution data are thus not sufficiently precise to document large changes in the extent of the ice through time.

The earliest known glacial deposits of the Cool Mode (Namurian of South America and eastern Australia) were positioned between ~60° and 40° latitude and the youngest (marine ice deposits of south-eastern Australia and Siberia) were located between ~70° and 60° in the Kazanian. On the other hand, low- latitude ice-rafting during the Asselian–Sakmarian phase of the glaciation, probably extended to ~35° in South America and the Arabian peninsula. This contrast argues against widespread glaciers in the late Permian and supports the concept of seasonal (high latitude) ice-rafting as discussed earlier. The final ice-rafting phase thus may have persisted for up to 15 m.y. after the glaciers melted in south-eastern Australia. A similar and more-or-less contemporaneous sea-level phase of rafting developed in Siberia but here evidence of prior glaciation is lacking.

Areas which were glaciated in the late Palaeozoic predominantly lay within 45° of the Gondwanan pole. The formation of large but apparently independently waxing and waning ice masses can therefore be attributed to the high-latitude positions of the continents. However, much of Gondwana (parts of Antarctica, South America and Africa) lay at latitudes above 45° before glaciation began, which raises the question why glaciers did not develop earlier. Crowell (1983) and Caputo and Crowell (1985) argue for a large early Palaeozoic loop in the Gondwana apparent polar path, which would mean that only South America occupied moderately high latitudes in the Silurian–Devonian. In this scenario, both the North Atlantic Ordovician–Silurian glaciation and the late Devonian to early Carboniferous ones in South America represent direct responses to polar-continental motions. The late Palaeozoic glaciation was the third in the sequence, developing as the pole traversed Gondwana from Africa into Antarctica (Crowell and Frakes, 1970). Recent palaeomagnetic evidence tends to support this model (Bachtadse, Van der Voo and Halbich,1987).

Fig. 5.2. Variations in climatic indicators during the late Palaeozoic Cool Mode. Column 1 shows the frequency distribution of age dates of glacial and ice-rafted deposits (IRD); column 2 latitudinal extent of evaporites (Laurussia = North America and central and western Europe); column 3 volume of evaporites; column 4 distribution by latitude of coals (ENA = North America and Europe; G = Gondwana; A = Asia); column 5 shows the latitudinal extent of carbonate sediments.

Glaciated Gondwana apparently lost its ice in the early Permian but this seems to have occurred while a large sector remained within 30° of the pole (much of Antarctica and southern South America and Africa), according to all reconstructions. The early Mesozoic (Triassic, Jurassic) saw Gondwana move rapidly in and out of the polar zone, leading to a progressive decrease in the proportion of high-latitude land (Schmidt, Cudahy and Powell, 1986). A reasonable interpretation would be that glaciation was not terminated everywhere but was restricted to Antarctica from the end of the Sakmarian; the resulting icebergs led to rafted deposits during later stages of the Permian in the adjacent Australian sector.

A summary of the late Palaeozoic glaciation by palaeolatitude is given in Fig. 5.1. The prominent feature of this compilation, based on ages as given in Hambrey and Harland (1981) and plotted on the Namurian and Sakmarian maps of Smith *et al.* (1981), is the marked decrease in mid-latitude glacial effects in the Stephanian. Glacial deposits of Stephanian age are known only to about 30° palaeolatitude, compared with about 35–40° in both the Westphalian and the Sakmarian. Together with chronological data (Fig. 5.2), indicating a disruption of the glaciation at this time, this information constitutes a strong case for a comparatively warmer Stephanian.

Other climatic indicators

Palaeontological and sedimentological data add to the picture of global climate during the late Palaeozoic Cool Mode. Useful among these are distributions of low-latitude cyclothemic sequences, evaporites, tropical and temperate carbonates, and the nature and diversity of fossil assemblages.

The use of oxygen isotopes in palaeotemperature analyses of the late Palaeozoic, unlike those of the Cenozoic, is complicated by extensive post-depositional alteration of carbonate and uncertainties about the isotopic composition of the oceans (Veizer, Fritz and Jones, 1986). Available data include those of Rao and Green (1982) (giving palaeotemperatures of 4°C to − 8 °C around Tasmania) and substantial work on carbonate and phosphatic fossils and on early diagenetic cements. The results of Rao and Green are comparable to $\delta^{18}O$ values generated during the Pleistocene glacial maximum, but there is uncertainty about the degree of early diagenesis versus variations in salinity of surface waters. Soviet workers have made palaeotemperature estimates based on the Ca/Mg ratio in fossil shell material (e.g. Yasamanov, 1980) but the theory behind the relationship is not well established (see Wilkinson and Given, 1986). For this reason we have chosen to treat the Ca/Mg results as perhaps indicative of trends. Their value lies in determining the broad chronology rather than the intensity of climates and, in this sense, these data support the general cooling in the Carboniferous and early Permian; one study (Dorofyeva, Davydov and Kashlik, 1982) suggests a major warming in the Stephanian.

Veizer *et al.* (1986) pointed out a major shift in $\delta^{18}O$ across the Devonian–Carboniferous boundary, from very light values (< − 4‰) to values closer to modern ones (− 1‰). Popp *et al.* (1986), having devised a method of cathodo-luminescence detection of diagenetic alteration, observed the same shift but placed it in the early Carboniferous (Visean). Both studies thus indicate an important cooling. Close stratigraphic control in the latter work dates the change as within the early Carboniferous. Popp *et al.* (1986) also calculated that a temperature difference of 10–15°C would be required to explain isotopic differences between subtropical North American reefs (Delaware basin) and temperate carbonates of the Russian Platform.

The use of oxygen isotope determinations in chert–phosphate pairs from Palaeozoic rocks permitted Karhu and Epstein (1986) to confirm early suggestions of generally warmer Palaeozoic oceans. Their data suggest subequatorial temperatures of ∼40°C in the Devonian as compared with ∼18°C in the late Carboniferous–Permian. The latter figure is somewhat cooler than the average subequatorial temperature at present (∼24°C), but the very high temperatures indicated for the Palaeozoic as a whole remain suspect and further work is required. Recent summaries of Palaeozoic

isotope data from brachiopods, conodonts and marine cements indicate that the average fall in temperature from the early to the late Carboniferous could have been as large as 15°C (or as small as about 3°C) (Hudson and Anderson, 1989; Lohman and Walker, 1989).

Rhythmic sedimentation

Gondwanan glacial deposits originating from nearshore or subaqueous environments frequently display rhythmicity on the scale of a few metres thickness. At least half a dozen such cycles are known from the Parana basin in Brazil (Rocha-Campos and dos Santos, 1981) and several dozen have been suggested for a sequence in Victoria, Australia (see Davis and Mallett, 1981). However, it is difficult to establish these lithological variations as true rhythmic cycles owing to poor lateral extent of individual units and limited palaeontological dating. There is, at best, only the suggestion of rhythmicity.

The most obvious cyclicity in the late Palaeozoic is that seen in the trangressive–regressive coaly sequences of North America and Europe. These cyclothems were attributed by Wanless and Shepard (1936) to glacial-eustatic control of sea level and, more recently, Crowell (1978) suggested that detailed stratigraphic studies of these sequences could perhaps provide insights into timing of poorly dated glacial events in Gondwana. Recent studies have already established that some rhythmicities fall within the range of periods attributed to the Milankovitch parameters, thus suggesting a relationship between orbital motions, variations in intensity of glaciation and sea-level changes.

In north-western Europe, Ramsbottom (1979) distinguished 31 large-scale cycles (mesothems), each bounded by unconformities, over the interval from the beginning of the Carboniferous up to the middle part of Westphalian C time. These transgressive–regressive cycles are on average about 2 m.y. in duration; most are composite and contain two or more transgressive pulses equated with cyclothems as originally defined. Mesothems have also been recognized in the Appalachian sequence of North America (Busch and Rollins, 1984) and applied to the 50 global late Palaeozoic cycles (earliest Carboniferous through Artinskian; Ross and Ross, 1985), again yielding a periodicity of about 2 m.y. Close examination of the late Westphalian–Sakmarian cyclothems and lesser cycles of the mid-continent USA led Heckel (1986) to suggest periodicities which approximate to the Milankovitch cycles with periods between 19 000 and 413 000 years.

It is not known whether rhythmic cyclothems characterize the remainder of late Palaeozoic time so well, nor can it be said that the effect is global for the Westphalian–Sakmarian. The effects of spasmodic basin subsidence and delta-switching constitute an unquantified component of cycles of this type. However, approximate synchroneity of transgressions and regressions

is indicated by Ross and Ross (1985) for much of the mid-continent USA, north-western Europe, the Russian Platform, the Moscow Basin and the southern Urals region, thus supporting a eustatic cause for the mesothemic cycles. There is no presently known Milankovitch cycle (nor any tectonic one) corresponding to a 2 m.y. cycle, and a link of mesothems with climate variation therefore is doubtful. However, since this figure could represent an approximate five-fold multiple of the 400 000 year eccentricity cycle, the relationship between the shorter-period cyclothems and Milankovitch cycles suggested by Heckel (1986) also might account for the apparent longer cycles.

If it is accepted that late Palaeozoic cyclothems have a glacial-eustatic cause, then one would expect a glacial effect to have been present throughout the time represented by the cycles. However, cyclothems of comparable magnitude extend over a longer period from the earliest Carboniferous to the early Permian (a few until at least the end of the Artinskian), while Gondwanan tillites occur only in the Namurian–Sakmarian interval. This suggests early Carboniferous (but not late Devonian) initiation of significant Gondwanan glaciation, and also an extension of glaciation almost into the late Permian.

The decreased glacial activity suggested for the Stephanian is not readily discernible in transgression–regression curves for the northern hemisphere late Palaeozoic, although Stephanian mesothemic transgressions appear to extend farther onto the shelf than in several earlier and later phases (Ross and Ross, 1985, Fig. 3). The 10 Stephanian mesothems also are rather closely spaced, having an average periodicity of about one million years, half as long as in the Westphalian and Sakmarian; the significance of this is not clear but it perhaps suggests that Milkanovitch effects may have been more pronounced on ice bodies which were relatively small or on those located in comparatively high latitudes.

The importance of cyclothems in climatic reconstruction also derives from the presence of coal in the sequences, at least since the Visean (Donetz Basin). However, the palaeoclimatic significance of coal depends on the environmental tolerances of the plants which make the coal and this is an important variable in the late Palaeozoic. Fig. 5.2 shows the palaeolatitudinal distribution of Phanerozoic coal deposits (Briden and Irving, 1964). The earliest, Carboniferous, coals consisted of the debris of ferns and related plants which, although widely distributed by latitude, were mostly concentrated in the low-latitude rainy belt (0–15° latitude). At the beginning of the Permian, the pattern of coal accumulation changed markedly and has remained to the present in a pattern of dominant middle- and high-latitude distribution. Reflecting their palaeolatitudes, Laurasian coals are mostly Carboniferous in age while Gondwanan coals are predominantly Permian (Artinskian and younger). Gondwanan coals also tended to be formed from

newly evolved *Glossopteris* and related gymnospermous plants which were seemingly more tolerant of the cooler, more seasonal climates to be found at high latitudes.

The initiation of coal formation was a consequence of the dispersal of land plants in the early Carboniferous, but increased humidity relative to the Devonian also played a role. Similarly, coal deposition ceased in the early Permian (Sakmarian) in much of North America, Europe and the Soviet Union, probably as a result of increasing aridity in low latitudes. In each of these three continental blocks the first large deposits of post-Devonian evaporites also appear in the Sakmarian (Zharkov, 1981). In contrast, the accumulation of Gondwanan coals spanned the interval from the latter part of the late Carboniferous through to the late Permian. Early Triassic and younger Mesozoic coals are also common in Gondwana.

The late Palaeozoic cold phase has not yet yielded abundant evidence of short-period cyclicity ($<100\,000$ years) other than in cyclothemic sequences. Post-glacial chemical sediments of North America, however, display these features in extraordinary fashion (Peterson, 1980; Anderson, 1982).

Other sedimentary indicators

The line of separation on continental reconstructions between indicators of humid climates (generally high-latitude coals, bauxites) and those of arid climates (low-latitude evaporites, redbeds) can be seen to vary from period to period. Fluctuations of this kind can be taken to represent changes in the global evaporation/precipitation state (Fig. 5.2). The first observation regarding the late Palaeozoic Cool Mode is that most evaporites were formed in low latitudes ($<30°$) throughout the Carboniferous, except for pre-Namurian ones in Siberia (to $\sim 45°$). Most Carboniferous coals also were deposited in low, rather than middle or high latitudes, a fact possibly related to some distinctive and as yet undetermined characterisitic of Carboniferous plant life. The abundant vegetation required to form these coals is an indication of at least local humid climates, but why they were so abundantly preserved in low-latitude zones, unlike the present, is unknown.

The spread over Gondwana during the Permian of unique floras previously adapted to middle- to high-latitude conditions, including generally wet environments, led to coals forming up to palaeolatitudes as high as $\sim 70°$ in the late Permian. Evaporite deposits extended to $\sim 30°$ palaeolatitude in Bolivia. The hemispheric distribution of evaporation/precipitation for parts of Gondwana immediately following the end of glaciation therefore attained a state similar to that of the present. Similarly, in the northern hemisphere, scant early Permian coals of Siberia extended equatorward to near 40° and northern hemisphere evaporites are known only up to $\sim 30°$

palaeolatitude (eastern Europe). In the late Permian, Gondwanan coals were laid down in palaeolatitudes as low as 28°.

Significant changes in the distribution of late Palaeozoic evaporites are seen in Fig. 5.2, which was constructed from the stratigraphic data of Ronov *et al.* (1984) and Zharkov (1981) plotted on Namurian, Sakmarian and Tatarian palaeocontinental maps of Smith *et al.* (1981). The pre-Namurian period displays relatively high-latitude deposition of evaporites (non-Gondwanan) as in the Devonian but there is a sharp restriction to palaeolatitudes less than ~ 33° from the Namurian onward. Only a slight latitudinal expansion of evaporite formation occurred up to the end of the Sakmarian. Also, Gondwanan evaporites extended to slightly higher latitudes in the Stephanian than in either the Westphalian or the Sakmarian, providing a further suggestion of warmer climates in Stephanian time. The Kungurian was the greatest era of Palaeozoic evaporite deposition, containing more than 32% by volume of all Palaeozoic salts (Zharkov, 1981). However, there was little spreading of evaporites to higher latitudes in the Kungurian and the data suggest that large volumes of low-latitude salts merely accumulated more continuously and at higher rates than at other times in the Palaeozoic on all continents except Europe.

During the Palaeozoic, total latitudinal extent of coal accumulation was greatest in the Westphalian (Ronov, 1976) but not on all lithospheric plates. Laurussia, formed of North America and western and central Europe, shows progressive restriction of coals to lower latitudes until deposition ceased there in the latest Carboniferous. Gondwana shows a strong shift of coal deposition to higher latitudes during the Permian. The Asian (Siberia, Kolyma) plate shows a similar trend of variation through the same interval but with a more or less mid-latitudinal spread. The greatest global change occurs in the Triassic, when coal distribution assumes a pattern typical of later Phanerozoic time. Continental docking probably affected coal accumulation on each land mass, through orographic effects and ecological pressures resulting from the mixing of floras.

Overall, the evaporite–coal relationships in the late Palaeozoic Cool Mode suggest pre-glacial global aridity, signified by the occurrence of evaporites to mid-latitudes in the Visean. Evaporites drew back sharply by Namurian time concurrently with an expansion of coal into lower latitudes. Together these data indicate widespread increases in humidity. The subtropical coals of Laurussia accumulated through the Westphalian before ceasing in the Stephanian, to be replaced by evaporite–redbed suites, including the first halite–glauberite associations to be seen in Phanerozoic time (Zharkov, 1984). The increase of middle- and high-latitude Gondwanan coals in the Permian was not accompanied globally by increased bauxite formation (Bardossy, 1979) nor by equatorward shrinking of the evaporite belt. Early Permian desert sands of Europe (e.g. Rotliegend; Glennie, 1987) indicate aridity at between 10° and 30° latitude in accord with this.

The global abundance, accumulation rate and distribution of carbonate sediments all give information bearing on late Palaeozoic climates. First, Carboniferous and Permian bioherms are largely confined to less than 20° latitude. The carbonate mass data of Ronov (1980), converted to accumulation rates of total inorganic CO_2 in sediments by Hay (1985), both show high values for the early Carboniferous and the early Permian, but are among the lowest values in the Phanerozoic for the late Carboniferous and the late Permian.

As pointed out by Hay, Palaeozoic figures are incomplete since the contribution of pelagic (oceanic) and continental margin sediments cannot be evaluated. Moreover, latitudinal distributions for both Gondwana and northern continents imply increased carbonate sedimentation, with expanded development in the early Carboniferous and early Permian (Fig. 5.2). Early Permian temperate carbonates accumulated to exceptionally high latitudes in Tasmania and Chile, although the latter must be viewed with caution as they may represent part of a displaced terrane. Similarly, carbonate rocks, ordinarily taken as representing moderate climates, must be interpreted with caution, until the distribution of temperate versus tropical carbonates is better understood.

Fossil soils provide an indication of the precipitation/evaporation balance at the time of formation. Unfortunately for our purposes, soil formation in the late Palaeozoic was bound up with the evolution and dispersal of early vascular land plants since the late Silurian. Hence, many Carboniferous soils are related to acidic ground water in forested terranes and Devonian ones tend to reflect alkaline waters (Retallack, 1986). Wright (1980), working on fossil soils of Wales, showed this in the transition from palaeokarst and hardpan calcretes (arid to semi-arid) to peaty soils formed in river flood plains during the Arundian age (~ 345 Ma). Sigleo (1983) found evidence of a late Pennsylvanian less humid interval in Appalachian soils. Lateritic soil profiles of the western Soviet Union (late Carboniferous) and north-east Asia (late Permian) are shown by Van Houten (1985) to be among those deposited in highest latitudes (to $\sim 60°$) during the Phanerozoic; these find their counterparts in Gondwana in Artinskian kaolinitic and lateritic palaeosols in central eastern Australia (Loughnan, 1975).

The earliest significant deposits of bauxitic weathering crusts occurred in the late Devonian to early Carboniferous interval (Bardossy, 1979). This production of lateritic bauxite, signifying local intensive weathering and hence humid climates in the Ukraine, USSR, was followed by generally lower levels of bauxite production through the remainder of the Cool Mode. Oolitic ironstone and sedimentary manganese, also indicators of humid climates, are virtually non-existent at this time (Van Houten, 1985).

Palaeontological indicators

Marine faunal and terrestrial floral provinces have been recognized for the early and late Carboniferous and for the Permian (Chaloner and Lacey, 1973; Chaloner and Meyen, 1973; Runnegar and Campbell, 1976; Ross, 1979; Ziegler *et al.*, 1981a). However, the climatic significance of Palaeozoic provinces is more often inferred from their palaeolatitudes than from their biological composition, in part because the assemblages correspond poorly to modern forms of known environmental tolerance. Diversity of an assemblage can give some measure of the relative equability of climates but, for the late Palaeozoic, Hallam (1984b) has shown that diversities at the generic level are of doubtful use for interpretations of climate in many cases. However, Permian invertebrate diversity, based on data from Waterhouse and Bonham-Carter (1975) and Ziegler *et al.* (1981a), show the hoped-for gradation from high values in low latitudes to low values in high latitudes (Hallam, 1984b, Fig. 7.4).

Changes through time have similar limitations, as for example the replacement of dominant pteridosperms and sphenopsids by tree ferns in non-coaly deposits near the end of the Euramerican Westphalian (Pfefferkorn and Thomson, 1982). Although a decrease in humidity is implied, there has been no quantification of the extent of aridity in the Stephanian. Similarly, although Phillips and Peppers (1984) found variable climatic wetness in Euramerican coal floras through the Pennsylvanian (wettest late Westphalian; driest early Stephanian), they inferred 'temperate' or 'subtropical' conditions mainly by reference to the palaeolatitudes of the occurrences. Since these terms have uncertain meanings on an earth of shifting zonal boundaries, palaeolatitude cannot be considered as a reliable indicator of climate.

During the periods immediately preceding the beginning of the Cool Mode (Visean to earliest Namurian A) and following the Visean glaciation in South America, global climates appear to have been relatively warm. On the basis of palaeobotanical studies, Raymond (1985) and Raymond *et al.* (1987) attribute this to deflection of equatorial surface currents into high latitudes by the newly amalgamated Pangaea (Ross and Ross, 1985).

The fossil floras of Mongolia (Durante, Dmitriyev and Pavlova, 1985) belong to two provinces (Angaran and Tethyan provinces). These workers ascribe variations in the floras through the late Palaeozoic to variable climates: (1) early Carboniferous non-tropical climates; (2) middle Carboniferous coal (Angaran) in the north and subtropical Tethyan climates in the south (including evidence from fusulinids), but signifying an overall cooling; and (3) considerable warming into the late Carboniferous. The early Permian featured gradual cooling at first together with evidence of a widespread Angaran flora, and the late Permian saw substantial warming. This

scenario provides further evidence of a lull in late Palaeozoic cooling in the late Carboniferous, although the changes cannot be quantified nor precise ages established. However, it matches the period of decrease of glacial sediments, as shown in Fig. 5.1.

Endemism of marine faunas shows great variability in the late Palaeozoic. Variations may be in part related to climate changes but the late Carboniferous and late Permian increases (Ziegler *et al.*, 1977) more likely reflect restricted seas due to global regressions (Hallam, 1984b).

In their summary of growth rings in fossil wood, Chaloner and Creber (1973) point out that Euramerican late Palaeozoic woods display little indication of growth rings, while Permian Gondwanan woods are characterized by well-developed rings. They attribute this difference to equable climates for the Euramerican low-latitude zone versus seasonal climates in high-latitude Gondwana. Again, however, there is not yet sufficient detailed study to allow either placement of the boundary between these two climate states on maps, or to determine how these conditions varied through time in the late Palaeozoic. One detailed study (Jefferson and Taylor, 1983) found that late Permian wood from Antarctica was indicative of a seasonal climate with warm temperate summers, but frost damage suggested freezing winters.

The early vertebrate faunas give some indications of climates through the late Palaeozoic Cool Mode. For subequatorial Euramerica, Olson (1979) infers persistent warm, humid climates and weak seasonality during the whole of the Carboniferous. The evidence from aquatic and terrestrial vertebrates from the USSR and Gondwana suggested to Olson that rainfall became more seasonally distributed in the early Permian. Climates in mid-latitudes in the late Permian may perhaps have been akin to modern temperate regimes, in contrast to the low latitudes, where aridity became increasingly common and intense.

Chronology of the Cool Mode

The most obvious feature of the late Palaeozoic Cool Mode is the existence of high-latitude glaciers on all sectors of Gondwana. These had their inception in elevated terranes of South America and eastern Australia by early Namurian time at the latest. Early glaciers grew into ice sheets at sea level and persisted for some 65 m.y. until the Sakmarian, and perhaps longer. Glaciers had their greatest extent, to ∼ 35° latitude, in the Westphalian and Sakmarian, with some amelioration indicated by less extensive glacial deposition in the intervening Stephanian. That glaciation was episodic and geographically variable is shown by repetitive coaly sequences of the northern hemisphere, which were most likely glacial-eustatically controlled, and which perhaps were initiated at the beginning of the Carboniferous.

51

After the Sakmarian, marine periglacial conditions continued for a further 15 m.y. at latitudes above ~70° in Australia; similar conditions developed at ~70° latitude in north-eastern Asia about the same time (and once earlier, in the middle Carboniferous) but apparently without associated glaciers. The late Palaeozoic Cool Mode thus can confidently be said to have had an age range of Namurian to Sakmarian, but may have had a greater range, from Tournaisian to Kazanian. Judging from penecontemporaneous coal cycles, glacial cycles appear to conform to Milankovitch orbital rhythms.

The beginning of the Carboniferous saw little change in environments in South America, which had already suffered a glaciation in the late Devonian, or in the latitudinal spread of evaporites to 45–50° in both hemispheres. Carbonate deposition continued at a high rate and to moderate latitudes from similar conditions in the Devonian. Remarkable changes in oceanic chemistry are reflected in the isotopes of oxygen and carbon at this time, the latter perhaps related to the increased diversity of land plants at the beginning of the Carboniferous. A marked increase in $\delta^{18}O$ suggests the beginning of a build-up of polar ice, in keeping with the initiation of northern hemisphere cyclothems at about this time.

The onset of the Cool Mode, though possibly dating from the late Devonian, for now is best recognized as taking place at the beginning of the Namurian. At about this time, the oldest known glacial deposits were formed, and while cyclothemic development continued, carbonate shelves were more restricted in both hemispheres (late Namurian), and evaporite accumulation suggests a marked contraction of the arid zones from latitudes as high as 50° to about 30°. The beginning of the glaciation was thus accompanied by increased wetness in mid-latitudes. It is possible that the former resulted from the latter through deflection of equatorial currents and hence the resulting provision of a moisture source. The spread of Laurussian coals into mid-latitudes also reflected this incease in wetness.

The growth of glaciers in the Namurian and development of new ice centres led to intense glaciation in the Westphalian. There was contraction of the low-latitude humid (peat) zones in both Euramerica and Gondwana. Asian high-latitude peat environments expanded slightly to extend as low as 35° latitude. There was massive deposition of coal, signalled also by a marked $\delta^{13}C$ increase, but no discernible change in evaporite accumulation. That is, the mid-latitudes continued fairly wet, even to the point of some formerly dry areas becoming humid peaty regimes. Shallow-marine carbonates were deposited at latitudes as high as 60° in the northern hemisphere but only to about 40° in Gondwana.

The evidence for Stephanian climates suggests a contraction of the ice, probably primarily from relatively low-latitude centres near the Arabian peninsula and in South America. This was paralleled by a slight expansion of the arid (evaporite) zone in Gondwana, extensive marine transgressions,

and less humidity as indicated by soils in the Euramerican cyclothemic sequences. Data on coals and carbonates are not specific for the Stephanian, and relative abundance, latitudinal spread, etc., therefore cannot be determined.

The beginning of the Permian saw a renewed increase in glaciation, a marked shift of Gondwanan and Asian coals to high latitudes, a slight equatorward contraction of the arid zone and relatively little change in carbonate deposition. It is likely that the Asselian–Sakmarian interval was comparable to the Westphalian in many respects, except that South American ice was less extensive. An obvious change was the drying out of Euramerica as testified by the transition from widespread peaty swamps to redbed–evaporite environments in the latest Carboniferous to earliest Permian.

The remainder of the early Permian featured marine ice-rafting in high-latitude Gondwana ($\sim 80°$, south-eastern Australia) and the onset of ice-rafting in the north-eastern USSR. The former suggests the presence of glaciers in adjacent Antarctica but the latter probably represents only high-latitude seasonal ice at sea level. The Artinskian and Kungurian also saw a tremendous build-up of both tropical and temperate carbonate sediments which extended to at least 75° latitude in Tasmania. This interval thus records the dying stages of the late Palaeozoic Cool Mode, but it is important to note that, first, the formerly glaciated terrane of Antarctica and terranes which experienced ice-rafting (Tasmania, north-east Asia) remained in high latitudes during the Triassic and thus should have continued to experience cold climates. Second, sea level, which seems to have fallen thoughout the Cool Mode, apparently dropped drastically at the end of the Permian (Hallam, 1984b; Holser and Magaritz, 1987).

The late Permian saw an expansion of Gondwanan and Asian humid zones to lower latitudes (coals, to $\sim 30°$ and 40° respectively), together with a withdrawal of carbonate sediments to latitudes more reminiscent of those at the height of the glacial intervals. Accompanying these changes was a marked decrease in $\delta^{13}C$, probably reflecting the decreased abundance of coal-forming environments. Although older studies on oxygen isotopes (e.g. Lowenstam, 1964) suggest higher palaeotemperatures for the late Permian, it is uncertain whether these signify a complete end to glaciation in the late Palaeozoic. Despite these uncertainties, it is apparent that marked climatic changes occurred before the end of the late Permian and therefore it would seem appropriate to continue to regard the Kazanian as the end of the late Palaeozoic Cool Mode.

The available evidence for the late Palaeozoic Cool Mode supports a picture of global climates generally cooling and becoming more humid than in the pre-Namurian, and of cold conditions probably more extreme than present.

Other factors bearing on climate

Up to this point, the late Palaeozoic Cool Mode has been discussed in terms of parameters of climate, i.e. temperature, aridity. We now describe additional factors which are significant for, but not necessarily descriptive of, climate and which show variation over the interval. These may well prove to have been instrumental in effecting change, and they include global and regional tectonic activity, volcanism, and variations in sea level and ocean chemistry.

Two tectonic events are of prime interest in relation to climate change at this time. The earliest was the Namurian deformation of northern Australia, western South America and possibly Victoria Land, Antarctica. Powell and Veevers (1987) in fact attributed the early Gondwanan glaciations to the formation of high-latitude uplifts resulting from this event. The Hercynian event affected both Europe and North America during the Carboniferous and, although there is no directly related glaciation in these low-latitude zones, the timing of the Hercynian orogeny, roughly coincident with global cooling, suggests a relationship. The Hercynian orogeny must also be considered when trying to unravel palaeobiogeographic factors (land bridges, mountain building etc.) from climatic effects on biotic diversity in the late Palaeozoic.

Volcanism, a variable long known as being related to tectonic activity, has been shown by Ronov (1976) to be abundantly represented in the continental rock record of the early Carboniferous and early Permian and much less so in the late Carboniferous and late Permian. The record of volcanism shows a positive correlation with carbonate rock abundance, and Budyko and Ronov (1979) related both to global climate through the atmospheric CO_2 connection. Interestingly, deposits of sedimentary chert (Hein and Parrish, 1987), which would be directly or indirectly related to silicic volcanism, show a progressive decrease though the Cool Mode.

Eustatic changes in sea level can also result from tectonic activity but the relationship is complicated by the effects of glacial eustasy. During the late Palaeozoic glacial phase, a major and apparently global regression in the Namurian (Mississippian–Pennsylvania boundary) was perhaps a result of the closing of the Iapetus Ocean or the Hercynian orogeny. Other global transgressions and regressions were pointed out by Veevers and Powell (1987); major regressions occurred just before the beginning of the Carboniferous and others in the early Visean, the Stephanian and the Sakmarian. The latter one was preceded by a major transgression. Some of these events no doubt correlate with mesothem unconformities in Euramerica, so that tectonic and climatic effects are intermixed in the sea-level record of the late Palaeozoic Cool Mode.

The possiblity of variations in the chemistry of the world ocean can be

seen from the abundance of non-skeletal aragonite in the late Palaeozoic and its paucity in warm modes earlier and later (Sandberg, 1985), and also in carbonate oolith abundance, which is extraordinarily high in the late Carboniferous (Wilkinson *et al.*, 1985). The carbon isotope record contains variations that are also likely to be significant. The $\delta^{13}C$ of carbonate rocks (summarized in Holser, 1984) was extraordinarily high in the late Palaeozoic, the highest in the Phanerozoic. In the Stephanian there was a sharp fall in $\delta^{13}C$, which possibly led to increased liberation of CO_2 to the atmosphere coinciding with apparent warming in the Stephanian. Also, Holser, Magaritz and Clark (1986) have documented a large positive shift of $\delta^{13}C$ of 5–7‰ in localities from Europe, North America and Asia, almost precisely at the Kazanian–Tatarian boundary. This period of anomalously high $\delta^{13}C$, which also coincides with the sedimentation of the Kupferschiefer base-metal deposits in eastern Europe, continued on to the Permian–Triassic boundary, at which time there was a substantial fall of about 3‰.

Accumulation of marine organic matter was remarkably low during the late Palaeozoic (Holser, 1984) (apart for moderate rates in the early Carboniferous) suggesting overall well-ventilated oceans with respect to oxygen. The lowest accumulation rates occurred in the late Permian. The time of accumulation of the giant phosphorite deposits in the Phosphoria Formation in western USA slightly predates this, falling in the age range Artinskian to Kungurian.

6

The Warm Mode: late Permian to middle Jurassic

Although the Mesozoic as a whole has been interpreted as a warm and arid interval, it now appears that such a generalization does not take account of significant variations within the Era. Much of the early Mesozoic appears to have been a warm time but extensive aridity developed only after the middle Triassic, and significant cooling occurred in the mid-Jurassic. A Warm Mode is here taken to extend from the end of Gondwanan and Asian ice-rafting in the late Permian (Kazanian) through part of the middle Jurassic (to the end of the Aalenian), after which ice-rafting was again common in high-latitude sites (Frakes and Francis, 1988). Following a cool, though arid, time from middle Jurassic to roughly the middle of the Cretaceous, Mesozoic warming resumed.

Oxygen isotopes

The general thermal state of the earth is deduced by the oxygen isotope method and by interpretations of climatically sensitive indicators. Evidence for the middle Jurassic part of the Warm Mode includes an abundance of isotopic information, most of it gained, however, at a time when techniques were imperfect. Most of such data are summarized in Bowen (1966) and have been reinterpreted by Stevens and Clayton (1971) and Hallam (1975). As stated in Frakes (1979), the problems of changing oceanic composition, vital effects, diagenesis and the practice of data averaging before presentation, all mitigate against acceptance of these data as reliable indications of early Mesozoic palaeotemperatures. Unaveraged data for this part of the geologic record suggest a cool mid-Jurassic in Australia ($\sim 18°C$ to $15°C$) and the Soviet Union ($\sim 15°C$ to $11°C$). Interestingly, the Australian temperatures remain low (down to $\sim 12°C$) until the Aptian stage of the Cretaceous, whereas the Soviet data suggest renewed warming beginning in the Oxfordian (see Frakes, 1979, fig. 6–7).

The late Permian to early Triassic, on several lines of evidence, was quite warm: the seasonal ice which characterized the Kazanian seems to have vanished; aragonite-secreting benthic faunas became dominant; early Trias-

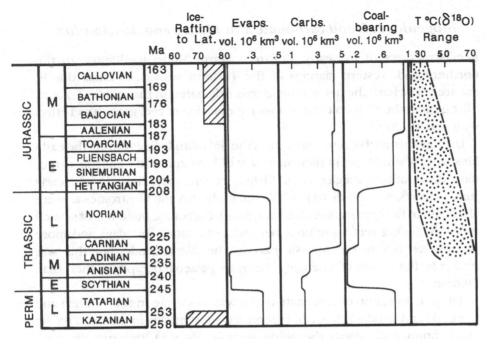

Fig. 6.1. Variations in climate indicators during the late Permian to mid-Jurassic Warm Mode. Column 1 shows the latitudinal extent of ice-rafting; column 2 volume of evaporites; column 3 volume of sedimentary carbonate rocks; column 4 volume of coal-bearing sediments; column 5 temperature range estimate from δ18O. Time scale from DNAG (Palmer, 1983).

sic faunas were extremely cosmopolitan. Extreme warm temperatures in fact may have existed in low latitudes if isotope work on biogenic phosphates and chert–phosphate pairs is accepted (Fig. 6.1). The Karhu and Epstein temperature curve (1986) depicts a strong warming at the beginning of the Warm Mode and a cooling thereafter which continues on into the early Cretaceous. We have picked the end of the Warm Mode on the occurrence of high-latitude ice-rafting, which is first seen in the middle Jurassic (Bajocian).

Karhu and Epstein, while acknowledging that the scale used for calculating their absolute temperatures may not be entirely reliable, put forward estimates of ~38°C for subtropical marine surface temperatures in the late Permian and Triassic. (On this method, these oceanic temperatures are matched by comparable extreme values only in the Cambrian, although still higher temperatures are suggested by Precambrian data.) Their results for the late Jurassic (Ethiopia, Scotland: ~20° latitude) approximate those from cool early Cretaceous marine carbonate material (15°C–20°C) and therefore it would appear that the warm excursion was not of long duration. The cause of such an extreme warming (perhaps not so extreme if scales for calculation are revised) poses a major problem in palaeoclimatology.

Biogenic reefs and carbonate abundance and distribution

Reef structures of the early Mesozoic Warm Mode are limited to the northern and western margins of the Tethyan seaway, because this is, apparently, where the hermatypic corals originated in the middle Triassic. Throughout the Mesozoic there was an expansion of reef growth as Tethys widened westward.

Despite this, reefs occur only within the 35° latitude lines during the early Mesozoic Warm Mode. In their summary of Mesozoic and Cenozoic carbonate distributions, Ziegler *et al.* (1984) record early Triassic and early Jurassic reefs, but middle to late Triassic bioherms and biostromes also are abundant in Europe (e.g. the Dolomite Alps). Carbonate build-ups of several types, including reef structures, became both more abundant and more widespread during the Jurassic part of the Warm Mode, but this was probably the result of changing Tethyan palaeogeography rather than climate.

The accumulation of carbonate rocks was moderate in the Tatarian and the early and middle Triassic but increased substantially in the late Triassic and continued so through the middle Jurassic (Fig. 6.1). Paradoxically, reefs and carbonates seemed to have increased in abundance in parallel with a cooling trend. Moreover, while both reefs and significant carbonate accumulation were largely Tethyan phenomena during the early stages of the Warm Mode, carbonate sediments were able to form along the Pacific margin of Pangaea in the early Jurassic (Pliensbachian), and these reached latitudes of more than 40° in Chile and northern India.

Variations in humidity

Evaporites

The Triassic is often cited as the time of greatest evaporite accumulation in the Phanerozoic. However, totals recently compiled by Ronov (1980), indicate that that distinction belongs to the Permian (particularly the early Permian), followed by the Jurassic ($\sim 27\%$ versus $\sim 12\%$ of the Phanerozoic total volume). A large quantity of evaporites was deposited in the Triassic, about 9% of the Phanerozoic total, but 80% of this vast quantity was formed entirely in the late Triassic; the early and middle Triassic accumulated salts at rates near the Phanerozoic average. Similarly, Jurassic evaporites are strongly concentrated in the late Jurassic and figures for the early and middle Jurassic are also near the Phanerozoic average. In addition to the late Triassic, another interval of major evaporite formation occurs in the early Mesozoic Warm Mode; this was the late Permian, with a total of about 5% of Phanerozoic salts, and about half of these were formed in the Tatarian part of the Warm Mode.

The Ronov summaries then indicate that there was widespread aridity in the Tatarian and in the late Triassic to a very unusual extent. Late Permian evaporites formed on all continents except Australia and possibly Antarctica, many in large intracratonic basins, and major late Triassic accumulations formed in rift basins now marginal to the North Atlantic, in part controlled by local climates (Hay *et al.*, 1982). An important feature of the distribution of Triassic evaporites is that they occur across the equatorial zone in Africa, the Arabian peninsula and south-eastern North America and in such a way as to suggest that the tropical rain belt did not exist. This problem was addressed by Robinson (1973), who carefully showed the influence of the Pangaean landmass on circulation patterns, particularly the Intertropical Convergence Zone (ITCZ). Quasi-permanent high-pressure systems would have been located over the Gondwanan and Laurasian segments which were configured such that the equatorial and subtropical zones received dry winds year round. Hence, an equatorial rainy belt was not developed during the interval from the late Permian to the late Triassic. However, this explanation does not account for the significantly smaller volumes of evaporites in the early and mid-Jurassic.

Late Triassic monsoonal areas would have developed where evaporative winds from the high-pressure systems crossed the Tethys and dropped their precipitation on the Tethyan margins (southern Asia, a coastal fringe in Somalia, and western India; Robinson, 1973; Parrish, Zeigler and Scotese, 1982). The intensity of late Triassic monsoonal conditions may have been controlled by Milankovitch orbital parameters (Rossignol-Strick, 1983; Olsen, 1986). Based on the distribution of evaporites (Fig. 6.1), it appears that in the early and middle Jurassic the monsoon would have reached to 40° latitude in Asia, and attained an even more exaggerated expansion to about 50° latitude in both hemispheres in the late Jurassic. The regions of highest latitude evaporite deposition are located along the western margin of Pangaea, partly in tectonized regimes and may represent exotic terranes.

Coals

The volume of coal-bearing sediments (Ronov, 1980) was comparatively low during the Warm Mode, representing only some 10% of such strata deposited since the Silurian. The rate of accumulation is about half the Phanerozoic average. Despite this, there are large variations in coal accumulation through the Warm Mode; most importantly there is a marked decrease during the Triassic. Correspondence with evaporite abundance is not perfect, however, as the late Triassic shows some increase in coal abundance during the period of major evaporite deposition. Generally, the abundance of coaly strata increases progressively from a low in the early Triassic through the Warm Mode and on into the late Jurassic.

Unfortunately, map compilations of climatically significant rock types by

epoch are not yet available for this part of the geologic record. From distributions by period, northern hemisphere Triassic coals appear to be restricted to higher latitudes in North America (to $\sim 50°$ or higher) than in Eurasia ($\sim 35°$), perhaps reflecting an expected large extent of west coast deserts. Jurassic coals occur farther south in both continents ($\sim 45°$and $30°$ respectively), indicating a decrease in aridity, as already documented in evaporite and coal volumes. Patterns in the southern hemisphere are complicated by the occurrence of coals in intermontane basins at $\sim 30°$ latitude in the Andes of South America and by seasonally variable (monsoonal?) climates along the western margin of the Tethys seaway in East Africa and India.

Soils

The late Permian to middle Jurassic Warm Mode is characterized by a general scarcity of bauxitic and lateritic soil profiles. During the Warm Mode, bauxites were restricted to areas at less than $30°$ latitude. As shown by Bardossy (1979) the interval lies between major peaks in bauxite abundance in the Devono-Carboniferous and the Cretaceous. However, Van Houten (1985) shows laterite distributions as extending from the tropics to over $60°$N in Asia and to about $60°$S in Australia in the late Permian and early Triassic.

Van Houten also shows that Asian late Triassic calcretes lie between $15°$ and $40°$ while laterites are rare but extend poleward to about $50°$ latitude. There is thus a suggestion of wet climates in high-latitude Asia during the late Permian to early Triassic, in contrast to the evidence for relative global dryness from bauxite (and evaporite) distributions. If similarly dated palaeosols of eastern Australia (Loughnan *et al.*, 1974; Retallack, 1977) are interpreted as products of deep weathering at $\sim 55°$ latitude, it would appear that the high mid-latitude zones of both hemispheres featured seasonally wet climates in the late Permian to early Triassic. Laterites of the early Jurassic are also widespread (again to about $50°$) but middle and late Jurassic laterites reach to only about $40°$ latitude.

Evidence of climates from palaeontology

Discussion of Warm Mode land plants as important indicators of climate must necessarily take account of major changes in floral character occurring at this time. Most notable among these is the disappearance of the Gondwanan *Glossopteris* flora late in the Permian and its eventual replacement during the Triassic by the *Dicroidium* flora. The latter was in turn replaced by new forms in the early Jurassic. Such significant and rapid shifts in composition make floral diversity an almost meaningless parameter for interpretation of Gondwanan climates. The Euramerican palaeofloristic

region is dominated by cycadophytes, ferns and conifers over broad regions in both the Triassic and Jurassic, and corresponding climates have been classified as warm temperate to subtropical, although with some variation in seasonality (Barnard, 1973). Growth rings in late Permian and early Triassic fossil wood from Antarctica display evidence of frost damage, indicating that low temperatures characterized the high latitudes for at least part of the year (Jefferson and Taylor, 1983).

Vertebrate animals, precursors to the dinosaurs, were widespread until extinctions affected most of them in the late Triassic, possibly through the mechanism of vegetation change (Tucker and Benton, 1982). They extended as far south as Antarctica and as far north as eastern Siberia, areas which at the time lay beyond 70° latitude. However, the environmental tolerances of these largely extinct organisms are unknown and they thus offer few clues about climate. Among invertebrate fossils there is little variation in diversity with latitude in the Triassic, suggesting broad zones of similar climate over the Earth. The early and middle Jurassic forms show variations in diversity, but these cannot be unequivocally assigned to climatic causes (Hallam, 1975; Frakes, 1979). Perhaps the most notable aspect of the Jurassic is a progressive warming suggested by a gradual shift northward of palaeofloristic province boundaries in the USSR (Vakhrameev, 1964; Krassilov, 1981); this has a counterpart in the marked 10° shift of the boundary between the Tethyan and the Boreal (invertebrate) Realms in the middle Jurassic between the Bathonian and the Callovian (Hallam, 1975).

Other factors possibly related to climate

Several recent compilations of evidence for eustatic sea-level change (Vail *et al.*, 1977; Hallam, 1984b; Haq *et al.*, 1987) differ in detail but agree on general trends as follows:

1 a slow, slight and punctuated rise (Tatarian to Norian; ~255–220 Ma);
2 a slow, slight and punctuated fall (Norian to Sinemurian; 220–200 Ma);
3 a long, slow and punctuated rise (Sinemurian to Kimmeridgian; 200–140 Ma).

The first rise began at the end of the Kazanian (i.e. at the beginning of the Warm Mode) and approximately contemporaneous with the termination of known ice-rafting of the previous Cool Mode. Ten rises and falls are recognized by Haq and his colleagues over episode 1, giving a mean cycle length of about 3.5 m.y. and a total estimated rise of about 100 m. If this represents eustatic rise due solely to melting of high-latitude land ice, the volume would have been about the same as that for ice presently locked up

in Antarctica.The apparent lack of late Permian to Triassic tillites in the geological record makes such an explanation doubtful. However, given the abundance of ice-rafting in the Kazanian, it seems likely that a somewhat smaller volume of land ice would have melted after the Kazanian and that this contributed to some unknown extent to the sea-level rise in the Tatarian–Norian interval.

The continued absence of early Mesozoic glacial and ice-rafting deposits constitutes a similar strong argument against glacial eustasy as the sole explanation for sea-level changes later in the Warm Mode. The sea-level record, however, does conform roughly to the suggested evolution of climate during the early part of the Warm Mode – warming through much of the Triassic followed by Jurassic cooling. Post-Sinemurian rises in sea level apparently took place despite continued slight cooling; these rises lasted until well after ice-rafting resumed in the Bajocian. The influence of sea-floor tectonics on eustasy is likely in this case.

Any relationship between sea level and indicators of global humidity is also uncertain. The large late Permian evaporite accumulations occurred late in a Phanerozoic low stand but the similarly abundant late Triassic ones were deposited in a period of high stand and extensive coastal onlap. Coal-bearing strata increased steadily in abundance throughout the Triassic and early and middle Jurassic, again suggesting a poor correlation with sea-level changes.

It is possible that the slow rise in sea level suggested for episode 3 (Sinemurian and later) was a result of increased sea-floor spreading and ridge volume as Pangaea fragmented, inasmuch as the oldest known sea floor related to this event is of mid-Jurassic age. This activity and related collisional tectonics on a global scale may have contributed not only to the level of the sea but also to climatic trends which emerged through the mechanisms of CO_2 emissions during volcanism and changed palaeogeography and topography, particularly their effects in high latitudes. A time lag of some 15 m.y. before the earliest ice-rafted deposits (central Siberia; Bajocian) suggests that any such changes proceeded slowly.

Black marine shales are common in the late Permian to middle Jurassic interval but thus far, few distinct times of concentrated black shale deposition have been recognized. A widespread black shale event occurs in the Tethys region at the Pliensbachian–Toarcian boundary and a less intense one in the late Pliensbachian. Significantly, both these units display positive $\delta^{13}C$ excursions (Jenkyns and Clayton, 1986), in parallel with oceanic anoxic events (OAE) of the Cretaceous. Other black organic-rich shales of note include immediately post-glacial deposits in Gondwanan intracratonic basins, including the Ecca Formation (late Permian of the Karoo Basin, South Africa). All shales discussed above were laid down during times of rising sea level.

The early and middle Jurassic, up until the Bajocian, was the time of greatest oolitic ironstone accumulation in Earth history (Van Houten, 1985). Triassic ironstones are virtually absent and there is a gradual decline in their abundance in the late Jurassic to Cretaceous. Their distribution in the early–mid Jurassic was to about 65° latitude in both hemispheres but the western Tethys platforms saw their greatest concentration. Like their Ordovician counterparts, Jurassic ironstones formed in conditions of rising sea level and slight global warming.

The early sea-level rise of the Warm Mode (episode 1) was accompanied by comparatively low levels of volcanism on the continents, while the fall in episode 2 (Norian to Sinemurian) corresponded to a strong late Triassic peak in volcanic activity as measured by Ronov (1980). Episode 3 again showed a coincidence of rising sea level and rare volcanism. These relationships suggest that early rifting in the Atlantic region may have contributed in some way to a eustatic fall which interrupted an otherwise nearly continuous rise in the level of the sea during the Warm Mode. The overall rise was crudely paralleled by a progressive increase in the abundance of chert, which continued on into the Cretaceous (Hein and Parrish, 1987).

The late Permian was the time of greatest recorded accumulation of sedimentary phosphate and is exemplified by the Phosphoria Formation of the western United States (Cook and McElhinny, 1979). This post-glacial episode was initiated as the sea began to rise and ice-rafting ceased to operate. The middle part of the Warm Mode featured relatively little phosphorite, but accumulation increased strongly to a major peak late in the Jurassic.

Within the Triassic period there was a reversal in the composition of invertebrate shell material from dominantly calcite to dominantly aragonite. The late Permian to early Triassic also saw the lowest $\delta^{34}S$ of Phanerozoic time (Claypool *et al.*, 1980). Railsback and Anderson (1987) draw attention to the possibility that increases in three variables led to an expansion of the aragonite-secreting organisms at this time. These variables were (1) dissolved sulphate; (2) surface temperature; and (3) the Mg/Ca ratio. Indeed, the sea-water sulphate reservoir was at its Phanerozoic peak between the late Permian and the middle Triassic, according to the model of Francois and Gerard (1986b). The relative contribution of seawater temperature to this calcite–aragonite changeover remains uncertain.

The $\delta^{13}C$ record for the Warm Mode shows a marked fall from the characteristically high values of the late Palaeozoic to a marked low ($\sim 1\%_0$) in the first half of the Triassic (Holser, 1984). Thereafter $\delta^{13}C$ increased irregularly until some time in the middle Jurassic, when values fell again. These trends fairly closely parallel, and probably can be explained by, the volume trends of coal-bearing strata (Fig. 6.1).

Chronology of the Warm Mode

The climate history of the late Permian to middle Jurassic Warm Mode is poorly defined owing to a scarcity of reliable temperature indicators. At the beginning there may have been an extraordinary heating of the Earth which terminated seasonal freezing in high-latitude rivers, lakes and seas in the Tatarian. The evidence from oxygen isotopes of chert–phosphate pairs and fish debris (Karhu and Epstein, 1986) suggests that a strong cooling trend followed. The limited isotope data from shelled organisms suggest that at the beginning of the succeeding Cool Mode (middle and late Jurassic to early Cretaceous), the mean global temperature was slightly warmer than today.

Perhaps the best available indication of temperatures comes from the progressive poleward shift of palaeofloristic boundaries from plant fossils in Asia and North America. However, this information may reflect precipitation rates as well as temperature, and moreover the data bear only on the Jurassic segment of the Warm Mode.

As regards precipitation history, evaporite and coal abundances and distributions indicate that the 'Pangea effect' created more extensive areas of low rainfall and less extensive monsoonal conditions in the late Triassic than in the late Permian. The early and middle Jurassic featured progressively more widespread humid climates, a situation which reversed itself only in the late Jurassic.

7

The Cool Mode: middle Jurassic to early Cretaceous

The Jurassic and Cretaceous have classically been considered very warm intervals, with low temperature gradients, temperate conditions at high latitudes and extensive evaporite deposition. However, there is a great deal of data that indicate that this is an over-simplification and, in fact, parts of the Earth were quite cool during this period. Reports of ice-rafted deposits in high-latitude regions at intervals during the latter part of the Jurassic and into the early Cretaceous (Bajocian to Albian) suggest that freezing conditions occurred near the poles and that temperate glaciers may possibly have existed (Kemper, 1987; Frakes and Francis, 1988, 1990). The equator-to-pole temperature gradient therefore appears to have been greater than previously considered. Marked seasonality also seems to be a prominent feature of climate during this interval. The designation of this late middle Jurassic to early Cretaceous interval as a Cool Mode (183–105 Ma, Bajocian to mid-Albian) is therefore based on the presence of at least seasonal ice in high latitudes.

In contrast, there are no reports of ice-rafted deposits for the latter half of the Cretaceous. The climate appears to have changed during the mid-Cretaceous to much globally warmer conditions. We have therefore split the Cretaceous Period into two different climatic modes.

Palaeoclimate information for the latter half of the Jurassic (from the Bajocian onwards) is rather sparser than for later times. The rarity of Jurassic ocean floor sediments and the consequent scarcity of ocean oxygen isotope data prohibit detailed temperature analysis for short-term intervals, as is possible for the late Cretaceous and Tertiary. Most Jurassic climate information is for continental areas. Land plant fossils are a valuable source of information about the Jurassic climate but their climate interpretations are rather broad since physiognomic analysis of leaf margins, which has yielded much detailed information for the late Cretaceous and Tertiary, is not possible for non-angiosperm Jurassic plants. More information, such as isotope data, is available for the early Cretaceous.

Fig. 7.1. *Palaeoclimatic belts for the late Jurassic. From Hallam (1985).
With permission of the Geological Society.*

Geological setting

During the late Jurassic all continents were assembled in the single land
mass of Pangaea, surrounded by a huge world ocean, Panthalassa, with the
Tethys Sea extending as a wedge into the eastern margin. The palaeogeogra-
phy was similar to that of early Jurassic and Triassic times, and the climate
was therefore probably not drastically different. This large continental area
and great expanse of oceans surrounding it led to extreme climate con-
ditions on land. Three main climate belts have been determined by Hallam
(1985) for the late Jurassic (Fig. 7.1): an equatorial arid belt extending across
the equator and up to latitudes of about 45–50°, a zone of seasonally wet
climates bordering this up to about 60°, followed by wet, temperate climates
at high latitudes.

From qualitative climate models based on analogies with present climatic
parameters, Parrish and Curtis (1982) proposed that, during the Triassic,
zones of coastal upwelling occurred on the western coasts of Pangaea and

Fig. 7.2. *Global palaeogeography and ocean circulation for the middle Jurassic (about 178–173 Ma) adapted from Haq (1984). Small arrows represent surface ocean currents and large arrows indicate possible source areas for warm saline bottom waters. Barron and Peterson (1989) proposed that circulation within Tethys was more likely to have been in a clockwise direction resulting in an eastward-flowing current in the northern part. The occurrence of coals (C) and evaporites (E) is plotted, plus areas of relatively high rainfall (hatched shading) and relatively low rainfall (dotted shading), based on Parrish and Curtis (1982). From Haq (1984). Copyright © 1984 by Van Nostrand Reinhold. All rights reserved.*

monsoon-related upwelling occurred on the margins of the Tethys Sea. Seasonal monsoonal conditions prevailed over Gondwana and the northern continents (though slightly weaker in the north). Interiors of the continental masses would have remained dry due to their positions within the subtropical dry belts (Fig. 7.2). The coupled atmosphere/mixed layer ocean models of Kutzbach, Guetter and Washington (1990) predicted that the continental climates were very seasonal, with strong summer and winter monsoonal circulations, hot summers, and winters cold enough for the formation of sea ice in winter.

Ager (1981) proposed that climatic phases and sea-level changes during the late Jurassic to early Cretaceous occurred in approximately 30 m.y. intervals, with warm intervals (Bajocian, Tithonian, Barremian/Aptian) following major transgressions. After a warm phase during the middle Jurassic (Bajocian, or slightly earlier according to the time scale of Haq *et al.*, 1987, Fig. 3), when extensive carbonate deposition occurred, there was a major transgression world-wide during the Callovian, related to an increase in the rate of sea-floor spreading. The peak of the rise was reached during

the Oxfordian. Carbonate deposition was extensive and reefs spread over North Africa, Europe and North America (Hallam,1975; Ager, 1981).

During the latest Jurassic the seas regressed and most of the sedimentary record is therefore non-marine, continuing on into the early Cretaceous. The latest Jurassic appears to have been arid, at least in low and mid-latitudes, as indicated by the extensive development of evaporites. However, the early Cretaceous was much more humid and evaporitic deposits diminished as coal deposits became more prominent.

The global climate was greatly affected by the change due to the break-up of the large Pangaean land mass and the development of seaways between continents (Fig. 7.2). Initially Africa and South America were joined to North America and Eurasia to block an equatorial current, but during the early Callovian the North Atlantic started opening at the southern end. After an initial stage of evaporite deposition in the shallow seas, barrier reefs and carbonate platforms developed along the margins. The central Atlantic continued to open throughout the Jurassic, with the South Atlantic opening much later. In the late Jurassic to early Cretaceous continental evaporites filled the initial rifts of the South Atlantic. After the Valanginian transgression two large basins formed but the high Falkland Plateau restricted deep-water flow into the South Atlantic until Aptian–Albian time.

The opening of the central Atlantic saw the development of an equatorial current. The current spanned the early ocean and entered the Pacific until the early Cretaceous, when movement of the Nicaraguan block restricted access to the Pacific and the Caribbean, and the Gulf Stream circulation was initiated (Gradstein and Sheridan, 1983). A westward-flowing circum-global current has been postulated from the evidence of foraminiferal migration patterns and homogeneous Tethyan faunas (Berggren and Hollister, 1974). However, numerical ocean modelling experiments for Cretaceous palaeo-geography suggest that although westward currents flowed through the tropical Pacific, circulation within the Tethys Sea may have been in a predominantly clockwise direction, producing an easterly flow along the northern margins (Barron and Peterson, 1989).

Ocean temperatures

Some temperature estimates from ocean oxygen isotopes are available for the Jurassic and Cretaceous. For the Jurassic the information is mainly from belemnite guards (possibly subject to some diagenetic alteration, as summarized in Stevens, 1971). From these data it appears that late Jurassic palaeotemperatures ranged from $\sim 10°C$ to $27°C$. Although temperatures varied greatly with location, a general trend shows warmer temperatures in the Oxfordian–Kimmeridgian followed by a cooling phase through the late Jurassic and early Cretaceous until the late Aptian warming.

Ocean temperatures have also been determined from foraminiferal oxygen isotopes of the north-west Pacific (Douglas and Woodruff, 1981). During the middle Jurassic low-latitude sea surface temperatures (SSTs) were high ($\sim 26°C$) but continued to increase through the late Jurassic to reach temperatures approximately the same as at present ($\sim 28°C$). Bottom waters were then about 17°C, giving a vertical temperature gradient of about 11°C. The isotope curve (Fig. 7.3) indicates a cooling phase with a 2–3°C drop in SST from the Berriasian to Barremian. In their isotope curve Douglas and Woodruff (1981) imply that the surface-to-bottom water temperature gradient remained constant throughout the Jurassic and Cretaceous, although the actual isotope measurements show a marked cooling of surface waters during the Valanginian to Barremian cooler interval.

Since temperature estimates are only available for this time from the Pacific, it is difficult to estimate the equator-to-pole ocean temperature gradient. There is no evidence of cold bottom waters resulting from polar ice caps. In contrast, it has been proposed that warm, highly saline water, which formed as a result of intense evaporation in low- and mid-latitude coastal settings, may have sunk to the bottom of the ocean. These warm saline bottom waters would have had a low oxygen content and would have contributed to deep ocean anoxia (Brass *et al.*, 1982).

The sea-level curves of Haq *et al.* (1987) and Hallam (1984b) indicate a general rise in sea level during the middle to late Jurassic, a slight drop during the latest Jurassic to early Cretaceous, rising to high sea levels in the mid-Cretaceous. Unfortunately the timing of the changes depends very much on the time scale used; the scale used by Haq *et al.* (1987) is very different from that used by DNAG (Palmer, 1983) with reference to stage boundaries (see Fig. 7.3). This presents many problems when trying to correlate climatic information.

Ocean faunal realms

Information about ocean climates has been obtained from changing distributions of invertebrate faunas. Two major faunal provinces existed in the northern hemisphere during the late Mesozoic: the northern Boreal Realm and the low-latitude Tethyan Realm. These were considered to correspond to cool-temperate and tropical/subtropical climates respectively (Stevens, 1971). A southern Austral Realm, believed to be absent at least during the Jurassic (Hallam, 1984b), has been recognized from the early Cretaceous onwards. The belemnites (from the Aptian onwards), foraminifera and bivalves (early Cretaceous onwards) define specific provinces, and the ammonites may also have been restricted to southern latitudes by the late Cretaceous (Doyle, 1988). This Austral zone corresponds to the northern cool-temperate realm (Stevens, 1971).

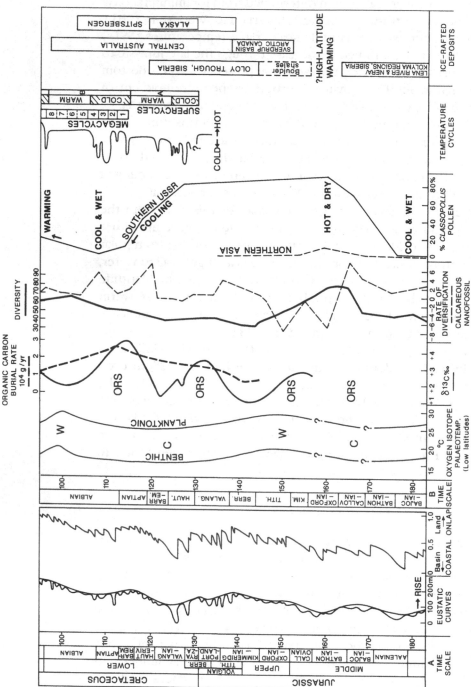

Fig. 7.3. The mid-Jurassic/early Cretaceous Cool Mode. Time scale A and eustatic and coastal onlap curves from Haq et al. (1987). Time scale B from DNAG (Palmer, 1983) – other curves are plotted against this. Oxygen isotope temperature curves for the North Pacific from Douglas and Woodruff (1981). δ13C curve from Weissert (1989). Organic carbon burial rate from Arthur et al. (1985). ORS represents organic-rich shales. Calcareous nanofossil diversity data from Roth (1987). Classopollis pollen curve from Vakhrameev (1981) (plotted according to stage). Temperature cycles for the Cretaceous from Kemper (1987). Ice-rafted deposits as summarized by Frakes and Francis (1988). For other sources see text. Curves plotted against DNAG time scale according to age rather than stages, unless otherwise indicated. C= cooler, W= warmer (authors' interpretation).

Kimmeridgian to Tithonian faunas (in particular ammonites, bivalves and belemnites) appeared to have been widespread across the mid to high latitudes in the southern hemisphere, forming the Indo-Pacific faunal province. At this time warm low-latitude waters would have been able to flow around the shelf margins of the relatively low-latitude Gondwanan continent, encouraging a homogeneous fauna (Stevens, 1981; Crame and Howlett, 1988). However, by the end of the early Cretaceous the Gondwanan continents began to disperse, creating new seaways. Centres of faunal endemism developed and faunas became much more restricted. A distinct high-latitude faunal province can be traced from southern South America through western Antarctica to eastern Australia and New Zealand (Crame and Howlett, 1988). Some families show distinct restrictions to southern high latitude regions, e.g. the belemnite family Dimitobelidae (Doyle, 1987), and links between Australia, New Zealand, Antarctica and Patagonia are shown by the distributions of specific groups of ammonites and bivalves.

The latitudinal restriction of these faunas suggests that climate, and in particular temperature, may have been an important control. Belemnite isotope temperatures indicate a poleward decrease in ocean temperatures (Stevens and Clayton, 1971) but other controls such as salinity, ocean currents and barriers may have been influential (Hallam, 1975).

Roth (1987) related the changes in the diversity of calcareous nanofossils to tectonic events and climate. During high stands of sea level, climate warming may have resulted in stratified oceans with poor mixing and the resultant low surface-water fertility would have caused high diversity. In contrast, at times of low sea level, climate contrasts increased, creating more vigorous ocean mixing and better ventilated waters with enhanced productivity. Diversity of nanofossils then became low. The largest evolutionary changes occurred during times of extreme environmental change. Roth related these changes ultimately to tectonics and changes in sea-floor spreading that then influenced climate and ocean productivity (see Chapter 11).

During the middle Jurassic the diversification rate of nanofossils was high, but decreased during the regressive phase of the late Jurassic, when the rate of extinctions was much higher than the rates of speciation. In the early Cretaceous diversification rates were high during the early Valanginian, early Aptian and middle Albian but declined in the Hauterivian and Barremian (Roth, 1987). These peaks in evolution rates correlate with anoxic events, times of high rates of organic carbon accumulation. In contrast, during regressions, such as in the early Tithonian, extinction rates were high as were speciation rates, as the Jurassic nanoflora was replaced by Cretaceous types. In general, the plot of nanofossil diversity through this interval (Fig. 7.3) indicates high diversity (which could be interpreted as resulting from high sea levels and warmth) during the Callovian and Oxfordian but

lower diversity (perhaps low sea levels and cooler climates) in the Bajocian and Bathonian, and in the Tithonian. During the early Cretaceous diversity was low but increased slightly during the Valanginian, dropped slightly during the Hauterivian and Barremian and then rose through the Aptian–Albian as climate warmed.

On a much smaller scale, Thomsen (1989) reported that Cretaceous calcareous nanoplankton production was seasonal, comparable to modern seasonal fluctuations in the Atlantic Ocean. Lower Cretaceous (middle Barremian) sediments in the central North Sea (the Munk Marl Bed) consist of alternating laminae of marl and calcareous nanofossils. The laminae were interpreted as due to seasonal variation in the production of nanoplankton. The laminae are preserved because they were deposited in oxygen-poor bottom waters devoid of bioturbating organisms.

A change in the nature of bryozoan faunas has been interpreted in terms of climate change by Walter (1989). In the Jura region of Europe three bryozoan faunas have been identified in Lower Cretaceous sediments. A marked change in the species composition of the faunas occurred during the late Valanginian, followed by another change back to the original conditions during the Hauterivian. These changes were interpreted by Walter to be due to a phase of cool climate, accompanied by a sea-level fall, during which the warmth-loving species were eliminated. The restoration of the faunas during the late Valanginian to Hauterivian records the return to warmer climates and higher sea levels.

The carbon record

Climate has a strong influence on productivity of ocean biotas, ocean temperature, anoxia and carbon burial, and thus on the isotope record. Several intervals of positive excursions of $\delta^{13}C$ occurred within this late Jurassic to Early Cretaceous Cool Mode. These include the Kimmeridgian (Renard, 1986), Valanginian (Cotillon and Rio, 1984) and the Aptian–Albian (Scholle and Arthur, 1980; Weissert, 1989) (Fig. 7.3). In contrast, the late Tithonian experienced a rapid decrease in sedimentation of organic carbon and also, to a certain extent, the Barremian. Organic-rich shales were deposited in the mid-upper Callovian (Oxford Clay, UK), the Kimmeridgian–Tithonian (Kimmeridge Clay in the UK, North Sea and western Siberia) and during the late Jurassic in the southern Andes, Falkland Plateau, western Antarctica and eastern Greenland (Hallam, 1987b). Organic-rich muds also occur in the Blake-Bahama Formation of Valanginian–Barremian age in the North Atlantic (Summerhayes, 1987).

There was a significant decrease in $\delta^{13}C$ near the Jurassic–Cretaceous boundary (Scholle and Arthur, 1980). The carbon isotope signal in pelagic carbonates from the Southern Alps in Italy shows a decrease from values of

$\delta^{13}C$ of about 2.07‰ for the Kimmeridgian to early Tithonian interval to about +1.26‰ for the late Tithonian to Berriasian (Weissert and Channell, 1989). Across this boundary there was also a change from silica-rich radiolarian sediments, platform carbonates and ribbon cherts to calcareous nanofossil limestones (Roth, 1986), lower sea levels and a drop in the Carbonate Compensation Depth (CCD). Accumulation rates of organic matter in the western North Atlantic also dropped from the late Jurassic to the late Tithonian to Berriasian. This change in the carbon budget was considered by Weissert and Channell to be the result of reduced surface water productivity, the cause of which was a decrease in the supply of the nutrients required by the planktonic biota for the carbon fractionating processes. Since the nutrient supply to the oceans, particularly phosphorus, is principally from continental run-off, this decreased supply was attributed to a change to much drier climate conditions around the borders of the Tethys and Atlantic oceans.

Weissert and Channell (1989) attributed the cause of the drier climate during the late Tithonian to Berriasian to lower levels of atmospheric CO_2, resulting from a decrease in sea-floor spreading rates (Sheridan, 1986), and increases in surface albedo. However, this implies that the climate was also somewhat cooler during this time, conflicting with other evidence for warmth. The oxygen isotope data of Weissert and Channell do not show a significant trend, although the isotopes became slightly heavier from the Kimmeridgian to mid-Tithonian, and then significantly lighter by the late Tithonian (which perhaps might be interpreted as a cooling followed by a late Tithonian warming).

The Aptian–Albian was the time of deposition of large quantities of black shales and organic-rich sediments accumulated in many oceans and in basinal settings, submarine highs and shallow shelf environments (Schlanger and Jenkyns, 1976; Weissert, 1989; Arthur *et al.*, 1990). This has also been defined as an OAE (ocean anoxic event) by Schlanger and Jenkyns (1976), a time of oxygen-poor bottom waters in which organic matter was preserved in large quantities. Although the OAE is defined as a broad interval of late Barremian to early Albian in age, discrete episodes of organic carbon deposition may have occurred during the early Aptian, early Albian and late Albian (Schlanger and Jenkyns, 1976; Scholle and Arthur, 1980), or alternatively, organic carbon deposition may have been more random (Summerhayes, 1981; Parrish and Curtis, 1982; Waples, 1983). A shift of $\delta^{13}C$ values of up to 4‰ have been recorded from several sites, possibly indicating an increase in productivity of surface biotas and increased burial of the isotopically light organic carbon, ^{12}C (Parrish and Curtis, 1982; Roth, 1986: Weissert, 1989).

Higher rainfall and the resulting increased nutrient supply to the oceans may have been an important factor in enhancing productivity; Weissert

estimated that the phosphorus flux, an important nutrient for organic productivity, increased by up to 10% and carbon burial increased by 20% during this time. Others have proposed that organic matter was preserved because oxygen levels in bottom sediments were unusually low due to slow mixing in warm ocean waters (deep water anoxia) (Bralower and Thierstein, 1984; see others referred to in Arthur *et al.*, 1990). The organic matter in the shales is dominantly of terrestrial origin (Tissot , Deroo and Herbin, 1979; Hallam, 1987b) suggesting that land floras were particularly productive at this time and the climate were favourable for plant growth.

Evidence of ice

Although the Mesozoic has been considered to be globally warm and icefree there are many reports of deposits formed by ice transport at high latitudes during the late middle Jurassic and early Cretaceous (Kemper, 1987; Frakes and Francis, 1988) (Fig. 7.4). The deposits consist of marine fine-grained shales or mudstones that contain outsized clasts composed of exotic rock types. Lack of evidence for the lateral transport of clasts by water currents suggests that the clasts were emplaced into the shales by dropping vertically from rafts. Due to similarities between these clasts and dropstones in modern glacial-marine sediments, the mechanism of icerafting seems the most reasonable explanation for the origin of the Mesozoic examples. The environment envisaged is not one of permanent glaciers, owing to the lack of recognized glacial tillites (Frakes and Francis, 1990), but of a periglacial setting with seasonal winter ice. Cold winters allowed the formation of ice on rivers and shorelines, the ice incorporating clasts from the banks and bases of the rivers and from beaches as it froze. The following summers were warm enough to cause the ice to break up, move out into the basin and drop the transported clasts as it melted.

Epshteyn (1978) has documented several reports of ice-rafted deposits in northern Asia. They range in age from Bajocian to Callovian and Berriasian to Barremian and occur along the arctic borderlands of Siberia, especially in the Lena River, Oloy Trough and Nera-Kolyma river regions. There are also other reports of pebbly mudstones in this region, although the authors have not discussed their origins. During the late Jurassic to early Cretaceous Siberia was situated at very high latitudes and, being part of a large continental mass, would have experienced extremes of seasonal temperatures. Epshteyn suggested that the climate became periodically cooler with lower winter temperatures and an increased duration of the cold season, allowing the formation of winter ice and boulder muds. These cycles may have been influenced by orbital parameters (Kemper, 1987). On similar evidence Brandt (1986) proposed that such cycles were glacially driven and caused by the waxing and waning of polar continental ice.

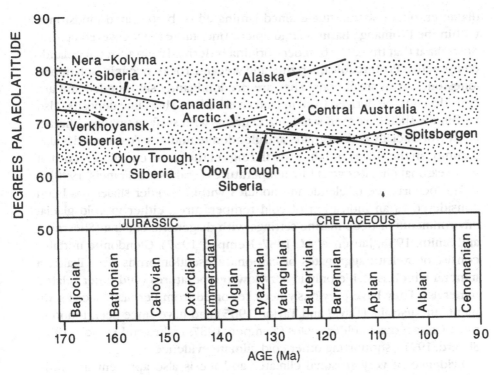

Fig. 7.4. *Palaeolatitude versus age distribution of late Jurassic and early Cretaceous boulder shales or ice-rafted deposits. The location of the ice-rafted deposits, and thus of seasonally cold climates, was between approximately 60–80° palaeolatitude. From Frakes and Francis (1988). Copyright © 1988 by Macmillan Magazines Ltd.*

In the northern hemisphere outsized clasts also occur within Upper Jurassic to Lower Cretaceous shales of the Sverdrup Basin in Arctic Canada (Embry, 1984; Kemper, 1987), which Embry has attributed to ice-rafting. In Alaska the 'Pebble Shale' of Hauterivian–Barremian age also contains erratic pebbles in a shale matrix but its origin has not yet been investigated (Bird, 1985). On Spitsbergen Dalland (1977) has reported outsized clasts in shales of early Cretaceous age. Similar deposits of Palaeocene age were considered by him to have been ice-rafted. Although Birkenmajer and Narebski (1963) proposed that the Tertiary clasts could have been carried by rafts of wood or kelp instead of ice because fossil wood is usually present in the shales, the restriction of all these reported boulder shales to high-latitude sites argues against this. If tree trunks and branches were responsible for the transport of clasts then it would be expected that boulder shales should be found at all latitudes.

At southern high latitudes early Cretaceous ice-rafted deposits have been reported from central Australia, situated at above 65°S during the early Cretaceous (Frakes and Francis, 1988). Large boulders, up to 3 metres in

diameter, occur within fine-grained laminated or bioturbated mudstones within the Eromanga Basin, a large epicontinental sea that covered central Australia at that time. The boulders, originally derived from adjacent uplands composed of Proterozoic quartzites and volcanic rocks, occur in isolation as lonestones or in small pods. They are predominantly subrounded and are considered to be derived from upland riverbeds or shoreline deposits. Also within the shales are sand-sized quartz grains with surface textures characteristic of glacially etched grains, and glendonite nodules. Fossil wood is also common in the shales and the growth rings in the wood contain a record of very seasonal climates with low temperatures (Frakes and Francis, 1990).

The occurrence of glendonite nodules within boulder shales has been considered as an indication of cold temperatures, either in cold glacial environments or deep marine settings with diagenetic cooling (Shearman and Smith, 1985; Jansen *et al.*, 1987; Kemper, 1987). Glendonite nodules consist of stellate aggregates of carbonate pseudomorphs after ikaite, a mineral which is stable only in very low temperatures (and other related minerals). They have previously been reported from Permian glacial sediments in Australia (David and Taylor, 1905) and have since been found in most Cretaceous boulder shales (Kemper, 1987; Frakes and Francis, 1988; Sheard, 1991), supporting other cold climate evidence.

Evidence of very seasonal climates and ice is also apparent in southeastern Australia in Victoria. During the early Cretaceous (Valanginian–Albian) this area, including the Otway and Gippsland Basins, formed part of a rift complex resulting from the separation of Antarctica and Australia. A sequence of continental fluvial sediments was deposited in the rift basins, burying plant fossils and dinosaur bones. Palaeotemperature estimates from oxygen isotope ratios of calcite in early-diagenetic calcite concretions (Gregory *et al.*, 1989) indicate that meteoric waters were depleted in ^{18}O, with mean $\delta^{18}O$ values approaching $-20‰$, suggesting that ice was present. Mean annual temperatures of less than 5°C and perhaps even as low as -6°C have been proposed by Rich *et al.* (1988; 1989), though on the whole the climate is considered to have been mainly seasonally cold.

Permanent glaciers may have existed at high altitudes and latitudes, although evidence for glaciers may have subsequently been eroded or not yet found (Frakes and Francis, 1990). Robin (1987) proposed that continental ice sheets could have been present at high elevations in East Antarctica at about 100 million years ago.

Even though seasonally cold temperatures were present at high latitudes during the mid-Jurassic to early Cretaceous, temperatures from oxygen isotopes of low-latitude oceans (Douglas and Woodruff, 1981; Fig. 7.3), indicate that the latest Jurassic oceans were very warm. This suggests that either polar cold temperatures were very localized and did not affect lower latitudes to any great extent, or that the cold phases occurred in fairly short-

term cycles (as suggested by Epshteyn, 1978, and Kemper, 1987) which cannot be resolved from the available isotope curves.

Kemper (1987) proposed that Cretaceous sedimentation was very cyclic and that cold periods lasting from a few thousand to 2 million years in length were separated by periods of warm, equable climate (Fig. 7.3). He suggested that the warm periods were much warmer than today to account for floras in polar regions but the cold spells were comparable to those of the Quaternary. It is proposed that these were related to orbital cycles. Rhythmic sequences in Cretaceous pelagic deposits have also been linked to orbital variations (Weissert, McKenzie and Hochuli, 1979; de Boer, 1982; Barron, Arthur and Kauffman, 1985; see review in Fischer, de Boer and Premoli Silva, 1990).

Climate significance of fossil floras

The distribution and thermal tolerances of land plants provides one of the main sources of climatic evidence for this time. The presence of fossil floras at high latitudes has led to the assumption that polar climates were uniformally warm (Jefferson, 1982; Douglas and Williams, 1982). The assemblages seem to consist of mainly ferns, cycadophytes and conifers, a fairly ubiquitous collection that occurred over a broad expanse of the globe and through a long time span, at least through the late Jurassic and early Cretaceous. The assemblage is characteristic of high-latitude areas and is found in Gondwanan continents of India, South America, Australia and Antarctica, and in northern regions too (e.g. Europe; Harris, 1975). The southern floras, however, appear to consist mainly of types ancestral to the modern southern vegetation, such as the podocarp and araucarian conifers. In contrast, the extinct conifer family the Cheirolepidiaceae dominated northern vegetation, although each hemisphere does contain elements of the other. This lack of variation has resulted in rather poor stratigraphic constraints and broad climatic interpretations for the floras, compared with the later Cretaceous.

In mid to low latitudes, particularly in the northern hemisphere, the vegetation was dominated by the Cheirolepidiacean conifers, which shed *Classopollis* pollen. These conifers were small bushes (Batten, 1974) or large forest trees (Francis, 1983) which grew on drained slopes, often near areas of evaporite deposition. Pollen of this type is most commonly found in association with evaporitic sediments and thus the presence of this conifer family has been used as an indicator of arid climates (Vakhrameev, 1981).

The percentage of *Classopollis* pollen found in sediments has been correlated with the intensity of warmth and aridity during the late Jurassic to early Cretaceous (Vakhrameev, 1981). In general, the abundance of *Classopollis* decreases away from the equator, probably due to wetter and

possibly cooler climates at high latitudes. The high-latitude regions remained areas of coal deposition rather than of evaporites, particularly in northern Europe and the USSR, and these coals have extremely low contents of *Classopollis* pollen.

The maximum abundance of *Classopollis* is found in Upper Jurassic sediments, which has thus been considered to represent the time of maximum aridity. A plot of the percentage of *Classopollis* in palynological assemblages through the late Jurassic (Fig. 7.3) has been used as a signal of climate change. A warm period in the Toarcian is indicated by an increase in *Classopollis*, followed by an abrupt decrease in amount to only 1 or 2% of pollen assemblages during the Aalenian to Bajocian. The latitudinal expansion of coals at the same time suggests that the climate became significantly cooler and wetter.

During the Bathonian and Callovian *Classopollis* increased in abundance again and reached a peak in the Oxfordian, when it seemed to constitute over 90% of many pollen assemblages (Vakhrameev, 1981). Evaporites also increased in quantity at this time, there were fewer clastics, and corals spread further north in the northern hemisphere (Frakes, 1979). This coincides with a large trangressive pulse, and it may be that the increase in area of shallow shelf seas provided greater areas for the formation of evaporites and the entrapment of pollen, rather than reflecting a great increase in aridity.

The abundance of *Classopollis* decreased slightly but was still high during the latter part of the Jurassic and Cheirolepidieacean conifers were still a dominant part of the flora in some areas (Francis, 1983). But by the early Cretaceous, when evaporite accumulation had decreased, these conifers formed only a minor portion of the vegetation (Batten, 1982). These floras, with a greater percentage of moisture-loving ferns, spread into lower latitudes as the climate became wetter.

The distribution and movement of floral zones has been interpreted as due to climate change during this time. Vakhrameev (1964) divided northern hemisphere floras into a warm-loving Indo-European and a high-latitude Siberian-Canadian province. These zones separated a mid-latitude arid region from subtropical humid climates to the south and temperate humid areas, possibly with winter temperatures below zero, to the north. The floral zones moved northwards across Eurasia during the late Jurassic and early Cretaceous, possibly reflecting a general warming trend. A late Jurassic warming was also proposed by Krassilov (1981) on similar evidence. However, a temperature decrease in eastern Asia from the Berriasian to the Albian was suggested by Krassilov (1973), based on the decline of cycadophytes. Many climatic interpretations of fossil floras have been based on the assumption that the cycadophytes were warmth-loving plants like their living relatives, the cycads, yet their morphology has been shown to be

different (Kimura and Sekido, 1975). Since they were common in Mesozoic high-latitude floras they were probably well adapted to living in high-latitude climates (Spicer and Parrish, 1986).

Many of the Jurassic and early Cretaceous fern, conifer and cycad floras occur both at low and high latitudes in both hemispheres. They have been interpreted as representing warm temperate climates in the past because of their composition (including cycadophytes) and latitudinal spread (Smiley, 1967; Douglas and Williams, 1982; Jefferson, 1982). However, reinterpretations of floras in Australia suggest that the climate was cool temperate (Dettman, 1986a; Drinnan and Chambers, 1986) and extremely seasonal to account for the occurrence of winter ice and forest floras (i.e. warm summers; Frakes and Francis, 1988; Rich *et al.*, 1988). These southern hemisphere floras, and possibly northern ones, may therefore represent warm summer climates, rather than year-round warmth.

In the past, floral boundaries, such as those delineated by Vakhrameev, have been drawn parallel to lines of latitude, on the assumption that climate changes across continents were influenced mainly by latitudinal tempera- ture gradients. However, the climate models of Sloan and Barron (1990) have shown that this kind of zonation was unlikely because of the large sizes of the continents and that continental interiors would have experienced more seasonal extremes that marginal areas.

The seasonality of late Jurassic to early Cretaceous climates has been recorded in the growth rings in fossil wood (Creber and Chaloner, 1985). Tree rings in Cretaceous conifer wood from Alexander Island, Antarctica (Jefferson, 1982) and the northern tip of the Antarctic Peninsula (Francis, 1986b) show definite annual boundaries, indicating that growth ceased in winter. This may have been due to a combination of low temperatures and low light levels at such high latitudes. In southern England, at palaeolatitudes of about 35°N, fossil forests grew in association with extensive evaporitic deposits during the latest Jurassic. Growth rings in the fossil wood and associated sedimentary evidence illustrate that the climate was of Mediterranean type, with warm, wet winters and hot, arid summers (Francis, 1984; 1986a). This allowed the formation of evaporites during the arid summer season and the growth of forest vegetation during the wetter and cooler parts of the year. During the early Cretaceous in the same region the climate remained seasonal though wetter, as reflected in the sediments (Allen, 1981) and in fossil wood (Francis, 1987).

Clay minerals and climate change

The increase in quantity of evaporite deposits during the latest Jurassic indicates that this was an interval of increased aridity, compared with the previous more humid climates. Large evaporite deposits formed in Europe

and central Soviet Asia (Hallam, 1984a). Clay mineral assemblages became devoid of kaolinite (Upper Kimmeridgian, France, Deconinck, 1987; Oxfordian to Kimmeridgian, Central Atlantic, Chamley *et al.*, 1983). A change in climate from humid to semi-arid during deposition of the Upper Kimmeridge Clay in Britain was also proposed by Wignall and Ruffell (1990), based on changes in clay mineral assemblages (decrease in kaolinite), a change from soft-ground to hard-ground faunas, and a change from dolostone to carbonate lithologies.

In the North Atlantic illite, chlorite and mixed-layer clays dominated Oxfordian–Kimmeridgian sediments but decreased gradually through the latest Jurassic and early Cretaceous as the proportion of smectite increased (Chamley, 1979). Chamley suggested that the smectite was derived from hydromorphic soils of low relief areas and that the climate of regions bordering the North Atlantic during the early Cretaceous was therefore warm, with strongly seasonal contrasts of humidity (a short intense humid season followed by a long dry season).

Changes in clay mineral suites in north-west Europe also reflect climate changes during the early Cretaceous (Sladen, 1983). From the latest Jurassic to the Valanginian, kaolinite increased in abundance and mixed layer minerals decreased, reflecting a change from hot, semi-arid climates to warm humid-temperate climates. Soils changed from alkaline and poorly leached to acid and well-leached podsols. During this interval the land areas in Europe changed from low relief to relatively high land, which may have influenced source area climate to a certain extent, and the expanding North Atlantic may have contributed to the development of more humid climates. However, from the Hauterivian to the Aptian, mixed-layer clays increased in abundance at the expense of kaolinite, reflecting lowered source area relief and perhaps climate warming. A change in clay mineralogy occurred at this time in the Tethyan region, changing from kaolinite-rich to smectite-rich sediments, also suggesting increased aridity (Hallam, 1984a; 1986b).

Seasonal climates

There are several sources of evidence that climates during the mid-Jurassic to early Cretaceous were very seasonal. Hallam (1985) suggested that seasonal climates were characteristic of mid-latitude zones, based on Robinson's (1973) model for the Triassic, which proposed that the large Pangaean land mass resulted in monsoonal effects (see also Parrish and Curtis, 1982; Parrish *et al.*, 1982), though this relates to seasonal rainfall. This seasonality is recorded in sediments and fossil wood from Europe (see above).

Climate models have consistently predicted very seasonal climates for the large continental land masses that existed during the Jurassic and early Cretaceous. Kutzbach and Gallimore (1989) and Kutzbach *et al.* (1990)

predicted hot summers and cold winters over the Pangaean land mass, the winters cold enough for the formation of sea ice in high latitudes. Global atmospheric modelling of the mid-Cretaceous by Barron and Washington (1982) predicted warm polar surface temperatures (15–19°C) and tropical surface waters similar to today. However, their model predicted seasonal extremes of temperatures over the large continental masses (USSR, North America, Gondwana), even if polar sea surface temperatures were assumed to be a minimum of 10°C. For example, the model predicted summer surface temperatures of about 17°C and winter temperatures of −23°C for Siberia, and 17°C summer and −13°C winter in central Australia. Strong monsoonal conditions were predicted for continental margins, particularly the northern margins of Tethys. Donn and Shaw (1977) had previously predicted low temperatures for the northern high-latitude areas from thermodynamic models.

Schneider, Thompson and Barron (1985) obtained similar results to Barron and Washington from modelling with mid-Cretaceous palaeogeography and ocean temperatures set at a warm 20°C. High-latitude continental interiors still suffered from winter freezing, probably because warm ocean temperatures reduced the strength of atmospheric circulation, providing insufficient ocean-to-land atmospheric heat transport.

Further climate-model experiments for the Cretaceous (and Eocene) by Sloan and Barron (1990) have shown that it would have been extremely unlikely that above-freezing temperatures could have occurred during winters in mid-latitude large continental interiors or high latitudes, regardless of warm ocean temperatues or the surface temperature gradient. Sensitivity experiments indicated that continental areas have little heat capacity or thermal inertia, so the seasonal cycle of temperatures is a response principally to the levels of insolation received. The amount of solar energy available in mid- and high latitudes during the winter would not have been sufficient to raise the surface temperatures above freezing. If the polar oceans were warm and the equator-to-pole temperature gradient low, atmospheric circulation would not have been strong enough to transport heat to continental interiors. The idea that the Cretaceous climate was thus 'equable' (reduced annual cycle of temperatures and lack of seasonal extremes) is therefore questioned.

The energy-balance climate models of Crowley *et al.* (1986) and Crowley, Mengel and Short (1987) also demonstrated that large continental land masses would have experienced large seasonal extremes of temperature in the past. They suggested that to maintain an ice-free earth (without permanent ice) it is necessary to have high summer temperatures to melt winter ice, rather than year-round warmth. They proposed two states for non-glacial climates: 'cool non-glacials' when winters were cold but permanent ice cover was prevented by warm summers, and 'warm non-glacials' with

higher average temperatures. Cycles reflecting these two states were proposed for this time interval by Epshteyn (1978) and Kemper (1987), and were suggested to have been orbitally forced. In general, the late Jurassic to early Cretaceous seems to have had a 'cool non-glacial' climate, though it remains to be seen whether evidence for glaciation might be found at high latitudes (Frakes and Francis, 1990).

Kemper (1987) proposed that the Cretaceous climate alternated between warm and cold periods, each of about 2 m.y. duration, constituting his 'megacycles' (see Fig. 7.3). These were suggested to have been caused by variation of the angle between the Earth's axis and the Earth's orbit. A higher order of cycles of warmth and cold, the 'supercycles', were superimposed on this, each warm/cold period lasting for approximately 20 m.y. From the compilation of curves in Fig. 7.3 it is apparent that climate shifted from relatively warm to relatively cool periods over intervals of approximately 10 m.y.

Chronology of the Cool Mode

The presence of seasonal ice in high latitudes of both hemispheres indicates that this period was cooler than the preceding Triassic to mid-Jurassic Warm Mode and the following late Cretaceous to early Tertiary Warm Mode, although not as cold as previous Cool Modes during which full-scale glaciation occurred. Compilation of palaeoclimate information suggests that there was an almost cyclic change in the intensity of the cool climate within this mid-Jurassic to early Cretaceous Cool Mode.

The Bajocian and Bathonian appear to have been relatively cool (seasonal ice, land and ocean floral evidence) but the climate started to warm during the Callovian. The Oxfordian, Kimmeridgian and Tithonian were the warmest times during this mode; oxygen isotopes show high temperatures, sea level was high, evaporites and carbonates were abundant and floras and clay mineral suites are indicative of warm climates. The latest Jurassic appears to have been particularly warm and arid. In the early Cretaceous climates became more humid (fewer evaporites and more coals, change in clay mineral suites, low temperatures from oxygen isotopes). The Valanginian and Hauterivian seem to have been quite cool, with extensive ice-rafting in high latitudes. Ice-rafting continued during the Aptian but there are many signs that the climate was increasing in warmth towards the late Albian peak.

8

The Warm Mode: late Cretaceous to early Tertiary

The period from the mid-Cretaceous (mid-Albian) to the mid-early Eocene (approximately 105 Ma to 55 Ma) was one of the warmest times in the late Phanerozoic. The average global temperature was probably about 6°C higher than that of today (Barron, 1983), allowing polar regions to be free of permanent ice. Temperatures were high enough to allow forests and vertebrates to live near both poles, particularly during the peak of warmth in the early Eocene. In contrast to the mid-Jurassic to early Cretaceous there is no evidence for seasonal ice (as ice-rafted deposits) in high latitudes from the Cenomanian to Maastrichtian, although there may have been seasonal ice present during the Palaeocene in northern high latitudes.

The mid-Cretaceous was a time of globally high sea levels and extensive areas of shallow shelf seas, favouring moderate climates and increased evaporation and precipitation (Arthur, Dean and Schlanger,1985).The oceans are believed to have been more stratified than at present, with little vertical mixing and the deposition of large quantities of organic-rich sediments in anoxic bottom waters. In particular, restricted circulation occurred in the developing Atlantic oceans, although by the latter part of the Cretaceous more open circulation was established and the deposition of black shales diminished. Sea levels reached their Mesozoic–Cenozoic peak in the late Cretaceous and then gradually dropped (Haq *et al.*, 1987). During the latest Cretaceous the seas withdrew, exposing larger continental areas and leading to the establishment of more seasonal, continental climates (Hays and Pittman, 1973).

However, refined palaeotemperature data and reinterpretation of the thermal tolerances of fossil plants indicates that there were distinct warmer and cooler phases within the warm trend, particularly highlighted at high palaeolatitudes. This includes the Cretaceous–Tertiary transition and its associated climatic perturbations. Although a cooling trend with a mean temperature decline of 0.2°/m.y. has been proposed for the late Cretaceous (Lasaga, Berner and Garrels, 1985) the subsequent warming in the early Tertiary indicates that this Warm Mode persisted beyond the Cretaceous. The major global cooling started only in the late early Eocene when the

Fig. 8.1. Ocean circulation and palaeogeography for the mid-Cretaceous from Roth (1986). Atmospheric circulation for the northern summer from Parrish and Curtis (1982). L= low pressure H= high pressure. Distribution of coals (C) and evaporites (E) from Parrish et al. (1982). Base map for 80 Ma from Smith and Briden (1977).

dynamics of ocean circulation changed and permanent ice appeared at the poles.

Sea-level fluctuations and circulation

During the late Cretaceous the large Pangaean land mass continued to break up, resulting in the development of new ocean gateways and new currents. By the Cenomanian the global equatorial Tethyan current was established, as was the proto Gulf Stream (Fig. 8.1). The eastern Tethyan gyre was interrupted in the Campanian as India moved northwards (Roth, 1986). Large equator-to-pole currents were circulating in the extensive Pacific

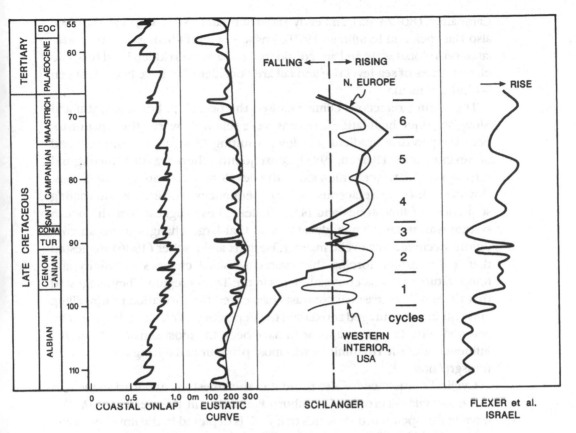

Fig. 8.2. *Sea levels during the late Cretaceous to early Tertiary Warm Mode. Time scale of Palmer (1983). Coastal onlap and eustatic curves from Haq et al. (1987). Sea level cycles for USA and Europe from Schlanger (1986). Schlanger interpreted six cycles during the Albian to Maastrichtian. Sea level cycles from Israel from Flexer et al. (1986).*

Ocean but circulation was far more restricted in the young Atlantic oceans. The South Atlantic developed restricted circulation with the Indian Ocean via the passage between Africa and Antarctica, and later included the region between Australia and Antarctica as the two continents rifted apart in the late Cretaceous (Veevers and Ettriem, 1988).

The late Cretaceous was a time of major sea-level transgressions and regressions (Schlanger, 1986; Haq *et al.*, 1987). Six major marine transgressive pulses have been recorded for Albian to Maastrichtian time, with a periodicity of about 5.5 ± 1 m.y. Summarizing cycles deciphered from Cooper (1977), Juignet (1980) and from the Western Interior Seaway of North America by Kauffman (1977), Schlanger (1986) defined six major transgressions: late Albian (100–95 Ma), Cenomanian to Turonian (95–90 Ma), Coniacian (90–87 Ma), Santonian to early Campanian (87–80 Ma), late

Campanian (80–73 Ma), and early Maastrichtian (73–67 Ma) (Fig. 8.2) (see also Hancock and Kauffman, 1979). These sea-level changes can be correlated on a global scale and are considered to have overridden local tectonic effects. Rates of sea-level rise and fall are considered to have been between 50 and 100 m/m.y.

The warm Cretaceous climate caused the oceans to become warm and 'sluggish'. Since high-latitude oceans were relatively warm, the equator-to-pole temperature gradient was low, resulting in weak atmospheric and ocean circulation (Barron, 1983). Consequently, there was little mixing and turnover of the water, which led to the development of anoxic conditions. However, many Cretaceous marine sequences consist of alternating sequences of limestones and black shales. These suggest that the ocean system was, in fact, highly dynamic and that large changes in ocean conditions occurred in a cyclic manner. Herbert and Fischer (1986) suggested that it was orbital forcing that caused seasonal changes in salinity and temperature in Cretaceous sediments in the Tethys region. They suggested that times of low seasonal contrasts prevented the formation of upwelling currents but encouraged anoxia and the deposition of black shales. The low- and mid-latitude regions appear to have been the most sensitive to orbital influence, and the rhythmicity was most pronounced during times of peak transgression.

Cyclical sequences are also present in the limestone and marl sequences in Cenomanian–Turonian Greenhorn Formation in the western USA. The oxygen isotope record indicates that $\partial^{13}C$ is depleted in the limestones and heavier in shales. This was interpreted by Barron *et al.* (1985) as a record of changing salinity and rainfall, rather than of temperature. They proposed that the periods when marl was deposited represented times of higher rainfall and run-off, bringing more detritus into the basin, and the limestones represented times of more arid climates. The oxygen isotope sequences from the Cenomanian limestone–marl cycles in the Chalk of Britain were interpreted as temperature records by Ditchfield and Marshall (1989), indicating that the marl layers were deposited in cooler waters (average 23°C) than the limestones (average 25°C). It should be noted that Hallam (1986a) believed that these marl–limestone cycles were produced by diagenetic processes.

Large-scale global sea-level changes were believed to have been tectono-eustatic in origin, produced by episodes of ocean-floor spreading at a time of intense mid-ocean ridge activity that caused a reduction in ocean basin volume (Hays and Pittman, 1973). After about 85 Ma spreading rates decreased, leading to the lowering of sea levels in the late Cretaceous (Haq, 1984). Extensive volcanicity was associated with this tectonic activity (Rea and Vallier, 1983; Vogt, 1989). In a summary of global volcanic events Arthur *et al.* (1985) showed that the Albian and the late Campanian to Maastrich-

tian were periods of relatively intense volcanic activity, with decreased activity during the intervening Turonian–Santonian. Since volcanic outgassing was an important source of CO_2 for the atmosphere, it might be expected that the peaks in volcanic activity led to times of significant climate warming. There was also an increase in hydrothermal activity at the Palaeocene–Eocene boundary (Vogt, 1979).

Ocean temperatures

Oceanic isotopic temperature data from several sources (oxygen isotope data from foraminifera from the equatorial Pacific (Douglas and Woodruff, 1981) and from belemnite isotope data from higher latitudes (Lowenstam, 1964; Savin, 1977) (Fig. 8.3) show a consistent trend of temperature changes throughout the late Cretaceous to early Tertiary. In general, this consists of a peak of warmth during the late Albian, followed by a slight cooling during the Cenomanian to Santonian, a slight warming in the Campanian and a further cooling in the Maastrichtian. Temperatures rose again during the early Tertiary to an early Eocene peak. Although temperatures declined for a while during the latter part of the Cretaceous (e.g. Arthur *et al.*, 1985), overall the temperatures were still high and the Earth still warm, compared with the late Tertiary.

A similar trend is also reflected in low-latitude shallow waters. Temperatures from oxygen isotope analyses of phosphate in late Cretaceous to early Tertiary fish debris from Israel range from 21 to 31°C. The Cenomanian and Turonian were warm (average 31°C), followed by relatively cooler times, though still very warm, in the Coniacian (28°C) and the Santonian (29°C). The lowest temperatures were recorded for the late Campanian (21°C). The early Eocene was again warm (29°C) (Fig. 8.3). The palaeolatitude of the Israeli sites was about 10° N throughout this interval (Flexer *et al.*, 1986).

The similarity of temperature trends in curves for low-latitude and high-latitude areas is striking (Fig. 8.3). However, there is an approximate but consistent 5°C difference between surface waters in low and high latitudes, which implies that these temperature variations are global trends, rather than local effects. The consistency throughout this interval also suggests that the equator-to-pole temperature gradient remained constant through the late Cretaceous to early Tertiary.

The parallel record of planktonic and benthic isotope data from the equatorial Pacific (Douglas and Woodruff, 1981; Savin, 1977) also show that bottom waters were consistently about 10°C cooler than surface waters (Fig. 8.3). Although ocean temperatures varied over time, the vertical temperature gradient remained stable. At higher latitudes the oceans were somewhat cooler. Isotope data from benthic foraminifera from about 65° S near the Antarctic Peninsula indicate that shelf temperatures ranged from

Fig. 8.3. Oxygen isotope curves for low-latitude surface and deep waters from Douglas and Woodruff (1981). Cross-hatched curve represents range in isotope values from benthic foraminifera from high-latitude Antarctic Peninsula (Barrera et al., 1987). Equatorial surface water temperatures from oxygen isotopes from fish fossils, Israel (Kolodny and Raab, 1988). Organic carbon burial rate, ocean anoxic events (OAE) and δ¹³C from Arthur et al., (1985). Temperature estimates from land floras for Alaska (Parrish and Spicer, 1988) and for south eastern USA (Wolfe and Upchurch, 1987). Records of high-latitude ice-rafted deposits on Spitsbergen from Dalland (1977). Curve of volcanism from Arthur et al. (1985).

about 4 to 10.5°C from the late Campanian to the Palaeocene (Barrera *et al.*, 1987). During the late Campanian to early Maastrichtian shelf temperatures were about 4–8.5°C, warming slightly in the mid-Maastrichtian (5.5–9°C), cooling in the late Maastrichtian (4–8.5°C) and then warming again in the early Palaeocene to 5.5–10.5°C (Fig. 8.3). Surface water temperatures are estimated to have been about 2.5–4.5°C warmer. Similarly, Pirrie and Marshall (1990) reported a decline in temperatures from 13.6°C in the Santonian–Campanian to 11.7°C in the Maastrichtian, based on oxygen isotope analyses from the Antarctic Peninsula region.

However, in some regions bottom water temperatures appear to have been quite variable. Temperatures as low as 5–7°C were recorded for the equatorial Pacific, suggesting that these bottom waters may have had a polar source (Haq, 1984). In contrast, bottom waters in the restricted Angola Basin (South Atlantic) were as warm as 22°C (Saltzman and Barron, 1982). It is possible that deep ocean waters had sources from both slightly cooler high-latitude regions and from warm mid- and low-latitude regions, where warm saline waters sank down into the deep ocean due to their high density (Brass *et al.*, 1982).

During the Palaeocene, intervals of warming and cooling in the oceans can be determined by the latitudinal movement of foraminifera. In cooler times warmth-loving foraminifera were restricted to latitudes between about 25° N and S. However, at about 60 and 62 Ma, during the early and latest Palaeocene, the same foraminifera migrated into much higher latitudes of up to about 55° N and S, suggesting that ocean waters warmed at these times (Boersma and Premoli-Silva, 1983). Oxygen isotope data (Shackleton, 1986a) indicate that ocean bottom waters were as warm as 12°C during the early Palaeocene, slightly cooler (10°C) in the late Palaeocene but then warmed considerably to a peak of 15°C during the early Eocene.

In contrast to the late Jurassic to early Cretaceous interval there are few records of ice-rafted deposits during the late Cretaceous to early Tertiary. Dalland (1977) reported clasts in fine-grained shales of Palaeocene age on Spitsbergen (Arctic Norway), which he interpreted as ice-rafted deposits. They are very similar to those of early Cretaceous age from high-latitude regions (JEF, pers obs. 1989). This suggests that high latitudes were, at times, cold enough for the formation of seasonal ice during the Palaeocene.

A change in climate may have been associated with biotic extinctions at the Cretaceous–Tertiary transition. There is still debate about whether there was a gradual disappearance of Cretaceous biotas (Kauffman, 1984; Officer *et al.*, 1987) or an extraterrestrial impact event which caused catastrophic extinctions on land and in the oceans (Alvarez *et al.*, 1984; Hsu, 1986). In the oceans, surface waters seem to have been the most affected, since planktonic microbiotas declined markedly whereas benthic biotas were hardly changed.

Organic-rich sediments

Mid-Cretaceous deposits are characterized by the occurrence of thick sequences of organic-rich shales, rich in hydrocarbons (see reviews of Schlanger and Cita, 1982; Brooks and Fleet, 1987; Arthur *et al.*, 1990). Deposition occurred mainly in the North and South Atlantic, but also in the Indian Ocean and on Pacific seamounts. Three main episodes of organic carbon deposition occurred in the Cretaceous: Aptian–Albian (117–100 Ma), a brief but world-wide event during the Cenomanian–Turonian (93–90 Ma) and a minor episode during the Santonian–Campanian. These have been correlated with OAEs (ocean anoxic events; Schlanger and Jenkyns, 1976), high sea levels and high values of $\delta^{13}C$.

The debate over whether high biological productivity produced large quantities of organic carbon or whether the prevalence of deep-ocean anoxia in warm oceans caused enhanced burial rates is discussed in Arthur *et al.* (1990). The warm climates of the late Cretaceous may have produced warm stagnant oceans with sluggish circulation (Barron, 1983; Barron and Washington, 1985). Such ocean waters, particularly bottom waters, would have had low oxygen levels and therefore would have been incapable of large amounts of oxidizing organic matter produced by ocean biotas, even if rates of biological productivity were not particularly high (de Boer, 1982). Episodes of strong anoxic conditions appear to have been global in extent at times (e.g. Cenomanian–Turonian OAE).

Alternatively, the rates of production of organic carbon by ocean biotas may have been enhanced at particular times during the Cretaceous. Productivity is related to the availability of nutrients (particularly phosphorus), which are more abundant in upwelling currents that carry the nutrients from the deep ocean to the surface for use by phytoplankton. This requires well-mixed, not stagnant oceans. Some contend that deposition of black shales was not global but related to regionally restricted circulation systems and to localized zones of upwelling related to wind and current circulation (Einsele and Wiedmann, 1982; Theide, Dean and Claypool, 1982; Parrish and Curtis, 1982: Barron, 1985).

The abundance of organic-rich deposits in the Cretaceous may well have been related to both these factors: high rates of productivity produced large amounts of organic carbon that, as it sank to the ocean floor, would have soon used up available oxygen. Oxygen levels may have initially been low due to the stagnant nature of the warm oceans and so a large portion of the organic carbon produced would have been preserved. Also, rising sea levels may have encouraged the development of upwelling currents on shelf areas, promoting high productivity and the deposition of black shales.

During the late Cretaceous there was a significant increase in ridge crest, plate margin and mid-plate volcanism, especially in the Pacific. A compila-

tion of data by Arthur *et al.* (1985) indicates two peaks in the intensity of volcanic activity: one during the Albian and a later one during the late Campanian to Maastrichtian, with a period of decreased activity between the two, centred at about the Turonian–Santonian (Fig. 8.3). This suggests that volcanic outgassing of CO_2 may have contributed to much warmer climates, particularly during the Albian and late Campanian to Maastrichtian. However, the increase in atmospheric CO_2 may have been compensated to some extent by increased productivity of biotas on land and in the oceans, and the subsequent burial of organic carbon.

During the late Cretaceous it appears that the drawdown of CO_2 as a result of high productivity may have been more than compensated for by the outgassing of CO_2 during volcanic activity, resulting in an overall net warming during this interval. However, during the Cenomanian to Turonian OAE Arthur, Dean and Pratt (1988) proposed that cooler climates may have prevailed. They calculated that high productivity and high burial rates of organic carbon (which produced the high values of $\delta^{13}C$ and black shales recorded for this time) would have depleted atmospheric CO_2 sufficiently to produce a cooler climate. In turn, a cooler climate would have led to a stronger latitudinal temperature gradient, stronger winds and more ocean upwelling, necessary to supply important nutrients for photosynthesis to ocean biotas (see Chapter 11 for discussion on this topic). At this time, levels of volcanic activity were relatively low and the production of CO_2 may have been insufficient to counteract the brief cooling, although perhaps sufficient to prevent the cool phase lasting any longer. Global sea levels were notably lower during this interval (Hancock and Kauffman, 1979; Fig. 8.2). The onset of cooler climates during the Cenomanian–Turonian has also been proposed by Jeans (C.V. Jeans, pers. comm., 1990) on isotope evidence.

There is a significant negative drop (up to 3‰) recorded in the $\delta^{13}C$ carbon isotope values at the Cretaceous–Tertiary boundary at many sites (Stott and Kennett, 1989). This drop is recorded mainly in planktonic samples; there is no corresponding drop in benthic isotopes. This indicates that the vertical carbon gradient, which is normally about 2–3‰, had been eliminated. Hsu and McKenzie (1985) suggested that the drop in planktonic $\delta^{13}C$ signalled mass mortality of ocean surface biotas and the consequent decline in productivity, resulting in a sterile dead ocean, named the 'Strangelove' ocean. They proposed that a meteorite impact and associated events, such as darkness due to clouds of ejecta, may have adversely affected ocean productivity, although Arthur, Zachos and Jones (1987) reported that the low productivity episode lasted for as long as 1.5 m.y. into the Palaeocene, rather a long time for the effects of an impact.

The decline in productivity would have caused a subsequent increase in CO_2 in the atmosphere, resulting in climate warming. A shift in oxygen

isotope values at DSDP Site 524, Cape Basin, South Atlantic, indicates that an interval of cooling at the boundary was followed by a warming during the next 50 000 years of the early Palaeocene (Hsu and Mckenzie, 1985), though a temperature change shown by oxygen isotopes has not been recognized at all sites (Arthur *et al.*, 1987). The oxygen isotope data of Barrera *et al.* (1987) records a drop in temperatures in polar oceans across the boundary and a warming in the early Palaeocene.

After the significant drop in $\delta^{13}C$ at the Cretaceous–Tertiary boundary a large peak in the carbon isotopes occurred during the late Palaeocene (Boersma *et al.*, 1979; Oberhansli and Hsu, 1986). A positive excursion from about 1‰ following the Cretaceous–Tertiary boundary to about 3.5‰ at about 60 Ma was recorded by Shackleton (1986a) and Stott and Kennett (1989). Both planktonic and benthic $\delta^{13}C$ values shift to the same extent, implying that organic carbon ^{12}C was removed from the ocean system through burial as organic rich shales or coals. The oxygen isotope data of Shackleton (1986a) indicate that the oceans cooled considerably while carbon isotopes were heavy (high $\delta^{13}C$) and were warmer in the early Palaeocene and earliest Eocene at the time when carbon isotopes were lighter.

In the early Eocene the carbon isotopes were much lighter (a drop of about 2.5‰). This decrease in $\delta^{13}C$ was considered by Rea *et al.* (1990) to be due to decreased biological productivity, ultimately the consequence of climate warming. An increase in atmospheric CO_2 from volcanic outgassing associated with tectonic activity caused the climate and oceans to warm. In turn, the equator-to-pole temperature gradient decreased and atmospheric circulation weakened, prohibiting upwelling of the oxygen and nutrient-rich waters required for high productivity levels.

Land temperatures

The late Cretaceous to early Tertiary was a period of vegetational change, due to the appearance of the flowering plants and their increasing dominance in the vegetation. Their great diversification during the mid-Cretaceous corresponded to the onset of this Warm Mode and the favourable climatic conditions may have been a crucial factor in their expansion.

Studies of global pollen distribution illustrate that palaeogeography and land connections determined floral provinces as much as climate in the Cretaceous (Srivastava, 1981; Batten, 1984). During the early Cretaceous there appear to have been two or three broad vegetational divisions, dividing the globe into essentially tropical/subtropical climates in the equatorial zone and warm temperate climates at higher latitudes. Srivastava (1981) proposed one pollen province in a central area including northern South America, northern Africa, southern Europe and the Middle East and

another incorporating the rest of the globe but separated from the central province by ocean. Others (in Crane, 1987) divide the northern hemisphere into a high-latitude Siberian–Canadian province and a lower latitude Euro-Sinian province, with the remainder of the equatorial and southern hemisphere region in the Gondwanan floral province (Fig. 8.4).

The major Albian transgression caused the isolation of land into four major provinces, which became well established until the end of the Maastrichtian. These included the *Aquipollenites* province in northern high latitudes and the *Normapolles* province in central Europe and North America, and two Gondwanan zones, including the *Nothofagidites* province encompassing the Gondwanan continents in the southern hemisphere (Fig. 8.4). The boundaries of these areas were constrained by palaeogeography and marine seaways and thus it was only as a result of a severe global regression in the late Maastrichtian that land connections were accessible for the floras to migrate and intermix, particularly in the northern hemisphere. Only the *Nothofagidites* province in Gondwana remained isolated and intact throughout the Tertiary.

A floral turnover appears to have occurred at the Cretaceous–Tertiary transition in some localities, although, as in the oceans, the magnitude of the change is variable from place to place and interpretations of data differ. The effect of a meteorite impact on land is proposed to have created dense clouds of dust or smoke that would have blocked out the sun, resulting in a decrease in photosynthesis and climate cooling. The presence of abundant soot particles at the boundary has been interpreted as evidence for widespread fires that would have destroyed vegetation (Wolbach *et al.*, 1988). However, fossil charcoal is common in most Cretaceous non-marine sediments and it is apparent that forest fires were a common occurrence in Cretaceous forests (Cope and Chaloner, 1980)

There do not appear to have been any consistent effects of the proposed bolide impact on terrestrial vegetation (Collinson, 1986), in that extinction rates appear to have been higher in low latitudes compared with high-latitude areas (Wolfe and Upchurch, 1986). But an increase in the percentage of fern spores occurs above the boundary at several localities (Tschudy *et al.*, 1984), which has been compared to the modern sequences of recolonization of vegetation in areas devastated by a dramatic event such as a volcanic eruption.

According to Wolfe and Upchurch (1986) the vegetation in western North America was not severely affected by temperature changes but instead reflects an increase in precipitation across the boundary. Rainforest-type vegetation became established there in the early Tertiary. In the vegetational changes evergreen types were affected more severely than deciduous plants. This suggests that the climate developed more seasonal extremes, although Wolfe and Upchurch propose that a low-temperature

A. CENOMANIAN

B. SANTONIAN–CAMPANIAN

C. PALAEOCENE

D. LATE PALAEOCENE–EARLY EOCENE

Polar broad-leaved deciduous

Broad-leaved evergreen

Paratropical rainforest

Tropical rainforest

Fig. 8.4. Floral provinces during the late Cretaceous to early Tertiary Warm Mode. (A) to (C) based on pollen zones from Crane (1987). (D) Mega-floral zones from Wolfe (1985). Base maps after Smith and Briden (1977). The vegetational provinces are rather broad as a result of the globally warm and uniform climate. From Crane (1987).

excursion took place at the boundary as a result of an 'impact winter'.

Vegetation zones remained quite simple during the Palaeocene (Fig. 8.4). A wide equatorial belt of tropical rainforest extended to latitudes of about 50°. Narrower zones of paratropical rainforest, broad-leaved evergreen forest and polar broad-leaved deciduous forests extended near to the poles (Wolfe, 1985). Only as the climate cooled during the Tertiary did the vegetation zones migrate to lower latitudes and become more complex.

In the past, late Cretaceous floras have been interpreted rather broadly in terms of temperate or tropical climates, supporting evidence from oxygen isotope data that the climate was warm and equable. However, analysis of physiognomy of fossil angiosperm leaf assemblages has been used as a tool to assess climate parameters in more detail (Wolfe, 1980, 1985; Spicer and Parrish, 1986). In North America megathermal vegetation (characterized by mean annual temperatures of 20–30°C) extended up to palaeolatitudes of 45–50° N and was characterized by open-canopy, broad-leaved evergreen woodland. Rainfall was possibly low or moderate in amount and distributed evenly throughout the year. Evergreen conifers were the largest and most dominant trees. Up to about 65° N broad-leaved evergreen woodland was established under a mesothermal climate (mean annual temperature 13–20°C). Growth rings in fossil wood are indicative of mild seasonality. In the north polar region the plants were predominantly deciduous, probably in response to long periods of low light levels during the long winter seasons. Comparisons with plant assemblages from other continents show similar characteristics (Wolfe and Upchurch, 1987).

A compilation of late Cretaceous and early Tertiary palaeotemperatures for North America at palaeolatitudes of 30° N (Wolfe and Upchurch, 1987), based on leaf physiognomy, suggests a general warming trend from the Albian to the early Tertiary with mean annual temperatures ranging from 21–29°C (Fig. 8.3). The mean annual temperature was approximately 21°C in the Albian and then rose gradually to a value of 25°C in the Santonian. Land temperatures declined to about 23–24°C during the Campanian and to a low of 22°C in the early Maastrichtian. The late Maastrichtian and early Palaeocene appear to have been as warm as the warm interval during the mid late Cretaceous (mean annual temperatures of about 26°C). A further relatively cool (though still warm) interval occurred in the late Palaeocene when mean annual temperatures dropped to 22°C prior to the peak warming of 29°C in the early Eocene. Latitudinal temperature gradients were lower than at present: a gradient of about 0.3°C/1° latitude was proposed by Wolfe and Upchurch (1987) for the Campanian and Maastrichtian.

Investigation of the late Maastrichtian environments in which dinosaurs lived in the Western Interior region of North America (Lehman, 1987) showed that a climatic transition occurred around 35° N latitude. To the south caliche-bearing palaeosols and fluctuating fluvial regimes suggest that

the climate was markedly seasonal and semi-arid. To the north the climate was much more humid. Sigleo and Reinhardt (1988) described palaeosols formed during Cretaceous weathering periods (Cenomanian to early Maastrichtian) in the south-eastern United States. They consider the palaeosols to have formed under tropical–subtropical climates, in warm, humid environments on well-vegetated piedmont surfaces. The region of Cretaceous laterite and bauxite, indicating humid climates, occurs mainly in palaeolatitudes between 5° S and 30° N (see Figure 16 in Sigleo and Reinhardt, 1988).

Temperate climates extended right up to the poles during the Cretaceous and early Tertiary, allowing the growth of forest vegetation at high latitudes (Creber and Chaloner, 1985). The plants were able to tolerate the rather extreme light regime that they would have experienced above the Arctic and Antarctic circles (several months of winter darkness and continuous summer sunlight) (Francis, 1990; Spicer and Parrish, 1990). In the north polar region physiognomic analysis of fossil leaves from Alaska indicates that the mean annual temperature during the latest Albian and Cenomanian was about $10 \pm 3°C$, then slightly warmer in the Coniacian up to a maximum of 13°C, followed by a cooling during the Campanian–Maastrichtian to mean annual temperatures of about 2–8°C (Parrish and Spicer, 1988). Distortion of leaf forms and the abundance of coals suggest that winter temperatures may have declined to freezing or below, preventing the complete decay of forest litter (Spicer, 1987). Growth of southern hemisphere plants under experimental conditions representing dark warm winters and dark cool winters indicated that many species of trees and shrubs, which are living relatives of Cretaceous and Tertiary types, survived best in cool dark conditions and suffered in the warm and dark (Read and Francis, in press). This suggests that the climate in polar regions was seasonally cool enough to encourage the growth of forest vegetation, and not annually warm, as generally believed.

Skeletal remains indicate that dinosaurs also lived on the North Slope of Alaska during the late Campanian–early Maastrichtian (Brouwers *et al.*, 1987). They lived in a deltaic environment among mild to cold temperate forests of mainly deciduous conifers and broad-leaved trees, and probably remained there year round, coping with the long winter darkness and low temperatures, rather than migrating to lower latitudes – lifestyles likened to those of Australian early Cretaceous dinosaurs (Rich *et al.*, 1989). In fact, the climate of the northern high-latitude regions during the latest Cretaceous may have been as cool as that of polar regions during the early Cretaceous, although there are no reports of ice-rafted deposits. Spicer and Parrish (1990) suggested that periglacial conditions did not develop at sea level in northern polar regions during the Maastrichtian but permanent ice was likely above 1000 m at 8° N.

In the southern hemisphere, temperate floras covered most continental

areas throughout the late Cretaceous. During the late Albian, temperate podocarp rainforests grew around the Antarctic Peninsula and Magellanes Basin area in Patagonia (Dettman, 1990). In the Antarctic Peninsula region, growth rings in fossil conifer and angiosperm wood (Francis, 1986b) and pollen analysis (Askin, 1983) indicates that the climate was warm to cool temperate, not unlike that of New Zealand or Tasmania today. In central Australia broad areas of low relief around the Eromanga Basin were also colonized by podocarp and araucarian forests. During the mid-Albian the vegetation changed due to the spread of warm temperate angiosperms and ferns, although podocarps remained dominant. In the southern rift valleys, formed in the early stages of the splitting of Australia from Antarctica, cool to mild temperate conditions prevailed and rainforests of podocarp and arau-carian conifers with tree ferns flourished in humid climates (Dettman, 1986b, 1990).

There was a transition from the high-latitude areas which had a high rainfall to drier climates north of southern South America. *Classopollis* pollen, from conifer trees often associated with evaporites and arid climates, is more frequent in assemblages from this northern region than further south (Dettman, 1986b). Tropical rainforest, mangrove vegetation and araucarian woodlands that grew in southern South America in the Palaeo-cene were replaced by more temperate vegetation as a result of climate cooling during the late Palaeocene and Eocene.

The distribution of coals and evaporites reflects the latitudinal zonation of the floras to some extent. Late Cretaceous and early Tertiary coals have been recorded from the northern hemisphere (Duff, 1987) and are distributed from high to low latitudes. During the Cenomanian coals were formed in mid- and high latitudes in the northern hemisphere and in high latitudes in the south. By the Maastrichtian coals formed from the north polar region to the tropics. Some evaporites formed in low- and mid-latitude areas (Parrish *et al.*, 1982). By the mid Eocene a coal-forming zone had developed in equatorial regions and, along with extensive high-latitude deposits, contri-buted to a greater abundance of coal at this time than in the Cretaceous, and possibly the cause of the peak in the $\delta^{13}C$ isotope values. Evaporites were then restricted to mid-latitudes (Parrish *et al.*, 1982).

Qualitative models of predicted rainfall patterns for these times (Parrish *et al.*, 1982) are, in part, supported by the geological evidence for zonal distribution of coals and evaporites. The models also predict high rainfall on continental margins bordering the Pacific Ocean, particularly western margins at high latitudes. In the Cenomanian the borders of the Tethys may have received high rainfall, probably of seasonal monsoonal type (Barron and Washington, 1982), but large areas in continental interiors, such as central Asia, were very dry. Very seasonal, continental-type climates may have been an important character of the late Cretaceous due to the large

size of the continental land masses, as has been discussed for the late Jurassic and early Cretaceous (Chapter 7). However, by the end of the Cretaceous the large continental blocks were breaking up; the opening of the North and South Atlantic resulted in additional high rainfall on adjacent continental margins and by the early Tertiary the movement of Australia away from Antarctica and the rest of Gondwana allowed high precipitation on its western shores.

Chronology of the Warm Mode

The compilation of palaeoclimate data indicates that the late Cretaceous to early Tertiary climate was generally very warm. Oxygen isotope data indicate that the Albian was the one of the warmest times. Ice-rafting in high latitudes had terminated by then and volcanic activity was abundant, possibly enhancing warming with the addition of CO_2 into the atmosphere. The middle part of the late Cretaceous was characterized by the deposition of large quantities of organic-rich shales and ocean anoxic events during the late Cenomanian to Turonian and Santonian to early Campanian. Warm anoxic oceans may have encouraged the burial of organic matter or, alternatively, large amounts of organic carbon may have been produced by high biological productivity in cooler, upwelling oceans.

The Coniacian and Campanian were also warm intervals, as indicated by the vegetation from high latitudes; this again correlates with increased volcanic activity. During the Maastrichtian there was a marked drop in sea levels and decline of temperatures on land and in the oceans. Freezing conditions may have been experienced in the northern polar regions. However, the climate warmed again during the early Palaeocene, cooled during the late Palaeocene and then warmed again to another very warm episode during the early Eocene.

9

The Cenozoic Cool Mode: early Eocene to late Miocene

The early stages of the Cenozoic Cool Mode started with the cooling in the early Eocene (55 Ma). From that time onwards the climate of the Earth gradually cooled from the Warm Mode of the late Cretaceous to early Tertiary to the cool glacial climates of today. Important changes which occurred during this phase include the enhancement of climatic zonation and the development of a thermally stratified ocean. Unlike the Palaeozoic record, the early part of the Cenozoic cooling is not recorded simply by the presence of ancient glacial deposits. In fact, any direct evidence of extensive glacial ice at the poles during the Tertiary is scarce, principally because the rocks in these regions are now covered with ice. However, the presence of ice-rafted debris in deep ocean cores provides positive evidence for the presence of at least seasonal ice.

The principal evidence for Tertiary cooling is documented in the oxygen isotope record of calcareous foraminifera from the oceans. This illustrates the decline in ocean temperatures and the build-up of ice at the poles (mainly the South Pole) from about 55 Ma onwards (Fig. 9.1). Climate evidence from fossil plant assemblages reflects the same cooling trend. Documentation of the Tertiary cooling is important because it illustrates the crucial transformation phase from a non-glacial to glacial state, as recorded by several geological parameters. This trend or transformation can be used as a model to determine the history of glacial build-up during former Cool Modes, such as those in the Palaeozoic, for which data are less reliable.

Superimposed on this major cooling phase are several shorter intervals of warming and cooling, such as the sharp decline in temperatures at the Eocene–Oligocene boundary and during the late Middle Miocene, and warming phases in the Eocene (possibly related to an enhanced greenhouse effect due to increased CO_2 levels). These warming phases are best documented by the spread of temperate vegetation to high latitudes close to the poles.

Although the continental floral and ocean isotope records both reveal the

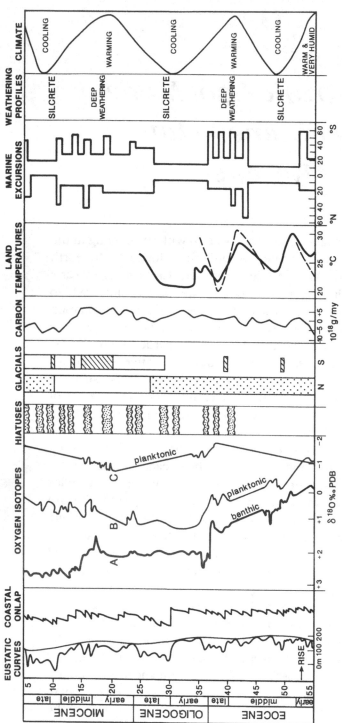

Fig. 9.1. The Tertiary Cool Mode. Eustatic and coastal onlap curves from Haq et al. (1987). Oxygen isotope curves after Shackleton (1984); A, B and C are South Atlantic deep water, mid-latitude eastern South Atlantic surface water and Equatorial surface water respectively. Global deep-sea hiatuses (dotted bars) from Barron and Keller (1982) and Keller et al. (1987). Record of glacial deposits in the northern (N) and southern (S) hemispheres; plain bar (N) represents ice-rafted deposits in the early Tertiary of Spitsbergen from Dalland (1977) and tillites in Miocene of Alaska from Denton and Armstrong (1969); patterned bar (S) represents presence of glaciers on Antarctic continent— oldest record at 29 Ma from Barrett et al. (1988); shaded bars represents tillites on South Shetland Islands, Antarctic Peninsula from Birkenmajer (1987). Carbon record of Shackleton (1987a) showing carbon isotope record of bulk sediment in DSDP sites 525 and 528, South Atlantic. Land temperature estimates from north-east Pacific floras (dotted line) compared with marine temperatures (solid line), adapted from Wolfe and Poore (1982). Latitudinal excursions of large tropical foraminifera from McGowran (1986). Weathering profiles for Australia and climate curve from McGowran (1990). Time scale of DNAG (Palmer, 1983).

dominant cooling trend, it is still difficult to resolve the temperature ranges that have been interpreted from them. Continental temperatures from plant assemblages are generally somewhat higher than those from ocean isotopes. Tectonic events also play an important role in the Tertiary climate history, particularly increasing topography due to uplift and changes in ocean circulation.

The dynamics of the Tertiary climate have been modelled (Barron, 1983) to examine whether changing palaeogeography, such as the uplift of the Tibetan Plateau or the expansion of the Pacific, was the principal cause of the cooling trend, as has been previously proposed (Frakes and Kemp, 1972; Barron, Sloan and Harrison, 1980). However, the modelling results suggest that the long-term cooling trend was probably the result of the interaction of a series of events, such as the opening of ocean gateways and the development of new ocean currents, and variations in atmospheric CO_2 concentration. The large amount of detailed data available for the Cenozoic has enabled several cycles of various orders to be recognized.

The onset of polar glaciation

Many of the global climate changes in the Tertiary are related to the isolation of Antarctica and the development of the Antarctic ice cap. Tangible evidence of ice is present in the form of ice-rafted debris and glacial-marine sediments in deep-sea cores. The presence of ice has also been proposed from the extrapolation of oxygen isotope data, which at times records the depletion of the light fraction of oxygen isotopes locked up in the ice or implies very cold surface water temperatures suitable for ice formation (Matthews and Poore, 1980).

At the onset of the cooling phase in the Eocene Antarctica was situated over the South Pole (Lawver, Sclater and Meinke, 1985). The Scotia Arc (comprising the Antarctic Peninsula and the Andean chain) formed a barrier to ocean currents around West Antarctica and the proximity of Australia and Antarctica also blocked ocean currents from being isolated in cold high latitudes. Fossil macrofloras and pollen assemblages show that the climate was warm–cool temperate (Kemp and Barrett, 1975; Francis, 1986b; Dettman, 1990) and oxygen isotopes reveal that both surface and bottom ocean waters were relatively warm: about 17.5°C and 17°C respectively in the mid-Early Eocene (Shackleton and Kennett, 1975). During most of the Eocene, bottom and surface water temperatures were only a few degrees different and both cooled gradually to temperatures of 9°C and 11°C respectively by the end of the Eocene. Up to this point there is no strong isotopic evidence for the existence of major ice caps and no record as yet of glacial deposition in ocean sediments during the Eocene.

However, evidence is accumulating for the presence of temperate, conti-

nental glacial ice on Antarctica as far back as the Palaeogene or even earlier (Hayes and Frakes, 1975; Barron *et al.*, 1981; Kvasov and Verbitsky, 1981; LeMasurier and Rex, 1982; Birkenmajer, 1987). Matthews and Poore (1980) interpreted the sharp enrichment of $\delta^{18}O$ in tropical surface waters at about 38 Ma (Eocene–Oligocene boundary) as a major increase in ice volume and a sea-level drop of at least 50 m, based on their assumption that Tertiary tropical sea surface temperatures were more like those of the Pleistocene than of today. This would imply that a significant amount of ice must have been present much earlier in the Palaeogene.

On land a subglacially erupted hyaloclastite in West Antarctica was considered evidence for an ice sheet at about 27 Ma (LeMasurier and Rex, 1982). Drewry (1975) suggested that small local highland ice masses may have been present in West Antarctica before the development of the continental ice cap. Ice fields with local, wet-based valley glaciers may have also built up on the windward side of the Transantarctic Mountains, although since these mountains were not uplifted until at least 55 Ma (Fitzgerald *et al.*, 1986), these valley glaciers can only have existed since the early Eocene. Wilson (1973) also speculated on the presence of temperate alpine glaciers in the McMurdo area 45–35 m.y. ago from the evidence of fossil cirques in Victoria Land.

The oldest glacial sediments reported to date were formed during the Krakow glaciation (Birkenmajer, 1987). On King George Island in the South Shetland Islands glacial-marine sediments occur beneath lavas dated at 49.4 Ma, so temperate glaciers on mountains may have been present even in the early mid-Eocene (Birkenmajer, 1991). Significantly, a drop in sea level is recorded on the curves of Haq *et al.* (1987) at precisely this time (Fig. 9.1). The onset of continental glaciation with an ice sheet at sea level in West Antarctica is considered by Hayes and Frakes (1975) and Birkenmajer (1991) to have been established by latest Oligocene. Robin (1987) suggested that by 40 Ma continental ice sheets would have been present on Antarctica above elevations of 1000 m, with perhaps a few valley glaciers reaching sea level.

Oxygen isotope data show a rapid enrichment of $\delta^{18}O$ at the Eocene–Oligocene boundary, at about 38 Ma. Oxygen isotopes in deep-sea benthic foraminifera from the tropical Pacific (Savin, Douglas and Stehli, 1975) and from the sub-Antarctic (Shackleton and Kennett, 1975) show that bottom waters cooled by as much as 5°C. This is interpreted as a major cooling of Antarctic surface waters and deep bottom waters, severe enough for the formation of the first sea ice around Antarctica (Mercer, 1983), although the cause of such an event is unknown.

By the middle to late Oligocene Australia had drifted far north of the Antarctic continent and oceanic circulation had developed between the two, beginning the isolation of Antarctica (Veevers, 1984). The Drake

Passage between Antarctica and South America was also beginning to open (Barker and Burrell, 1977), allowing ocean currents to circulate around Antarctica with no connection to lower latitude warm currents. Direct evidence that glaciers had reached sea level was found in Oligocene sediments (26 Ma) in the Ross Sea area as ice-rafted debris and glacial-marine mudstones (Hayes and Frakes, 1975). This record has now been extended back to 29 Ma (Barrett *et al.*, 1987) or even 31 Ma (Harwood, 1987) after the recovery of glacial diamictites in McMurdo Sound, Ross Sea.

The environment of 29 Ma was envisaged to be similar to that of today with continental glaciers calving at sea level (Barrett *et al.*, 1987). However, the discovery of a leaf of southern beech (*Nothofagus gunnii*) interbedded with the glacial mudstones (Hill, 1989) indicates that there were still some vegetated areas on nearby parts of the Antarctic continent, so the glacial evidence cannot be interpreted in terms of a full-scale ice sheet. Palynological assemblages in Ross Sea cores also illustrated that a low-diversity cool temperate flora with *Nothofagus* grew in this region until at least the end of the Oligocene (Kemp and Barrett, 1975; Truswell and Drewry, 1984).

During the early Miocene the circum-Antarctic current became well established and Antarctica was isolated in a cold polar zone. Up to this point calcareous sediments had been deposited very close to the continent but at this time the establishment of cold surface waters displaced the calcareous sediments, allowing siliceous biogenic deposition. For a short interval low-latitude surface waters warmed significantly.

In the mid-Miocene (12–14 Ma) another rapid enrichment of $\delta^{18}O$ (Fig. 9.1) is interpreted as the build-up of the permanent East Antarctic ice cap (Savin, 1977), possibly related to the previous warming phase in that the warmer waters may have caused increased precipitation for the build-up of ice. This was followed by a great expansion of the ice cap during the late Miocene (5 Ma) (Shackleton and Kennett, 1975). The Antarctic Convergence shifted northward, as did the zone of ice-rafting, and the area of siliceous biogenic sedimentation increased. The ice reached its greatest extent into low latitudes and covered a larger area than the ice cap at present.

While the Antarctic ice cap was forming in the south, the Arctic ocean remained ice free. The report of ice-rafted dropstones in sediments from Svalbard (Dalland, 1977) suggests that at least seasonal winter ice may have been present during Eocene times. Fresh water flowing from the Arctic borderlands may have frozen to form seasonal winter ice (Clark, 1982). Ice-rafted debris indicates that ice was present in the central Arctic Ocean at about 5 Ma in the late Miocene. Tillite sequences from the Wrangel Mountains in Alaska suggest that intermittent glaciation occurred there from the late Miocene (about 10 Ma onwards) (Denton and Armstrong, 1969). After the Miocene there followed a period of global low sea level and

climate cooling, during which the ice cap developed on Antarctica and the Mediterranean Basin was isolated prior to its desiccation.

Ocean records

Detailed oxygen isotope records are available for most of the Tertiary and have provided a good record of changing ocean temperatures and changing salinities induced by the build-up of continental ice. In general, the isotope record is similar for all the major oceans, including the Southern Ocean (Shackleton and Kennett, 1975), South Atlantic (Williams *et al.*, 1983; Poore and Matthews, 1984), North Pacific and Equatorial Pacific (Douglas and Savin, 1973; 1975) and the Indian Ocean (Oberhansli, 1986). Isotope information is summarized in Miller, Fairbanks and Mountain, (1987).

Following low $\delta^{18}O$ values in the earliest Eocene, both surface and bottom waters had higher values of $\delta^{18}O$ for the rest of the Eocene. Since there is little geological evidence for significant quantities of polar ice in the Eocene the isotopic change has been interpreted as cooling of ocean waters, rather than build-up of ice (Oberhansli and Hsu, 1986). Early Eocene isotopes signal bottom water temperatures of 11–15°C. If the isotopic value for Eocene sea water is considered to be the same as that present during glacial times (Matthews and Poore, 1980) a temperature of 15°C can be assumed. A temperature of 11°C results if the ocean is considered to be ice free. In either case, these temperatures infer that high-latitude surface waters would be too warm for ice formation. Bottom waters in mid-latitudes may have been warmed due to the sinking of warm, highly saline surface water, but high temperatures (up to 17–21°C) have been recorded from even high-latitude planktonic foraminifera.

During the middle Eocene (52–40 Ma) $\delta^{18}O$ values increased in two steps, at the Early to Middle Eocene boundary and in the late Middle Eocene. By the end of the Eocene the temperature of bottom waters was about 6°–10°C and high-latitude oceans are thought to have been ice free. The most significant increase in $\delta^{18}O$ values occurred at the Eocene–Oligocene boundary (~ 38 Ma). Values fell below the threshold of 1.8‰, which implies that the temperature of high-latitude surface waters was cold enough for ice formation. The isotope drop is synchronous in both benthic and planktonic foraminifera, indicating the growth of ice, rather than simply a temperature decline (Miller *et al.*, 1987). The presence of ice has also been inferred from synchronous isotope decreases at 28–31 Ma and in the North Atlantic at the Oligocene–Miocene boundary at 24–25 Ma. Apart from these two intervals the Oligocene isotope record is one of only slightly declining ocean temperatures.

The next major cooling occurred in the middle Miocene at 13–15 Ma. A synchronous drop in both benthic and planktonic isotopes records the

intensification of polar glaciation. However, the pole-to-equator surface water gradient increased during the middle Miocene from 6°C to 12°C. High-latitude surface waters cooled but low-latitude waters warmed slightly (Kennett, 1977), possibly as the result of continental break-up around the Tethyan seaway and consequent disturbance to the circum-equatorial current. Increasing $\delta^{18}O$ values for equatorial surface waters have been interpreted as indicating a slight drop of ocean temperatures to 22–24°C at some time during the mid-Miocene (Shackleton, 1986a), although this unusual cooling is not apparent if the isotope values are calculated on the basis of a glaciated ocean composition (Matthews and Poore, 1980).

After a short warming phase in the early Late Miocene a terminal Miocene cooling occurred at about 5–6 Ma, causing a major phase of ice growth and intensification of glaciation. Global sea levels dropped markedly. This caused the Mediterranean Basin to become isolated from ocean waters and led to its desiccation (Hodell, Elmstrom and Kennett, 1986). About 1 million cubic metres of evaporites were precipitated, lowering the salinity of the oceans by about 2‰ (Cita and McKenzie, 1986).

The carbon record

Carbon isotope records from the Indian, South Atlantic, Pacific and Southern oceans indicate several periods of ^{13}C enrichment during the Tertiary (Oberhansli and Hsu, 1986; Miller and Fairbanks, 1985). As discussed in Chapter 8, the most marked enrichment during the Tertiary was during the late Palaeocene to early Eocene (Shackleton, 1986a; Arthur *et al.*, 1990). However, in the early Eocene a sudden depletion of $\delta^{13}C$ was recorded in most oceans (Oberhansli and Hsu, 1986), the cause of which is unknown but may relate to a decrease in ocean productivity. From the mid Early to late Middle Eocene the vertical carbon gradient increased significantly (Arthur, 1982), possibly due to high productivity in surface waters.

During the Oligocene and Miocene three cycles of increased $\delta^{13}C$ values have been detected in the North Atlantic (Miller and Fairbanks, 1985). Peak values are recorded at 35–33 Ma (early Oligocene), 25–22 Ma (Oligocene to Miocene boundary) and 18–14 Ma (Early to Middle Miocene boundary). Similar cycles have been noted in the Atlantic, Pacific and Indian oceans so these variations are considered to represent global changes in the $\delta^{13}C$ content of ocean water. The $\delta^{13}C$ values increase by about 1‰ and occur with a periodicity of about 7–12 m.y. Miller and Fairbanks (1985) showed that the increases in $\delta^{13}C$ corresponded to increases in $\delta^{18}O$ (related to sea-level drops and ice build-up) during the early Oligocene (~ 36 Ma), near the Oligocene–Miocene boundary (~ 25 Ma), and in the late Early Miocene (~ 18 Ma). However, a $\delta^{18}O$ increase in the middle Oligocene was not marked by a significant carbon excursion, causing Miller and Fairbanks to

Fig. 9.2. The Monterey Carbon Isotope Excursion of Vincent and Berger (1985). Carbon and oxygen isotope data plotted against time. The records are from three genera of foraminifera from different depths: BF benthic, SPF shallow planktonic and DPF deep planktonic (see Vincent and Berger, 1985, for species names). M1 and M2 mark the beginning and the end of the 'Monterey Carbon Isotope Excursion' and A1–A2 marks the episode of Antarctic cooling and ice build-up, which occurred during the carbon excursion. From Vincent and Berger (1985). Copyright © by the American Geophysical Union.

conclude that there was no simple correlation with sea-level change and the carbon record for this time.

However, a correlation between the carbon record and climate was suggested by Vincent and Berger (1985) for the latest Early Miocene. A marked excursion in $\delta^{13}C$ for this time has been reported from several sites. Vincent and Berger recorded this signal in benthic and planktonic foraminifera from the latest Early Miocene of the Indian Ocean, the excursion starting at about 17.5 Ma and lasting for 4 m.y. (Fig. 9.2). They attributed this to the increased extraction and burial of organic carbon, resulting from a phase of increased biological productivity and carbon fractionation in the oceans. This process would have lowered atmospheric CO_2 levels and cooled the climate (reverse greenhouse). In turn, this would have acted as a positive feedback, the cooler climate allowing more vigorous ocean upwelling and nutrient supply, enhancing productivity. There is a strong correlation in the carbon and oxygen isotope records at this time. The $\delta^{18}O$ record indicates that cooling and ice build-up occurred during the carbon peak, about 2 m.y. after the carbon excursion began.

Vincent and Berger (1985) called this the 'Monterey Carbon Isotope Excursion' since large quantities of organic matter were deposited around the edges of the northern Pacific, including the Monterey Shales off Califor-

nia. There is evidence for strong coastal upwelling in these areas at that time, the result of strong zonal winds and a strong temperature gradient due to polar cooling. The cycle may have eventually been broken when the supply of nutrients was exhausted and productivity rates dropped. The level of CO_2 in the atmosphere would have then increased, halting the cooling trend. This is displayed in the oxygen isotope record of Vincent and Berger (Fig. 9.2).

Ocean faunas

The cooling trend and increase in latitudinal temperature gradient throughout the Tertiary is reflected by the migration of microfossil zones through the oceans. McGowran (1986, 1990; McGowran and Beecroft, 1986) recorded several extratropical excursions of large foraminifera from the Indo-West Pacific Province. The boundaries of this zone, now situated at 35° N and 35° S, expanded at times in the Eocene to 60° N and 45° S. The excursions can be correlated with oxygen isotope evidence for global warming and transgressions, indicative of warm phases during the late Palaeocene to early Eocene, late Middle to latest Eocene, and the late Oligocene to middle Miocene, interspersed with cooler periods in the early Middle Eocene, Early and early Late Oligocene (see Fig. 9.1).

Peak faunal diversity is recorded for dinoflagellates in the early Eocene, for nanofossils in the early–mid Eocene and later in the late Middle Eocene for planktonic foraminifera (Oberhansli and Hsu, 1986). Diversity then dropped during the rest of the Tertiary. After the early warm period three cooling phases caused shifts in microfossil zones (Haq, 1982). In the middle Eocene (46–44 Ma) high-latitude nanofossil assemblages moved to temperate and tropical zones in response to cooling. A major faunal turnover took place towards the end of the Eocene (Benson, Chapman and Beck, 1984) when warmth-loving microfossils were replaced by more cold-tolerant types, including ostracods and foraminifera. This has been related to the influx of cold bottom waters and the development of the psychrosphere (thermally stratified ocean). Another incursion of high-latitude foraminifera and nanofossils into lower latitudes took place in the middle Oligocene (32–31 Ma) in response to a cooling phase almost as extreme as that at the end of the Eocene, according to Haq (1982).

In surface waters four planktonic provinces that existed in the Eocene were expanded to six separate zones by the Oligocene as colder sub-Antarctic and Antarctic regions became established, signalling the development of a more differentiated latitudinal temperature gradient (Sancetta, 1979). Globigerinids became restricted to mid- and low-latitude sites and nanofossils also moved equatorward. Keller (1983) suggested that species variation in planktonic foraminifera during the mid Eocene–early Oligocene

were indicative of low frequency cool–warm cycles of about 0.5–1.0 m.y. frequency. However, during the late Early to Late Oligocene the cycles were of about 2.5–3.0 m.y. duration.

The discovery of many extinction events and microtectite layers in Eocene–Oligocene boundary sequences led to the speculation that such environmental and climatic changes were due to an asteroid impact (see references in Pomerol and Premoli-Silva, 1986; Corliss and Keigwin, 1986). However, refined studies have now shown that microfossil extinctions occurred in a step-like manner throughout the late Eocene and that the magnitude of extinctions was not higher at the boundary itself (Keller, 1986). The changes were more likely a response to bottom water cooling that occurred at this time. Many sensitive species of foraminifera were eliminated in the early Palaeocene and those that remained were able to tolerate ocean changes later on. This step-wise extinction pattern in the late Eocene to early Oligocene is also reflected in the record of terrestrial mammals and plants (Collinson, Fowler and Boulter, 1981; Wolfe, 1985).

Tectonic events

Many of the most important events in the Tertiary cooling trend are linked to changes in ocean circulation and the opening of ocean gateways, and to palaeogeography and topography. Tectonic events which affected ocean circulation and climate include the collision of India with Asia at 50–53 Ma (McGowran, 1978; 1990) and the northward movement of Australia from Antarctica from ~95 Ma (Veevers and Ettriem, 1988). Deep circulation may have commenced between Australia and Antarctica by the late Eocene but the circum-Antarctic current did not fully develop until later.

In the North Atlantic the Iceland–Faroe sill subsided during the late Eocene. Arctic waters were then able to mix with those of the North Atlantic in the earliest Oligocene due to the opening of the Svalbard–Greenland Strait (Schrader *et al.*, 1976). The primitive stages of the North Atlantic Deep Water may have developed from overflow from the Norwegian Sea in the late Eocene, injecting a flow of high salinity water into the middle depths of the Central and South Atlantic (Johnson, 1985). The Gulf Stream strengthened shortly after, due to the elevation of the Panamanian Isthmus and the closed link between the Pacific and Atlantic oceans, and the more latitudinally oriented currents were then replaced by warm surface currents crossing the ocean, taking warm, wet air into the northern high latitudes to help develop the northern ice cap (Berggren, 1982). In Europe the Alpine Orogeny in the late Eocene caused the restriction of ocean circulation between Tethys and the Atlantic due to the collision of the African and European plates (Biju-Duval *et al.*, 1976). The circum-equatorial current that previously circulated through Tethys was then greatly reduced, isolating the proto-Mediterranean ocean.

Within the ocean sediment record for the Tertiary there are many global unconformities or hiatuses, which represent either the erosion of sediments due to the increased erosive strength of cold bottom currents or non-deposition/dissolution of sediment during times of low productivity and weak circulation. Keller *et al.* (1987) and Barron and Keller (1982) documented several hiatuses during the Tertiary: seven global deep-sea hiatuses during the Miocene, one in the early Pliocene and six during the latest Middle Eocene to Late Oligocene (Fig. 9.1). Hiatus intervals during the Miocene correspond to times of rapid enrichment of $\delta^{18}O$, carbonate dissolution, cool planktonic microfossil assemblages and sea-level drops, all related to enhanced polar glaciation and bottom water formation (Barron and Keller, 1982).

In the Palaeogene, however, the hiatuses may have been caused by any of three mechanisms: sediment starvation in deep basins during sea-level rise or coastal onlap (McGowran, 1986), increased bottom water formation due to global cooling (and sea-level falls), or major changes in flow paths of currents due to new basin configurations (Keller *et al.*, 1987). For example, the late Middle Eocene hiatus is marked by a change to cooler water assemblages of faunas and coincides with shallow circum-Antarctic circulation. The Eocene–Oligocene boundary event corresponds to the development of cold bottom waters and the ocean psychrosphere, and the Oligocene–Miocene event due to bottom water erosion, probably as a result of the opening of the Drake Passage and changes in circum-Antarctic circulation. Most hiatuses are associated with major changes in ocean faunas.

Continental climates

Latitudinal variation in the distribution of fossil plant assemblages is a useful indicator of Tertiary climate trends. Wolfe and Poore (1982) summarized the variation of inferred mean annual temperatures of various plant assemblages through the early Tertiary as a climate curve (Fig. 9.1). Intervals of warming and cooling appear to have occurred in a cyclical manner, each warm-to-cold interval lasting for about 10 m.y. Plant evidence shows that periods of warmth occurred in the late Palaeocene–early Eocene, late Middle Eocene and possibly the latest Eocene. Cool phases developed in the early Middle Eocene (46–50 Ma) and the early Late Eocene (37–41 Ma). These intervals correspond approximately to climate changes recognised in the ocean faunas.

For the warmest phase during the Palaeocene to early Eocene Wolfe (1985) divided the vegetation into five zones. The vegetational pattern was dominated by a wide zone of tropical rainforests (mean annual temperatures of 25–30°C), which extended to latitudes as high as 50°. Beyond this, paratropical forests covered the zone from 60–65° and broad-leaved evergreen vegetation extended as high as 70° (Fig. 9.3). The polar regions were

Fig. 9.3. Vegetational zones for selected times during the Tertiary. Zones from Wolfe (1985) on base maps of Smith and Briden (1977). (A) Late Palaeocene to early Eocene, (B) Eocene cool intervals (early Middle and early Late Eocene), (C) late Middle Eocene, (D) Early Miocene. Note the equatorial contraction of the tropical rainforest, paratropical zones and broad-leaved evergreen zones from the warm late Palaeocene–early Eocene to the cooler Eocene intervals. During the warmer late Middle Eocene the same zones moved poleward again. By the early Miocene these zones were situated much nearer the equator and the broad-leaved deciduous zone expanded considerably in the north. From Wolfe (1985). Copyright © 1985 by the American Geophysical Union.

covered by dense forests of conifers. This was the time of maximum spread of warmth-loving vegetation.

During cool intervals the boundaries of these floral zones shifted towards the equator. Paratropical rainforest (mean annual temperature 20–25°C) replaced tropical rainforests in mid- to low latitudes. Broad-leaved evergreen forests were restricted to latitudes less than 50°. A very wide zone of mixed coniferous forest developed in the mid- to high latitudes and on highland areas in mid-latitudes. This coniferous vegetation was prolific at high latitudes during the middle Eocene and formed dense forests of large trees which would have produced a large quantity of organic carbon. The Yellowstone Petrified Forests in America (Dorf, 1964) and the fossil forests of the Canadian High Arctic (Basinger, 1987; Francis and McMillan, 1987; Christie and McMillan, 1991) testify to the warmth of polar climates at this time.

During the following warm interval in the late Middle Eocene the floral zones expanded poleward but not to such high latitudes, reflecting the overall cooling trend. In mid-latitudes a new vegetation type developed across North America and Asia, that of semi-deciduous tropical/paratropical forests, which were able to tolerate a distinct dry season. These included the floras of the Mississippi Embayment and the fossil flora preserved in the Green River Formation, Wyoming (Wolfe, 1985).

In Europe the humid tropical vegetation remained but the vegetation responded to climate cooling during the middle Eocene. Tropical and subtropical floras with palms in southern England were replaced by floodplain vegetation of conifers and ferns. Macrofossil and palynological analyses have shown that these changes occurred in two steps over a period of about 15 m.y., rather than resulting from a single dramatic climate change (Collinson *et al.*, 1981).

Interpretation of European pollen assemblages highlighted Eocene climatic fluctuations similar to those proposed by Wolfe. Hubbard and Boulter (1983) considered that European winters during the Eocene were probably mild but the summer and mean annual temperatures varied widely. In contrast, the Oligocene floras indicate that the climate was then much cooler with strong fluctuations in the severity of the winters, sometimes below freezing, but with more constant summer and mean annual temperatures.

In southern high latitudes southern beech (*Nothofagus*) rainforest dominated the vegetation. *Nothofagus*, podocarp and araucarian conifers, plus a few other early angiosperms, flourished in Antarctica throughout the Eocene in warm temperate climates with high rainfall (Kemp and Barrett, 1975; Dettman, 1990). Growth rings in fossil wood from the tip of the Antarctic Peninsula (Francis, 1986b) and leaf analysis of fossil floras from the South Shetland Islands (Zastawaniak, Wrona and Birkenmajer, 1985) indi-

cate that the climate in this area (about 60° S) was warm to cool temperate.

The severe cooling interval over the Eocene–Oligocene transition and the evolution of a less equable climate led to the development of vegetation with high mean annual temperature ranges (Wolfe, 1985). Broad-leaved deciduous floras characterized by low mean annual temperatures but high mean annual temperature ranges (i.e. cold, seasonal climates) tolerated high-latitude sites. Evergreen vegetation shifted to latitudes lower than 35° and tropical vegetation to less than 15° in the Oligocene (lower latitudes than comparative zones today). Pollen assemblages from the early–mid Oligocene of Canada illustrate that the high-latitude climate was by then very cool (Norris, 1982). Mixed coniferous forests also grew in low-latitude sites at high altitudes, such as in the newly formed Himalayan zone (Wolfe, 1985).

However, forest vegetation still persisted in Antarctica during the Oligocene, even though the ice sheets were locally well developed by this stage (Zastawaniak *et al.*, 1985; Birkenmajer and Zastawaniak, 1990). The southern beech and conifer forests may have been of tundra type on the continent but rainforests with large trees still grew on the relatively lower latitude South Shetland Islands.

During the cooler climates of the Miocene the vegetation became highly diverse. In the mid-Miocene warm interval broad-leaved evergreen vegetation expanded into higher latitudes up to 45° N along the Pacific coast of northern USA, in western Europe and in Japan. Mixed conifer forest was able to grow at latitudes as high as 75–80° N in the Arctic (Hills, Klovan and Sweet, 1974). The first dry, savanna-type vegetation appears to have developed in low latitudes in the early to mid-Miocene and grasslands in mid-latitudes slightly later (Wolfe, 1985). In the late Middle Miocene temperatures dropped again, although high latitudes remained relatively warm, especially the Arctic region, which was still free of perennial ice at this time. Broad-leaved deciduous forest remained in Siberia until about the late Miocene when it was replaced by mixed coniferous forest. Tundra-type vegetation does not appear to have developed in polar regions until late in the Miocene (Antarctica) or even the Pliocene (Arctic).

Climate change on land is also illustrated by the variation in soil types. In the Palaeocene–Eocene laterites and bauxites occurred in a fairly wide zone up to 45° latitudes, suggesting that the zone of seasonally high precipitation expanded to mid-latitudes (Frakes, 1979). They became less common in the Oligocene during a period of drier climates. McGowran (1979) correlated the occurrence of laterite and deep weathering to periods of warm and humid climates, and the formation of silcrete to cooler, more arid climates in Australia during the Tertiary. These weathering types correspond to times of warming and cooling in the oceans (Fig. 9.1) and from this McGowran (in press) proposed warm–cool climate cycles of approximately 20 m.y.

duration (10 m.y. between warm–cool). These cycles are of similar ampli-
tude to those proposed for the late Jurassic to early Cretaceous.

Chronology of the Cool Mode

Following the late Palaeocene to early Eocene peak of warmth for the
Cenozoic, the climate cooled in a series of sharp changes towards the
Pleistocene glacial environments. The cooling trend began during the
middle Eocene at about 50–55 Ma. Oxygen isotope data show that both
bottom and surface ocean waters cooled synchronously. Two moderate
enrichments of $\delta^{18}O$ occurred, near the Early to Middle Eocene boundary,
and in the late Middle Eocene; these have been interpreted as global ocean
temperature drops, either with or without the formation of ice. Although
there is no evidence for major ice cap or sea ice formation at this time, there
may have been continental ice in West Antarctica (Birkenmajer, 1991) and
seasonal winter ice in the Arctic regions (Dalland, 1977).

Enrichments in ocean $\delta^{18}O$ at the Early–Middle Eocene boundary and late
Middle Eocene suggest the oceans cooled markedly at this time, possibly
due to the development of shallow circum-polar circulation around Antarc-
tica. Bottom water temperatures had cooled to 6–10°C. However, about late
Middle Eocene time some large tropical microfossils expanded their ranges
to higher latitude zones (McGowran, 1986) while other microfossils seem
to indicate a latitudinal shift related to cooling. In contrast, land vegetation
zones expanded poleward, suggesting a climate warming. The degree of
warming suggested by the plant migration was not as large as that of the late
Palaeocene to early Eocene so this latter one was perhaps weakened by the
cooling climate indicated by ocean temperatures.

A dramatic climate change occurred at about 38 Ma (Eocene–Oligocene
boundary), marked by a strong enrichment of $\delta^{18}O$, particularly in benthic
foraminifera, indicative of deep-sea cooling and a clear sign of the build-up
of ice on Antarctica. At this time a psychrospheric ocean developed as south
polar ice caused the formation of cold bottom waters, replacing the more
homogeneous warm ocean structure of previous times. A major faunal
turnover occurred in the oceans and many species migrated to lower
latitudes. The proposal that an extraterrestrial impact caused environmen-
tal changes and faunal extinctions at this time is now regarded as unlikely
(Corliss and Keigwin, 1986).

During the Oligocene temperatures decreased only slightly and this
seems to have been a cool and fairly equable time. During the mid- to late
Oligocene the Drake Passage opened, allowing the isolation of Antarctica in
cold polar waters. Ice-rafted debris in the Ross Sea indicates that glaciers
reached sea level by about 30 Ma (mid-Oligocene), although fossil plant

evidence shows that Antarctica was by no means totally covered with ice at that stage.

The early Miocene appears to have been relatively warm but a striking enrichment of $\delta^{18}O$ at about 15–13 Ma during the Middle Miocene has been attributed to a massive build-up of ice on East Antarctica and a marked cooling. Latitudinal temperature gradients had become much stronger and the southern polar regions cooled markedly, although plant evidence suggests that the northern high latitudes were still warmer than the south. Low latitudes became noticeably arid at this time. At 6 Ma there was a shift in carbon isotopes to lighter values, probably as a result of a drop in sea level and increased input of organic carbon from sediments on land. After a slight warming in the early late Miocene the end of the Miocene was subject to another phase of intense cooling (at about 5 Ma). Ice-rafting occurred in lower latitudes, siliceous biogenic activity increased, and there was a pronounced drop in sea level which may have led to the desiccation of the Mediterranean Basin.

In summary, important features of the Tertiary Cooling Mode include the rather intermittent but marked, 'step-like' cooling phases superimposed on the general cooling trend. These have been related to specific events, mainly tectonic in origin and resulting in changing ocean circulation patterns. The carbon cycle appears to have been a strong influence on changing atmospheric CO_2 levels and warming or cooling phases. The change from a rather homogeneous warm ocean to one with cold bottom water circulation was an important event. High-latitude regions seem to have been far more sensitive to climate change than lower latitudes during the Tertiary, indicating that high-latitude regions of the past may contain the best clues to previous climate changes. Finally, cooling and warming phases were quite well represented by the migration patterns of ocean faunas and land plants.

10

The late Cenozoic Cool Mode: late Miocene to Holocene

The late Cenozoic was a time of major palaeoclimatic and palaeoceano-graphic events (Fig. 10.1). The latest Miocene was the most critical time: there was a global cooling of ocean surface waters at middle and high latitudes; a northward migration of cold Antarctic surface waters; an expansion of the Antarctic ice sheet; a major sea-level fall; isolation and desiccation of the Mediterranean Basin; and a lightening in carbon isotopes. In the early Pliocene, there was a warming and a major sea-level rise, followed by late Pliocene climate deterioration and gradual intensification of northern hemisphere glaciation leading to late Pleistocene climate changes visible in the prominent 100 k.y. cycle. The late Cenozoic thus has the most complete record available for defining global change during development of a Cool Mode, especially as related to orbitally driven glacial cyclicity.

Summary of marine climates from oxygen isotopes

Oxygen isotopic ratios in foraminifera from late Cenozoic deep-sea sediments are primarily a function of the $\delta^{18}O$ composition and temperature of the ambient sea water. The abyssal waters of the ocean experience relatively small changes in temperature, thus the $\delta^{18}O$ in benthic foraminifera is mainly a function of the isotopic composition of the bottom water, which in turn is controlled largely by global ice volume. However, the effect of diagenesis was emphasized by Killingley (1983) who noted that the long-term decrease in ocean temperatures from the Palaeocene to the Miocene, derived from oxygen isotopes, might reflect progressive recrystallization of calcite and therefore all palaeotemperature estimations for the early Tertiary remain speculative. Horrell (1987) considered the possibility of fluctuations in $\delta^{18}O$ content during geological time as being due to hydrothermal circulation at oceanic spreading centres, and Olausson (1988) argued that $\delta^{18}O$ values in pre-Quaternary sequences have been influenced by water circulation through oceanic crust. Oxygen isotope determinations there-

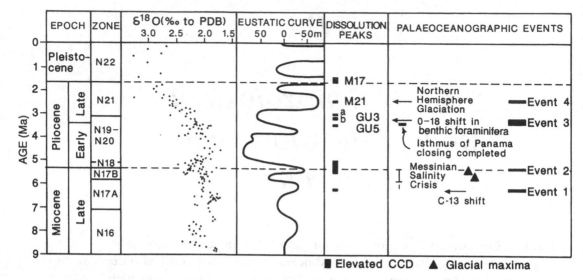

Fig. 10.1. *Summary of some palaeoclimatic changes of the late Neogene. The oxygen isotope composition of benthic foraminifera in DSDP Site 588 from the South Pacific, sea level eustatic changes and dissolution peaks are redrawn from Malmgren and Berggren (1987). Palaeoceanographic events 1 to 4 are as follows. Event 1 (at 6.2 Ma) is time of global cooling, major sea level fall, isolation of the Mediterranean Sea from the global ocean and the depletion of carbon isotopes. Event 2 (at 5.3 Ma) is the time of restoration of the marine connection between the Mediterranean Sea and the world ocean as an effect of global warming and sea level rise at the Miocene–Pliocene boundary. Event 3 (at 3.4 to 3.2 Ma) is associated with cooling and increased bottom water production at high latitudes. Event 4 (at 2.4 Ma) is known as a time of initiation of the northern hemisphere ice sheets. From Malmgren and Berggren (1987). Copyright © 1987 by the American Geophysical Union.*

fore must be applied to palaeoclimatology with caution but are most reliable where synchronous changes and marked reversal can be demonstrated from widely separated areas.

Following the sub-Antarctic cooling which began at ~9 Ma (Ciesielski *et al.*, 1987), the latest Miocene and early Pliocene (6.2–4.5 Ma) in the south Pacific were marked by increased $\delta^{18}O$ values, indicating cooling and glacial development in the southern hemisphere (Kennett, 1986). In order to resolve the details of this latest Miocene climate deterioration and glacial expansion, Keigwin (1987a) correlated a high-resolution stable isotope record from the North Atlantic with a second record from the south-west Pacific and identified two periods of glacial expansion centred at 5.2 and 4.8 Ma. Each event had a duration of ~15 k.y. and the 0.6‰ isotope signal suggested a minimum of 60 m amplitude for the latest Miocene sea-level change. Data from the Mediterranean (Gudjonsson, 1987) show coolings at

~4.8 and 4.2 Ma. Nikolayev, Blyum and Nikolayev (1986) published a reconstruction of the late Miocene thermal structure of ocean water masses from oxygen isotope data. They conclude that the surface layer (0–50 m) was ~1.7°C colder than it is now.

During the middle Pliocene a 0.4‰ enrichment in benthic $\delta^{18}O$ indicates an increase in global ice volume and cooling of bottom waters. During this interval (3.4–2.4 Ma) surface waters warmed by 2–3°C in middle latitudes of the southern hemisphere. The late Pliocene (2.6–2.4 Ma) increase in benthic $\delta^{18}O$ is commonly attributed to the development of large ice sheets in the northern hemisphere.

The Pliocene oxygen isotope record from the South Atlantic provides evidence for the temperature history from 4.6 to 1.7 Ma (Williams, Hodell and Vernaud Grazzini, 1985; Weissert and Oberhansli, 1985). The two major isotopic events in both the planktonic and benthic oxygen isotope records from several DSDP sites are centred around 3.2 and 2.6 Ma. There is general agreement that the benthic enrichment at 3.2 Ma seen in cores from around the world did not result from a large, permanent increase in global ice volume, but rather from a circulation event which included a decrease in bottom water temperatures (e.g., Prell, 1984a; Shackleton *et al.*, 1984; Weissert and Oberhansli, 1985; Kennett and Hodell, 1986). Hodell, Williams and Kennett (1985) suggested that this event resulted from a pulse-like increase in the formation of Antarctic Bottom Water (AABW).

The isotope event at ~2.6–2.4 Ma is shown by parallel $\delta^{18}O$ enrichment in both planktonic and benthic foraminifera, suggesting an increase in global ice volume. This is believed to have resulted from the initial rapid build-up of northern hemisphere ice sheets and coincides with changes in several other proxy climate indicators (e.g. ice-rafted debris, deep-sea calcium carbonate abundance). Shackleton *et al*. (1984; DSDP site 552A) determined that a North Atlantic bottom water mass similar to that of today first formed at ~3.5 Ma, and that major ice-rafting began at 2.5 Ma, with a maximum at 2.4 Ma.

Late Neogene climates

The late Neogene was a time of significant changes in the global climate system as evidenced by a wide range of proxy data (changes in living biomass, ocean and atmospheric circulation and ice volume). To summarize briefly, severe cyclic glaciations and cooling during the latest Miocene to earliest Pliocene (6–4.8 Ma) were followed by distinctive warming and rise of sea level during much of the early Pliocene (Keany, 1978; Keigwin, 1979; Haq *et al.*, 1987; Mercer, 1987), which in turn were followed by cooling during the middle to late Pliocene. An increase in $\delta^{18}O$ began at ~3.4 Ma and reflects cooling of deep water and/or an expansion of ice volume on

Antarctica (Prell, 1984a; Hodell, Williams and Kennett, 1985; Elmstrom and Kennett, 1986; Kennett, 1986). A further significant $\delta^{18}O$ enrichment occurred at ~ 2.5 Ma and has been interpreted as an indication of the onset of northern hemisphere glaciation (Shackleton *et al.*, 1984; Prell, 1984a; Louberie and Moss, 1986). Since then, major glacial–interglacial oscillations have intensified in northern hemisphere ice sheets, as is evidenced by a general increase in the magnitude and variability in $\delta^{18}O$ (Shackleton and Cita, 1979; Moore *et al.*, 1981).

As our knowledge of Neogene climatic history increases there is evidence for greater than previously expected variability of the global climatic system, especially fluctuations in ice volume and its distribution in high latitudes. During this time the Antarctic ice sheet has been a permanent feature, although oxygen isotopic evidence indicates changes in ice volume.

Climates from oxygen isotopes

Oxygen isotopic records for the Cenozoic show an increase in $\delta^{18}O$ towards those of the present, reflecting gradual cooling and increased glaciation. This trend did not proceed uniformly, but was punctuated by times of rapid change that represented sudden transitions from one climatic state to another. These transitions, according to Kennett and Hodell (1986), represented threshold events that incorporated strong positive feedback mechanisms which pushed the climatic system towards a new steady state. During the Neogene, two such step-like changes occurred: in the middle Miocene between 16 and 13 Ma and in the late Pliocene at ~ 2.5 Ma. Several events of smaller magnitude have been recorded: at ~ 10 Ma; during latest Miocene (5.6–5.1 Ma); and in the middle Pliocene at 3.4–3.2 Ma (Fig. 10.2).

On the basis of correlation between estimated short-term changes in sea level and the Miocene oxygen isotope record, Moore, Loutit and Greenie (1987) suggest that the polar cap built up and melted back to nearly the same volume several times in the mid- to late Miocene. The apparent correlations suggest that ice-volume changes have affected sea level at least as far back as 16.5 Ma.

Foraminiferal $\delta^{18}O$ changes reflect temperature and global sea-water compositional (ice volume) changes. Benthic foraminiferal values can be used to identify times of glaciation if it is assumed that cold ($< 2°C$) bottom waters indicate that high latitudes were frigid enough to support large ice sheets (Miller *et al.*, 1987). Covariance between planktonic and benthic $\delta^{18}O$ signals from a site is used to identify changes due to sea-water compositional changes and hence fluctuations in global ice volume (Shackleton and Opdyke, 1973). Positive covariance indicates times of ice sheet expansion, whereas negative covariance implies decreased continental ice volume. Synchronous changes have also been interpreted as concomitant temperature change in bottom water and low- to middle-latitude surface water (e.g. Savin, 1977).

Fig. 10.2. Record of the benthic oxygen isotope and carbon δ¹³C variations from DSDP sites 591 and 591B (south-west Pacific) and biostratigraphic zonations of site 591, plotted against age. Adapted from Kennett (1986).

From oxygen isotopes, the global climatic trends in the late Neogene are summarized as follows. The late Miocene $\delta^{18}O$ changes record mid- and high-latitude cooling and pulses of ice sheet expansion and contraction. The latest Miocene and earliest Pliocene (6.2–4.5 Ma) were marked by enhanced cooling and glaciation, with glacial events being most intense during 5.5–5.35 Ma, and at 5.2 Ma and 4.8 Ma (Keigwin, 1987a). Interglacial conditions prevailed between 4.8 and 4.1 Ma. After 4.1 Ma Pliocene climate began to deteriorate and at 3.4–3.2 Ma benthic $\delta^{18}O$ enrichment documents a net increase in global ice volume and cooling of bottom waters. Between 3.2 and 2.5 Ma conditions were stable. At ~2.5 Ma a series of $\delta^{18}O$ enrichments occurred and has been interpreted as establishment of northern hemisphere mid-latitude ice sheets (Keigwin, 1987b). During coolings, surface waters warmed by 2–3°C in southern middle latitudes and the vertical temperature gradient in the ocean increased.

Latest Miocene events

The latest Miocene climatic history has long fascinated students, and involves possible relationships between Antarctic glaciation and Mediterranean isolation (Frakes, 1979; Hodell *et al.*, 1986; Cita and McKenzie, 1986; Muller and Hsu, 1987). Keigwin (1987a), combining stable isotope stratigraphy and magnetostratigraphy from the North Atlantic and the southwest Pacific, has been able to document brief pulses of glaciation, based on covariance of $\delta^{18}O$ in benthic and planktonic foraminifera. The two maxima are dated at 5.2 and 4.8 Ma. Each event had a duration of ~15 k.y. and lowered sea level by at least 60 m. From the amplitude of the oxygen isotope signal, Keigwin *et al.* (1987) and Hodell and Kennett (1986) concluded that variability of latest Miocene ice volume was about one third that found in the late Pleistocene. Such small signal amplitude places restrictions on our ability to separate ice volume and temperature effects, especially from benthic foraminifera where the entire range of $\delta^{18}O$ variability could be accommodated by temperature changes of 2–3°C. The two events were inferred by Keigwin (1987a) to reflect glaciation, although an equally valid interpretation would be that surface and deep waters cooled simultaneously by 2–3°C. Records of glacier expansion in Patagonia (Mercer, 1983), brief intervals of ice-rafting in South Atlantic cores (Ciesielski and Weaver, 1983) and eustatic lowering of sea level in the latest Miocene support the ice-volume interpretation.

The Messinian salinity crisis was a catastrophic event which led to the desiccation of the Mediterranean Sea and the deposition of 10^6 km^3 of evaporites between 6.5 and 4.8 Ma. Weijermars (1988) suggested that tectonic activity, together with a glacioeustatic lowering of sea level, caused the isolation and subsequent desiccation of the Mediterranean. The rapid removal of calcium carbonate and calcium sulphate in the Mediterranean

Basin would have made the open ocean to some degree unsaturated, causing increased dissolution of deep-sea carbonates in the late Miocene (Thunell, Williams and Howell, 1987). In the early Pliocene the marine connection with the Atlantic was restored and the Messinian event terminated; deep-sea carbonate accordingly increased in abundance.

According to Muller and Hsu (1987), between 6.2 and 5.7 Ma the Antarctic ice sheet expanded rapidly and global sea level dropped, leading to partial water flow restriction to the Mediterranean. Glacial and interglacial conditions alternated and several layers of evaporite were precipitated. At 5.7 Ma Antarctic glaciation reached a maximum and at this point the Mediterranean Basin was completely isolated from the Atlantic. Evaporites accumulated at the following times: near 5.7, 5.5–5.4, and 5.25–5.15 Ma, the last of these just preceding the suggested deglaciation of the West Antarctic ice sheet. Flooding of the Mediterranean Basin between 5.1 and 4.9 Ma was probably caused by tectonic activity leading to opening of the Gibraltar Straits. The consequent southward flow of warm and saline Mediterranean waters is documented by fluctuations in $\delta^{18}O$ values in the South Atlantic (McKenzie and Oberhansli, 1985). This warmer water mass may have triggered the deglaciation of the West Antarctic ice sheet at about this time.

Hodell and Kennett (1986), on isotope data, also suggest that the isolation and desiccation of the Mediterranean during the latest Miocene may have been controlled by glacioeustatic sea-level oscillations related to ice-volume fluctuations on Antarctica. At 5.5 Ma there was an expansion of ice on Antarctica and the resulting eustatic fall may have severed the final connection between the Atlantic and Mediterranean, thus leading to deposition of the lower evaporite unit of the Messinian. A temporary ice-volume decrease between 5.3 and 5.2 Ma caused the intra-Messinian transgression that temporarily refilled the Mediterranean and interrupted evaporite deposition. Between 5.2 and 5.1 Ma further ice growth led to deposition of the upper evaporite unit. A rapid increase in $\delta^{18}O$ at 5 Ma marks a glacial retreat and marine transgression, which permanently terminated evaporite deposition (Hodell *et al.*, 1986).

Pliocene events

Oxygen isotope values continued to decrease during the early Pliocene, reaching a minimum during the interval between 4.5 and 4.0 Ma. However, the most depleted $\delta^{18}O$ values over this time are not significantly different from modern values, suggesting similar temperature and ice volume to today. During the middle Pliocene, heavy isotopes at \sim 3.4–3.2 Ma reflect an ice-volume increase and/or cooling of bottom waters. The enrichment at 3.2 Ma was a temporary feature in most planktonic and benthic records (Prell, 1984a; Keigwin, 1986) but is permanent in a deep-water site (DSDP 518) underlain by AABW (Hodell *et al.*, 1985). Prell (1984a) suggested that

the 3.2 Ma event was a brief, temporary increase in ice volume followed by a permanent cooling of bottom water. During the late Pliocene (2.6–2.4 Ma) a further step-like increase in $\delta^{18}O$ has been interpreted as due to the onset of northern hemisphere glaciation (Shackleton *et al.*, 1984; Backman, 1979). By 2.4 Ma the extent of the influx of ice-rafted debris to the North Atlantic and the range of $\delta^{18}O$ are similar to those observed at maxima during the middle Pleistocene (Shackleton *et al.*, 1984; Keigwin, 1986).

The middle Pliocene (3.3–2.5 Ma) is characterized by widespread ice masses in both hemispheres and by strong isotopic differences between glacial and interglacial stages. Near the base of the Gauss chron, low-amplitude changes in the benthic isotope curve for the South Atlantic are replaced by high-amplitude fluctuations and by a positive shift in the average $\delta^{18}O$ composition. This change at ~ 3.3 Ma documents the most fundamental change in Pliocene glacial history. A 0.4‰ positive shift in the average mean of the benthic values, coupled with a large standard deviation, reflects an increase in the difference between positive glacial and negative interglacial times and also documents the increase in southern hemisphere ice volume (Weissert *et al.*, 1984), as known from expanded glacial-marine sedimentation around Antarctica (Ciesielski, Ledbetter and Ellwood, 1982).

The transition from a warm early Pliocene to a cooler middle Pliocene is not reflected in the planktonic oxygen-isotope record. Assuming that the $\delta^{18}O$ value of the middle Pliocene was more positive by 0.4‰ than in the early Pliocene as a result of an increase in ice volume, the contrasting planktonic values suggest a 2°C warming of South Atlantic surface waters (to 30° latitude) at the time of ice sheet growth (Keigwin, 1979; Vergnaud Grazzini, 1980; Weissert *et al.*, 1984).

Neogene ice sheets

Shackleton and Kennett (1975) proposed an expansion of the Antarctic ice sheet in the late Miocene to a size 50% greater than present, on the basis of $\delta^{18}O$ enrichment. This figure was challenged by Woodruff, Savin and Douglas (1981) who argued that the maximum late Miocene ice volume was probably similar to that of the present day, and by Mercer and Sutter (1982) who estimated the late Miocene Antarctic ice volume to have been half or a third that of the late Pleistocene and suggested that additional ice was present in the northern hemisphere.

The history of Neogene continental ice is complex. Mercer (1983) found evidence that Patagonian glaciers extended beyond the Andean mountains for the first time between 7 and 5.2 Ma. Denton *et al.* (1984) suggested that the Transantarctic Mountains were overridden by an extensive ice sheet at least twice, before and after the middle to early late Miocene. The oldest sand-sized ice-rafted debris in cores from the south-east Argentine Basin is dated at 5.57–5.35 Ma. The latest Miocene glaciations were followed by a

warming during the early Pliocene (Keany, 1978; Ciesielski *et al.*, 1982), and Ciesielski and Weaver (1974) postulated early Pliocene surface water temperatures 10°C higher than present, which should have caused at least a partial deglaciation of the West Antarctic ice sheet. By 4.35 Ma, however, the frequency and amount of accumulation of ice-rafted debris at South Atlantic sites had increased significantly, suggesting an increase in the stability of the West Antarctic ice sheet.

Several arguments support the view of early development of northern hemisphere glaciation. Mercer and Sutter (1982) suggested that during the late Miocene, south-east Alaskan glaciation extended to sea level, in synchroneity with ice advances in Patagonia. The idea is supported by widespread cooling evident from oxygen isotopes, and by marine regression. From these events, a major late Miocene ice-volume increase in Antarctica is inferred.

Earlier growth of ice bodies is inferred for the Norwegian–Greenland Sea (4.3 Ma; Eldholm *et al.*, 1986), Iceland (~9.8 Ma; Mudie and Helgason, 1983), Alaska (6.3 Ma; Armentrout, 1983) and elsewhere (e.g. Leinen, 1985). Early growth of northern glaciers thus was irregular and regionally controlled, including very early ice-rafting in Kamchatka (possibly since the early Eocene, and pack ice intermittently present in the Arctic Basin since 13–10 Ma (Clark *et al.*, 1980; Herman and Hopkins, 1980; Zubakov and Borzenkova, 1983).

Thick late Cenozoic ice formed in the Soviet High Arctic and ice-rafting extended southward to 50° N (Laukhin, 1986); traces of ice-rafting are known in sediments from Sakhalin (Serova, 1985) and are accompanied by abundant glendonite crystals, which are taken as indicators of cold water conditions (Kemper, 1987). During the late Oligocene to early Miocene, latest Miocene to earliest Pliocene and late Pliocene climate cooling was extreme and ice-rafting reached to about 45° N in the Pacific part of the USSR.

The second half of the Oligocene to the earliest Miocene is identified in the far east of the USSR as the coldest time, including glaciation in Kamchatka (Fotyanova and Serova, 1987). Pre-Pleistocene glacial strata are widespread in Alaska, where the earliest evidence of glaciation is about early middle Miocene (Plafker, 1971; Pewe, 1976). The oldest recognized glaciation in Alaska is that along the north flank of the Brooks Range and is represented by erratic boulders (Hamilton, 1986). Mountain glaciation probably began about 13–10 Ma in the St Elias Range and other ranges surrounding the Gulf of Alaska according to Pewe (1976). Apparent tillites from the Wrangel–St Elias Mountains are interstratified with lava flows of late Tertiary age (tillites dated by K–Ar method are 10.5, 10.2, 10.1, 8.9, 8.5, 3.7 and 2.8 Ma) (Hamilton, 1986; Denton and Armstrong, 1969). During this time glaciers repeatedly formed ice shelves and discharged icebergs into the ocean and a mudstone (Yakataga Formation) accumulated offshore. Late

Miocene is the most likely age for the lower Yakatanga Formation (Armentrout, 1987). Large glaciers also existed on the flanks of the Wrangel Mountains at about 10 Ma according to Denton and Armstrong (1969). One of the most extensive glaciations is the Brown Glaciation on the north flank of the Alaska Range, which may be 2.7 m.y. old (Pewe, 1976). The first evidence of alpine glaciations in Iceland, Alaska, and western Canada has been noted by Mudie, de Vernal and Head (1990).

It was in the late Pliocene that the northern hemisphere ice sheets grew to significant size; this has been dated at 3.2 Ma (oxygen isotope shift; Shackleton and Opdyke, 1977). However, recent data suggest that large northern ice sheets did not accumulate until about 2.5 Ma. Prell (1982) presented oxygen isotopic evidence from equatorial regions which suggests that the 3.2 Ma oxygen isotopic event represents only a brief excursion in ice volume. Shackleton *et al.* (1984) found that high-amplitude oxygen isotope frequencies extend back only to about 2.4 Ma. The age of inception of ice-rafting observed in cores from high northern latitudes varies from region to region, but such material older than 2.5 Ma possibly represents glaciation not related to major ice sheets (Margolis and Herman, 1980; Poore, 1981; Warnke, 1982). Distinct surface water cooling associated with late Pliocene glacial development is known from the Southern Ocean (Kennett, 1985), the Mediterranean Sea (Thunell, 1979) and the Atlantic (Thunell and Beleya, 1982). Preliminary results from the Ocean Drilling Programme from around Antarctica point towards synchronous climate behaviour in both hemispheres. About 2.4 Ma production of diatoms in Antarctic surface waters decreased sharply, presumably in response to increased sea ice extent (Kerr,1987).

The oxygen isotopic record from DSDP site 548 (north-east Atlantic) shows that northern hemisphere glaciation developed gradually, through a series of cycles. This result contrasts with other records, which indicate constant palaeoclimatic conditions preceding the sudden, step-wise, development of glaciation at 2.5–2.4 Ma (Louberie, 1988). The results from site 548 indicate that the development of northern hemisphere ice sheets could be regarded as due to a progressive climatic deterioration rather than a geologically sudden climatic event (see Fig. 1 in Louberie, 1988).

Recent international correlation of Quaternary glaciations in the northern hemisphere (Sribrava, Bowen and Richmond, 1986) show that continental glaciation existed well before the Pleistocene, and there are many localities with glaciation of Miocene and Pliocene age (Fig. 10.3). In particular, in the USA, the ages of some Pliocene glaciations are well established; Cordilleran and Laurentide ice sheets were present as early as ~ 2.14 Ma and mountain glaciation by ~ 3.1–2.8 Ma (Richmond and Fullerton, 1986). Alpine deposits of several Pliocene glaciations have been identified in mountain ranges in the western USA and Alaska. Drift of single Laurentide

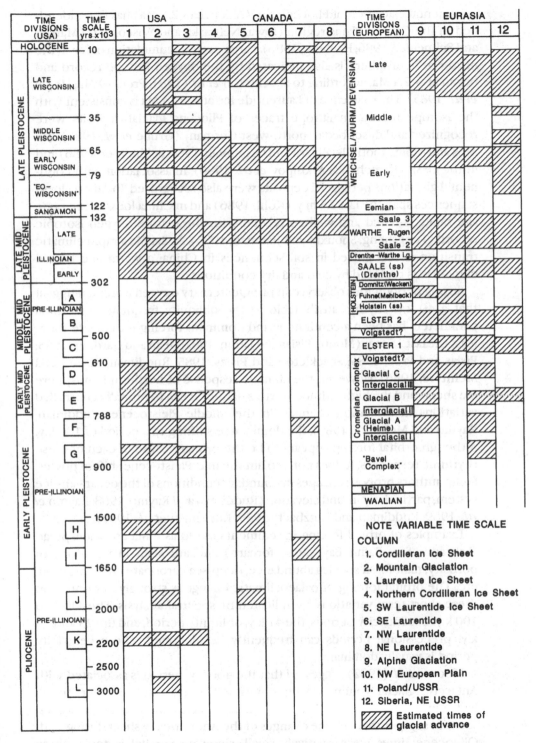

Fig. 10.3. *Correlation of glacial events of the northern hemisphere during the last 3 m.y. From Sribrava et al. (1986). Copyright © 1986 by Pergamon Press plc.*

continental glaciation of Pliocene age (older than 2.1 Ma) has been identified in the Central Plains region (Iowa, Nebraska and Wisconsin) (Matsch and Schneider, 1986; Hallberg, 1986). Large-scale accumulation of ice in the northern hemisphere is also indicated by the oxygen isotope record and began at ∼ 2.4 Ma, according to Shackleton *et al.* (1984; Prell, 1984a; Hodel *et al.*, 1985). The existence of Laurentide ice at ∼ 2.14 Ma is consistent with the isotope record. In Europe traces of Pliocene glacial deposits were recognized and described in north-west Germany (Grube *et al.*, 1986), the Alps and Alpine foothills (Billard and Orombelli, 1986; Schluchter, 1986) and in the USSR (Kamchatka; Arkhipov *et al.*, 1986). In association with continental glaciations periglacial deposits were also recognized. In Austria loess sequences span the last 2.5 m.y. (Kohl, 1986) and in China loess of the same age range is dated and well correlated with the marine oxygen isotopic record (Tungsheng, Shouxin and Jiaomao, 1986). At 2.4 Ma a rapid climatic transition is recognized in soil sequences in China; a warm and humid climate was replaced by cold and dry conditions.

Strong periodicities observed in pre-Quaternary oxygen isotopes suggest that a 40 k.y. rhythm attributable to the obliquity frequency of orbital parameters has been a continuous and dominant forcing for climate for at least the last 15 m.y. (Moore, Pisias and Dunn, 1982; Dean and Gardner, 1985; Pisias and Prell, 1985; Shackleton and Pisias, 1985; Ruddiman, Raymo and McIntyre, 1986). However, the climatic response of northern hemisphere ice sheets and the high-latitude North Atlantic Ocean to orbitally controlled variations in insolation changed in the middle Pleistocene (Ruddiman, Raymo and McIntyre, 1986), the dominance shifting to a period of 100 k.y. Although orbital forcing appears to be the primary cause of each of these rhythmic responses, it does not explain the mid-Pleistocene shift of power. Some authors propose changes in boundary conditions of the ocean–air–ice system, particularly in land elevation (Ruddiman and Raymo, 1988; Raymo *et al.*, 1989; Ruddiman and Kutzbach, 1989; Ruddiman *et al.*, 1989).

Examples of orbital forcing in sediments as old as 9 Ma are widespread (Labrador Sea, Baffin Bay, Japan forearc) and are seen in variations of porosity, nanofossil species abundance, deep-sea carbonate concentration, etc. The orbital forcing of palaeoclimatic change is strongly imprinted on these short-term variations, as indicated by spectral analysis. The 413 and 100 k.y. eccentricity periods, the 41 k.y. obliquity period, and the 23 and 19 k.y. precessional periods can frequently be identified. The eccentricity period is usually dominant.

Moore *et al.* (1982) suggested that the 400 k.y. period is associated with Antarctic glacial volume changes during Pliocene and Miocene times. Isotopic data presented by Leonard, Williams and Thunell (1983), however, seem to suggest that volume changes of the Antarctic ice sheet during mid-Oligocene times were relatively small, since no parallel isotopic phase

relationships were confirmed between the benthic and planktonic forami-
nifera. The wealth of palaeontological, sedimentological, and oxygen-iso-
tope evidence of deterioration of palaeoclimatic conditions in the deep-sea
environment before 2.4 Ma appear to reflect distinct temperature variations
rather than the waxing and waning of major ice sheets in Antarctica or in the
northern North Atlantic region (e.g. Shackleton and Opdyke, 1977; Ledbet-
ter, Williams and Ellwood, 1978; Keigwin and Thunell, 1979; Kennett *et al.*,
1979; Hodell, Kennett and Leonard, 1983). These temperature oscillations
appear to have occurred in coherence with orbital elements, and the 400
k.y. eccentricity period played a dominant role. The many indications of
climatic deterioration at ~ 2.8, 3.2 and 3.6 Ma cannot, at present, be
associated with major glacial build-up.

 Assuming that deep-sea carbonate cycles were caused by global events, as
the periodicities suggest, and assuming that variation in supply of terrige-
nous clastics was the main cause of the cycles, then we need a cyclic driving
mechanism that could pulse terrigenous clastics with the appropriate
periodicity to produce the cycles. The only likely mechanism for this is
climate change. While orbital cycles have been suggested as the main
driving forces behind short-term climatic change (Hays, Imbrie and Shackle-
ton, 1976), the relationships between orbital cycles, climate change, and
cycles of sedimentary climatic indicators imply periodicities that are more
regular than are actually observed. In addition, different climatic indicators
may have different periodicities. This suggests that the primary record of
climatic forcing has been modified by effects such as tectonic movements,
sediment bioturbation and ocean circulation, and that the resultant records
are complex products of primary climatic signals with superimposed local
effects (Dean and Gardner, 1986).

 Assuming that there is a connection between variations in the orbital
parameters and climate, what are the connections between climate and
sediment composition? One of the best established links is that between
global ice volume, the oxygen isotopic composition of the ocean, and the
oxygen isotopic composition of both planktonic and benthic foraminifera.
Studies of cyclic fluctuations in the oxygen isotopic composition of forami-
nifera from Quaternary deep-sea sediments have related these cycles to
variations in global ice volume and hence to changes in global sea level
(Shackleton, 1967; Matthews and Poore, 1980).

The role of carbon

According to present understanding (Shackleton, 1987a) carbon isotope
measurements in deep-sea sediments may be used to obtain the history of
the mean $\delta^{13}C$ content of dissolved carbon in the ocean, of $\delta^{13}C$ gradients
within the ocean, and of the globally averaged $\delta^{13}C$ content of the carbonate
accumulation on the sea floor. Carbon isotope reconstructions provide

information on the changing global budget of organic carbon accumulation and allow comparison with climatic events associated with these changes. The fundamental question in understanding long-term cycle changes is whether changes in the global organic carbon reservoir are controlled mechanically (e.g. Kennett, 1986) by the effect of global sea-level fluctuations on the accumulation and erosion of organic-rich sediments, or geochemically (Shackleton, 1987a) through atmospheric and oceanic oxygen levels.

Shackleton (1987a) suggested that a major post-Miocene decrease in organic carbon storage was climatically controlled (ocean deep-water cooling by 10°C during the late Tertiary) and was accompanied by a 20% decrease in the oxygen content of the atmosphere. Raymo, Ruddiman and Froelich (1988) proposed a scenario for the influence of increased mountain building during the last 5 m.y. on the ocean geochemical cycle. The increase in chemical weathering resulting from intensified uplift can lead to depletion of atmospheric carbon dioxide either by consumption of CO_2 during weathering reactions or by influencing the oceanic carbon cycle. They argued that carbon dioxide concentration in the atmosphere may be controlled not only by the *balance* between sea-floor spreading rates, land–sea fraction, and atmospheric feedbacks on weathering rates (Lasaga *et al.*, 1985; Volk, 1987), but also by relative *rates* of sea-floor spreading (CO_2 production) and continental orogeny (CO_2 consumption). Only recently has more effort been made to understand palaeoceanic chemical variability during the last 60 m.y. (Delaney and Boyle, 1988).

In particular, it has been realized that geochemical cycles are complexly linked (such as to nutrients and global atmospheric CO_2 and carbon cycles and those of sulphur and trace metals such as iron etc.), and simple scenarios for geochemical variations cannot explain these complex system interactions. Fluctuations in the $\delta^{13}C$ of the ocean's carbon pool are seen as records of change in the input/output ratio of organic and inorganic carbon (see for discussion Berger and Mayer, 1987; Shackleton, 1986a, b, 1987a; Sundquist, 1987). Differences between oceans are explained in terms of basin-to-basin fractionation (Shackleton, 1985; Keigwin and Boyle, 1985), whereas differences in the $\delta^{13}C$ of shallow and deep water are linked to the nutrient concentration in the deep ocean (Broecker, 1982; Berger, 1985; Shackleton and Pisias, 1985; Sundquist, 1987). Based on these ideas, fluctuations in $\delta^{13}C$ seen in the Neogene record can be interpreted as follows. If the benthic and planktonic signals are parallel, the cause of fluctuations is due to interaction with an external reservoir. If changes are different for benthic and planktonic, variations in productivity are involved. If changes are asymmetric between ocean basins, the cause is inter-ocean exchange.

Palaeoclimatic change contributed to many of the observed $\delta^{13}C$ changes. Global cooling led to greater aridity and reduction in size of the continental

biomass, whereas global warming enchanced the accumulation of organic carbon in continental reservoirs by expansion of the biosphere due to greater moisture supply (Berger, 1982; Kennett, 1986). For example, the palaeobotanical record indicates widespread changes in vegetation during the late Neogene. The general increase in desert and grassland area in Africa, Australia, Asia and Europe is a manifestation of the overall trend of biomass destruction on land due to climate changes (intensified glaciation and palaeoceanographic changes, for example, the closing of the connection between the Indian and Atlantic Oceans led to dry climate in the Mediterranean). The decrease of 1.5–2.0‰ in ^{13}C values since the Miocene may have resulted directly from these changes (Fig. 10.2). Also, the short-term (10^4 years) cycles in $\delta^{13}C$ are thought to be related to alternations of the glacial and interglacial climates (Sundquist, 1987; Curry and Crowley, 1987). Cycles seen in the record of benthic and planktonic foraminifera at Site 552A (Keigwin, 1986) may represent the effect of orbital variations on tropical precipitation, and hence biomass, as proposed by Keigwin and Boyle (1985) for the late Quaternary.

Broad changes in the Neogene $\delta^{13}C$ record were caused, according to Kennett (1986), by transfer of organic carbon between continental and oceanic reservoirs. These transfers were caused by marine transgressions and regressions on the continental margins. During sea-level high stands, increased storage of organic matter would occur in reservoirs on the continental shelves, whereas during low stands erosion of these reservoirs would result in transport of ^{12}C-rich organic matter to the deep sea. Barron (1987), following this argument, assigned times of increased $^{13}C/^{12}C$ ratios in deep-sea benthic foraminifera to times of higher global sea level, and abrupt negative shifts in this ratio to rapid sea-level falls. Palaeoclimatic changes reinforced $\delta^{13}C$ changes by influencing the size of the continental biomass.

The $\delta^{13}C$ depletion at 6.7–5.5 Ma in the latest Miocene is recorded worldwide and was followed by generally low values during the Pliocene. The exception was the early Pliocene (5–4 Ma) when an increase occurred which can be correlated with a marine transgression. In New Zealand the Blind River latest Miocene carbon isotope event shows a marked shift in both carbon ($-1.0‰$) and oxygen ($+0.9‰$) between \sim6.65 and 6.45 Ma, followed by a rapid carbon shift ($-0.8‰$) centred on 6.3 Ma (Edwards, 1987). The permanent part of this trend is about $-1.0‰$ carbon and $+0.45‰$ oxygen, indicating a significant long term decrease in productivity, in temperature (about $-1.4°C$), and in sea level (about -15 m), perhaps caused partly by increased polar ice accumulation.

Permanent depletion of carbon isotopes in foraminifera may have begun slightly earlier (\sim6.7 Ma) but maximum values were reached at \sim6.2 Ma (Hodell and Kennett, 1986). The shift has been ascribed to change in rate of

organic carbon burial resulting from change in upwelling and ocean fertility by Bender and Keigwin (1979) and to regression and change in deep-ocean circulation patterns and fertility by Berger (1982, 1985).

The lowering of the carbonate compensation depth throughout the oceans during the late Miocene had an origin similar to that of the carbon isotope shift, having resulted from an increase in carbonate and organic matter entering the deep sea from the continental shelves during the latest Miocene sea-level fall (Ciesielski *et al.*, 1982; Kennett, 1986).

Climates from marine organisms

The last 10 m.y. of climate history in the Atlantic was characterized by important changes in the marine biota. Variations in assemblages suggest strong cooling at ~1 m.y. intervals (9.5–5.5 Ma), followed by major episodes of cooling at ~4–2.8 and ~2 Ma. Several equatorward and poleward migrations of surface water masses in the Atlantic (essentially 1 m.y. periodicity) were delineated by Poore (1981) using numerical abundance of cold- and warm-water species of planktonic foraminifera. Cooling events were identified at 9.5, 8.5, 7.5 and 5.5 Ma, with a subdued cold event at 6.5 Ma; warming events were identified at 9.8 and 7.6 Ma. Somewhat later, the Arctic faunal province was first established during the time of initiation of northern hemisphere glaciation (3–2.4 Ma), and this was followed by continued provincialization of planktonic faunas. Wei and Kennett (1986) and Malmgren and Berggren (1987) show a large degree of correlation between evolutionary changes of Neogene planktonic foraminifera and palaeoceanographic and climatic changes. North Atlantic calcareous plankton biogeography during the Pleistocene is essentially a continuation of water-mass migration across ~20° of latitude (equivalent sea surface temperature oscillations of 12°C). It has been estimated that at least 11 such oscillations have occurred in the last 0.6 m.y. and perhaps as many as 20 during the past 1.2 m.y. (Berggren and Olsson, 1986).

Around the margins of the Atlantic, mass extinctions began in late Pliocene time (around 3.5 Ma) and ceased during early Pleistocene time. This pattern of extinction suggests that climatic cooling was the dominant agent. In the Mediterranean, there was a major extinction event at ~3.2–3.0 Ma. Additional marked extinction here and also in the western Atlantic region probably coincided with another pulse of cooling at 2.5–2.4 Ma (Stanley, 1985). The marine section from the US Atlantic coastal plain contains information about climate evolution during the last 4 m.y. (Cronin, 1988). Pliocene marine sediments deposited during a period of high sea level between 4.0 and 2.8 Ma contain increasing percentages of tropical and subtropical ostracodes, indicating gradual warming of climate. After maximum warming at ~3.2 to 2.8 Ma sea level fell and the climate cooled significantly.

At ∼3 Ma a sudden climatic deterioration took place that affected both the marine fauna and the flora of Iceland. The warmth-loving molluscs were replaced gradually by boreal and later by Arctic species (Gladenkov, Norton and Spaink, 1980). Molluscs indicate a water temperature, at ∼3 Ma, at least 5°C higher than present but cooling led to a fall in water temperatures by 2–3°C by 2.5 Ma. The cool form, *Portlandia arctica*, appears in glacial marine sediments along with the first foreign ice-rafted material at ∼2 Ma (Einarsson and Albertsson, 1988).

There is considerable evidence for a warm interval at ∼2 Ma, which conflicts with suggestions of a cooling at about this time. Houghton and Jenkins (1988) presented a calcareous nanofossil record from the Celtic Sea indicating warm climatic conditions and high sea level at about 2.1–1.9 Ma. A 2.1 Ma warming was also suggested from palaeoenvironmental studies on planktonic foraminifera from the Mediterranean Sea (Thunell, 1979). These warmings are of the same age as one in the Arctic reported by Carter *et al.* (1986) and Funder *et al.* (1985) and also coincident in age with the first known Laurentide ice from North America (2.14 Ma). Either warming and glacial build-up were concurrent, with warm oceans serving as a moisture source for ice formation, or the warming represents a subsequent deglaciation. High sea-level records between 2.1 and 1.9 Ma support the latter case (Jenkins *et al.*, 1985).

Pacific coolings took place at the Miocene–Pliocene boundary, between 3.3 and 2.4 Ma, and at ∼1.6 Ma. In the tropical Pacific, quantitative study of late Cenozoic silicoflagellates shows that the greatest amplitude of fluctuation in relative palaeotemperature occurred in the late Miocene. The warmest intervals occurred in the middle Miocene and in the late Pliocene to Quaternary. A Miocene–Pliocene boundary warm peak was followed by a cold period at ∼4.5 Ma and a further warm peak (3.8–3.4 Ma). A sea-level rise in the Pacific correlates with the tropical warming from ∼5.2 to 5.1 Ma (Bukry, 1985).

The distribution of Neogene molluscan faunas in Japan indicates cooling during the latest Miocene and warming in the early Pliocene. However, the Pliocene fauna consists of temperate and partly subarctic species, some of which are new immigrants from the north (Chinzei, 1987). At ∼2 Ma warm species were introduced into the Japan Sea. A northern fauna is present in Hokkaido during the Pliocene, while in northern Honshu it is known only during a short interval at ∼3 Ma, indicating cooling at this time. A major pulse of mollusc extinction due to climate deterioration has been reported from central Japan (Matsuoka, 1987) and was centred at 3.0–2.5 Ma. Matsuoka presented arguments indicating global extinctions due to cooling in the late Pliocene, and a gradual decrease of sea-water temperature during the period 3.8–1.6 Ma is recorded from central Japan by Ibaraki (1987), on the basis of changes in planktonic foraminifera assemblages.

Low-latitude diatom response to the late Pliocene cooling at 2.4 Ma is characterized by numerous extinctions and evolutionary first appearances of diatom taxa. The North Pacific features migrational events with cold water forms penetrating equatorwards along the California coast, and warm water and transitional forms extending their range poleward along the Japanese coast (Burckle and Abrams, 1986). Migrational events are dominant over extinctions. A similar scenario is proposed for the Southern Ocean where a cooling trend was interrupted at ~ 3.3 Ma by a warm interval lasting until ~ 2.5 Ma. At 2.4 Ma many Antarctic diatoms migrated northward, suggesting a northward movement of the Polar Front. Burckle (1987) examined diatoms in Pliocene sediments from the Atlantic sector of the Southern Ocean as well as from the Pacific and south-east Indian sector. Results show that circum-polar diatom productivity was generally high in the early and middle Pliocene, with sea ice some 3–10° south of the present limit and the Polar Front well south of its present position. In southeast Indian sector sites close to Antarctica there is a sharp increase in relative abundance of.resting cells of the ice-related diatom *Eucampia antarctica* at ~ 2.4 Ma, possibly reflecting increased sea ice and/or iceberg activity.

Major oceanographic singularities, including the polar fronts, were no doubt affected by late Cenozoic climate changes. Ciesielski and Weaver (1983) have been able to reconstruct the history of Polar Front migration during the last 4 m.y. from radiolarian assemblages from the South Atlantic. Fig. 10.4 displays the position of the data site relative to the Polar Front. From 4.1 to 3.86 Ma there were distinct warm intervals. The first was interrupted by cooling at ~ 3.96 Ma. Also a more extensive cooling occurred at ~ 3.8 Ma (full extent unknown due to hiatus). This may correspond to the Patagonian glaciation (3.59 Ma; Mercer, 1983) and to significant cooling in the northern hemisphere.

In New Zealand, a number of long-ranging warm-water molluscs and benthic foraminifera have their last known occurrences at ~ 3.1 Ma (Edwards, 1987), indicating the onset of cooling and/or water-mass restructuring. The late Pliocene record from the Blind River marine section contains a number of poorly dated events indicative of climatic deterioration. They include regressive strata with cool palynofloras, a number of minor planktonic microfossil events and, probably, New Zealand's oldest glacial strata (the Ross glaciation).

Morley (1987) interpreted variations in radiolarian assemblages from the north-west Pacific (34–42° N) as occurring in response to climate changes during the last 3 m.y. Mixed subtropical and subarctic faunal assemblages dominated all three sites between 3.0 and 2.7 Ma. By ~ 2.4 Ma the siliceous faunal assemblages at all sites resembled closely those existing at present. Between 1.6 and 1.1 Ma the relative abundance of polar assemblages increased gradually in response to a cooling trend throughout the north-

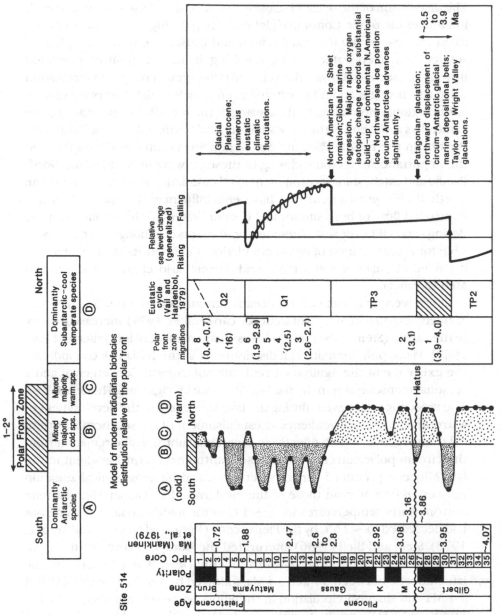

Fig. 10.4. Fluctuations of the Antarctic Polar Front Zone at Site 514 relative to its present location, based on radiolarian assemblage characteristics. Fluctuations are correlated to eustatic cycles and selected palaeoenvironmental events and episodes. From Ciesielski and Weaver (1983); Ludwig, Krasbennikow and Wise (1983).

west Pacific. At ~1.6 Ma a major change from low-amplitude variations to high-amplitude variations took place.

Ocean circulation

The circulation of the oceans is closely linked to the global climate and each influences the other. Continued global cooling during the late Neogene led to an increase in the thermal gradient and caused the final decoupling of surface and deep-water masses, resulting in the creation of restricted upwelling (see Baldauf and Barron, 1990). Neogene changes in circulation were also brought about by tectonic events such as the disruption of low-latitude global circulation by the closing of the Indo-Pacific passage and the Paratethys at the end of the early Miocene, and by the uplift of the Isthmus of Panama in the middle Pliocene. Scholl and Stevenson (1987) emphasized the importance of tectonic closing of these seaways in isolating the North Pacific and establishing its prominent gyral circulation pattern. Vigour of the North Pacific gyral circulation has been influenced by the tectonically controlled flow of both surface and deeper waters to the Pacific from the Beringa and Okhotsk Seas. These tectonic events are thought to be responsible for reorganization of ocean circulation and intensification of meridional wind circulation and a more arid to semi-arid climate of Africa and Mediterranean.

There were major effects on oceanic circulation as a result of the late Miocene expansion of the Antarctic ice cap (Frakes, 1979), increased aridity in the Sahara (Stein, 1986), and the Messinian salinity crisis (Muller and Hsu, 1987). Planktonic foraminiferal distributions in the Indian Ocean indicate the existence of the Agulhas current, the Subtropical Convergence and a seasonal monsoon system by the late Miocene (Wright and Thunell, 1988). The most important event during the late Miocene was the development of a strong temperature gradient and establishment of the southern Subtropical Convergence, which effectively separated subtropical circulation from the circum-polar current. An increased latitudinal thermal gradient in the late Miocene produced surface circulation and biogeographical zonation not very different from those of the modern Indian Ocean. High-latitude bottom water temperatures were ~5°C in the middle to late Miocene but had decreased to ~2.5°C by the Pleistocene (Vergnaud Grazzini and Letolle, 1978). Stein and Bleil (1986) interpreted the isotope record at Site 141 to show intensified advection of North Atlantic Deep Water (NADW) in the late Miocene between 5.7 and 5.4 Ma, followed by a reduction in NADW at 5.4 Ma and sluggish circulation in the Pliocene between 4.5 and 2.75 Ma. Intensified NADW production as indicated by sea-floor erosion took place at 2.75 Ma. Murray (1984) found a similar age for increased NADW in the North Atlantic. Stein, Sarnthein and Suendermann (1986) used a numerical model of the Atlantic Ocean for simulating the palaeo-deep water circulation during the late Neogene.

Despite several limitations of the model, simulated circulation results are encouraging. The calculated deep-water temperatures in the north-east Atlantic decreased from ∼9°C at 3.5 Ma to ∼5.4°C at 2.4 Ma. These results indicate that there was almost no temperature difference between deep (2600–4000 m) and bottom (>4000 m) water masses during the early Pliocene and also suggest enhanced circulation during times of climatic change rather than during the time of glacial maximum.

In the late Miocene the latitudinal temperature gradient in the Pacific was ∼12°C (Loutit, Kennett and Savin, 1984). Late Miocene planktonic foraminiferal faunas show a warming of low-latitude surface waters and northward migration of cold Antarctic waters into the Indian Ocean. The climatic warming in the early Pliocene (Hodell and Kennett, 1986) most affected the subtropical province and there was less distinction between the transitional and subpolar provinces (Wright and Thunell, 1988). This may have been due to slightly warmer high-latitude surface waters.

The relatively warm and stable conditions of the middle Pliocene were ended by northward migration of the Polar Front from ∼2.8 to 2.6 Ma (Fig. 10.4). Periods between 3.1–2.85 and 2.45–2.3 Ma were marked by significant bottom and deep water production and intensified circulation of bottom waters in the south-west Atlantic (Turnau and Ledbetter, 1989). Also, Abelmann and Gersonde (1988) indicate major cooling and changes in global deep-water circulation patterns near 2.3 Ma on the basis of radiolaria distribution in the Southern Ocean.

A continuing trend toward a modern deep-water stratification during the late Pliocene was a direct response to the formation of extensive northern hemisphere ice sheets (Turnau and Ledbetter, 1989). It was manifested by production of Mediterranean outflow water and a deepening of NADW in the South Atlantic. Prior to the late Pliocene, during the early Miocene warm climatic phase, global thermohaline circulation was strongly influenced by the influx of warm saline water from the Tethyan area into the northern Indian Ocean (Woodruff and Savin, 1989). During the late Miocene to Pliocene, deep-water circulation changed from deep water influenced by warm saline waters to one dominated by cold deep waters generated in high polar latitudes. This change was related to a closure of the Tethys and growth of the polar ice caps.

Extensive deposition of aeolian sediment in offshore equatorial Africa (Stein, 1985) and in North Pacific pelagic sediments (Rea and Janecek, 1982) also indicates intensified atmospheric circulation at roughly the same times. According to Stein this was caused by an increased temperature gradient between the north pole and equator due to extended northern hemisphere glaciation. This supports the idea that prior to the late Pliocene glaciation, in the late Miocene, sizeable ice masses may have been present in high northern latitudes (Mercer and Sutter, 1982; Mudie and Helgason, 1983; Laukhin, 1986).

Antarctic climate history in the Neogene

Pre-late Miocene global ice volume may have been more variable than previously assumed. For example, siliceous microfossils recovered from the Sirius Formation indicate open water and restricted ice in the Wilkes and Pensacola basins of Antarctica during the late Eocene to early Oligocene (40–33 Ma), late Oligocene to earliest Miocene (28–23 Ma), middle Miocene (17–14 Ma), and Pliocene (5–2.5 Ma) (Harwood, 1986). This history is supported by concurrent episodes of high sea level and light $\partial^{18}O$ isotopic values. In addition, there is evidence for ice volumes significantly greater than at present (Denton *et al.*, 1984). An increase in siliceous productivity at ~5 Ma indicates an intensification of upwelling associated with the Antarctic Convergence that may be related to growth of the West Antarctic ice sheet and/or extension of the East Antarctic ice sheet.

Harwood proposed the following scenario of climatic events in East Antarctica during the last 7 m.y.; late Miocene (7–5.1 Ma) general cooling and major ice-volume increase, followed by early Gilbert (5.1–4.2 Ma) deglaciation and marine invasion of the Dry Valleys. The late Gilbert (4.2–3.2 Ma) was a time of ice-volume increase and cooling, which was in turn followed by a warm interglacial with marine invasion during the Gauss (3.2–2.5 Ma). The following period of ice accumulation from 2.4 to 1.8 Ma is known as the Queen Maud Glaciation.

Ice-rafting was significant at 6.9 Ma in the south-west Atlantic Ocean and may have been in response to a short late Miocene warming, before the latest Miocene glaciation. Bornhold (1983) indicates that accumulation rates of ice-rafted debris remained relatively low until 4.15 Ma, after which they increased markedly.

The West Antarctic ice sheet may have become unstable and collapsed in the early Pliocene, according to Hughes (1982). Isostatic sinking of the West Antarctic continental shelf under the ice load, combined with glacial erosion of the interior, created a crustal depression which led to instability in the ice sheet. Collapse would have flooded the Southern Ocean with icebergs and might have triggered the cooling in high southern latitudes from 3.7 to 3.5 Ma.

Early Pliocene (4.5–3.5 Ma) fossiliferous marine sediments from the Marine Plain in the Vestfold Hills (East Antarctica) indicate reduced ice volume and warmer interglacial climate than the present (Pickard *et al.*, 1988). Harwood (1986) speculates that Pliocene deglaciation was sufficient to form a seaway into the Wilkes and Pensacola basins, with water temperatures of 2–5°C. Terrestrial vegetation established itself during this warm interval at latitude 86° S (Webb and Harwood, 1987; Carlquist, 1988). Later, the ice sheet thickened and eventually overrode the Transantarctic Mountains. These basins are covered by 3 km of ice at present.

Climates from floras

Major vegetational changes during the late Neogene included development of steppe in western North America (Pliocene), taiga formed in the Arctic at ~5 Ma and tundra at ~3–2 Ma (Wolfe, 1985). Late Neogene aridity in Australia and the Sahara and the Namib regions of Africa continued to expand in step with global climate deterioration (Kennett and Von der Borch, 1986; van Zinderen Bakker and Mercer, 1986).

Wolfe (1985) concluded that the origin and spread of low-biomass vegetational types during the Neogene resulted from two major factors. First, in polar areas temperatures continued to decrease and taiga and tundra types of vegetation gradually spread. Secondly, at low latitudes temperatures continued to increase. The resulting increase in the equator-to-pole gradient led to intensification of the subtropical high-pressure systems and summer drought along the western sides of continents. Low-latitude warming led to additional water requirements for plants and the low-latitude savanna thus increased.

The trend toward vegetation types reflecting cooler conditions was not uniform and there is some evidence for brief returns to warmer climate, with vegetation growing in high latitudes. Funder *et al.* (1985) indicated that 2 m.y. ago, forest tundra grew ~2500 km further north than at present in Greenland and the Canadian Arctic. Also Webb and Harwood (1987) reported *in situ* tree/shrub stems and roots of Pliocene and possibly early Pleistocene age from the Transantarctic Mountains. However, some form of tundra has probably existed around the Arctic Basin since the middle Miocene (Matthews, 1987). The flora of Iceland changed substantially after ~3 Ma; the conifers and temperate deciduous forests were replaced by floras similar to present ones, with dwarf *Betula* and *Salix* (Einardsen and Albertsson, 1988). At least eight glaciations occurred in Iceland in the interval from 2.0 to 0.7 Ma.

Hooghiemistra (1984), in studying a palynological record from Colombia spanning the last 3.6 m.y., recognized 26 glacial cycles, with the most prominent change occurring at ~2.8 Ma. Also, Fuji (1988) recovered sediment cores from Lake Biwa containing pollen records of climate for the past 3 m.y. Pollen analysis shows cold stadials in the earlier part of the record similar to the late Quaternary; however, lack of a more refined time scale prevents any more detailed conclusions about climate.

In New Zealand the dominant late Miocene flora of southern beech *Nothofagus* (as indicated by *brassi* pollen assemblages) was replaced during the latest Miocene to late Pliocene by *fusca* beech-dominated pollen assemblages (Nelson *et al.*, 1988). A gradual decrease in palynoflora diversity up the sequence probably represents cooling conditions. The change to *fusca*-type pollen assemblages (*Nothofagus fusca* zone) indicates the onset

of temperate climates. Further cooling is indicated by the extinction of *Acacia* during the late Pliocene to early Pleistocene.

Quaternary climatic events

We live in the Quaternary period which has been assigned a beginning at 1.8 Ma. During this time Earth climates have undergone dramatic fluctuations, with periodic glaciation of lands at high latitudes and high altitudes and accompanying changes in sea level, biosphere productivity and chemistry of the atmosphere and ocean. In particular during the last 0.6 m.y. the Earth experienced major environmental changes as climate oscillated between glacial and interglacial states. Significant variations have taken place in ice volume, sea levels, temperatures, atmospheric CO_2 concentration, animal and plant geography and patterns of oceanic and atmospheric circulation. Deep-sea sediments show that over a range of periods from 1 to 400 k.y. climatic spectra are dominated by three narrow band signals that are highly coherent with orbital forcing. Variations in the annual cycle of incoming radiation associated with changes in precession (19 and 23 k.y.), obliquity (41 k.y.) and eccentricity (100 k.y.) confirm the external forcing functions of the global climate system (Milankovitch cycles). According to Imbrie *et al.* (1984) approximately 80% of the observed climatic variance in these bands is linearly related to and presumably forced by these variations in the seasonal and latitudinal radiation patterns. Current efforts are focused on explaining and modelling the physical and chemical feedback mechanisms which make Earth's climate so sensitive to these small variations in the annual radiation budget.

A century of work with glacial deposits has shown that reliable chronologies cannot be achieved for the continents because of incomplete and undatable sequences (Kukla, 1977; Bradley, 1985). In contrast, the marine record is more often continuous; oxygen isotope composition of the shells of foraminifera in oceanic sediments preserve an almost complete record of fluctuating ice volumes and temperature. Records from many cores taken throughout the world ocean show very similar trends and contain the same periodicities in ice-volume fluctuations, as deduced from $\delta^{18}O$ (e.g. Martinson *et al.*, 1987), and in sea surface temperatures reconstructed from foraminiferal assemblages (CLIMAP, 1976, 1981; Hays *et al.*, 1976) (Fig. 10.5). The oxygen isotope record is a powerful stratigraphic tool that also represents a climatic standard; an accurate Pleistocene time scale for deep-sea cores is a basic prerequisite for comparisons of variations of incoming insolation with proxy-indicators of climate. Radiometrically dated horizons have been used to maximize coherencies between orbital and isotopic signals at the three frequencies (41, 23 and 19 k.y.). The resulting cross-spectrum has significant coherencies not only at those frequencies but also at 100 k.y., indicating that ice volume is driven by eccentricity as well as by precession and obliquity (Imbrie *et al.*, 1984).

Fig. 10.5. *Chronology of the Pleistocene oxygen isotope record from various ocean basins. Detailed oxygen isotope records are correlated using nanofossil biostratigraphy and palaeomagnetic stratigraphy. The Composite and Type isotope record reveals that the mid-Pleistocene change in climatic regime was a gradual transition that lasted from 900 to 600 k.y. BP and not a sudden shift as previously postulated. Figure assembled from Williams et al. (1988) and McKenna et al. (1988). With permission of Elsevier Science Publishers.*

The oxygen isotope record from DSDP site 610 in the North Atlantic (Jansen and Sejrup, 1987) shows the mode of climate variability for the last 2.4 m.y. Prior to 0.9 Ma the isotopic variability record is smaller in amplitude (but not in frequency) compared with the late Quaternary, indicating smaller ice-volume and climatic fluctuations, and higher average sea level. Still earlier, large fluctuations in polar planktonic faunas and oxygen isotopes between 2.4 and 2.3 Ma indicate a period of even greater climatic variability. Data from deep-sea cores indicate that large scale climatic shifts occurred at 1.3 and 0.8 Ma. Beginning at ~1.3 Ma the climate was punctuated by short, cold pulses with durations of up to 30 k.y. Watts and Hayder (1984) suggested that the longer cold periods which begin at ~0.9 Ma could have been triggered by relatively minor changes in any of several parameters that determine the ice sheet response to orbital parameter variations.

Oxygen and carbon isotope analyses of benthic foraminifera from DSDP site 517 on the Rio Grande Rise provide an expanded record of climatic trends during the entire Pleistocene (Vergnaud Grazzini *et al.*, 1983). This

period is characterized by more than 14 climatic cycles. Three separate intervals can be recognized. Glacial extreme $\delta^{18}O$ values earlier than 1.8–1.6 Ma were not isotopically heavier than present-day calculated equilibrium values. Between 1.6 and 0.9 Ma the values for glacial extremes increased progressively to reach values that were 1.5‰ heavier than present equilibrium values. During the last 0.9 m.y., glacial values remained nearly 1.5‰ heavier than today. This last phase might be related to a greater stability in the ice volume of northern hemisphere ice sheets. However, a composite isotope record for the last 1.88 m.y. (Williams *et al.*, 1988) shows that the mid-Pleistocene change in climate regime was a gradual transition that started at 0.9 Ma and ended at 0.6 Ma and not a sudden shift from one climate mode to another. The transition between climatic states involved a high degree of instability and a complicated mixture of frequencies.

Orbital theory of climate change

According to the astronomical theory of palaeoclimates, the long-term variations in the geometry of the Earth's orbit are the fundamental cause of the succession of Pleistocene ice ages. Spectral analysis of climatic records of the past has provided substantial evidence that a considerable fraction of the climatic variance is driven in some way by insolation changes caused by orbital forcing. Three variable elements of the Earth's orbit influence the total solar radiation received in its geographical distribution. These are the eccentricity, the obliquity of the ecliptic, and the longitude of perihelion measured from the vernal equinox (which itself is in motion). The total energy influx (insolation) depends only on eccentricity, if integrated over the globe over a year. The total insolation changes due to eccentricity vary only by less than 0.2%, and a major task therefore has been to show that this small insolation change could lead to cyclic glaciation of the Earth. The dominant term in the series expansion for eccentricity has a period of 413 k.y. and 8 of the next 12 terms range from 95 to 136 k.y. At low resolution they appear as a single broad peak of 100 k.y. cycles (Berger, 1977, 1984, 1987).

The geographic and seasonal distribution of insolation depends on obliquity and precession index $e \sin \omega$ (where e is eccentricity and ω is the longitude of perihelion). The spectrum of obliquity is dominated by components with a period of 41 k.y., although periods of 54 and 29 k.y. are not negligible. The vernal equinox precesses as the pole moves over the sky with a 25.5 k.y. period due to lunar perturbations. Variation of $e \sin \omega$ gives rise to a double spectral peak with periods of 23 and 19 k.y., which appears as a single broad peak at 21.7 k.y. The obliquity and precession variations lead to comparatively high-amplitude swings in the latitudinal distribution of incoming radiation as a function of season. For example, at 9 k.y. BP, when the Earth was last closest to the Sun in the northern hemisphere summer,

summer radiation was increased by 7% and winter radiation was decreased by the same amount, apparently with significant consequences for the strength of the global monsoon (Prell and Kutzbach, 1987).

It has been established that the climatic system is partly driven by insolation changes at the frequencies of variations in obliquity and precession. However, there has been difficulty with explaining how weak orbital variation at 100 k.y. frequency could be related to the dominant 100 k.y. frequency in ice-volume fluctuation. The question remains as to how the climate effects of the eccentricity cycle are amplified. The most often cited candidates are ice–albedo feedback, carbon dioxide and aerosol changes and non-linear effects (e.g. isostatic rebound, calving of icebergs into proglacial lakes, and bedrock deformation).

The palaeoclimate record, established through study of deep-sea piston cores, has been compared with the planetary changes of Earth motion including the three basic elements (obliquity, eccentricity and precession). The variations of insolation due to eccentricity are the smallest and yet the climatic record shows the dominant cyclicity close to 100 k.y. (according to Berger, 1977, insolation variations range from 0.17 to 0.014%). These variations are too small by a factor of ten to cause glacial–interglacial cyclic oscillations of climate.

A Walsh spectral analysis of oxygen and carbon isotopic variations during the last 1.7 m.y. reveals periodicities in orbital frequency bands (Tiwari, 1987). Besides the three orbital spectra cycles, additional periodicity is seen in a range of 57–60 k.y. throughout the Pleistocene. This periodicity corresponds to the combined effects of obliquity and precession as predicted by Berger (1977) and demonstrates an additional link between climate and incoming solar radiation modulated by orbital variations.

Imbrie *et al.* (1984) used techniques of cross-spectral analysis to determine the coherency spectrum, phase spectrum and accuracy of stacked oxygen isotope records with orbital forcing (see Berger and Pestiaux, 1984, for a discussion of spectral techniques). They have established that at each of the main orbital periods that are present in a 780 k.y.-long record (100, 41, 23 and 19 k.y.), coherencies between orbital and isotopic signals exceed 0.9 and are thus highly significant statistically. As much as 85% of the isotopic variance in each of the four main frequency bands is forced by orbital variation.

Berger, Guiat and Kukla (1981) used a simple qualitative multivariate model which translates insolation changes due to orbital forcing into the climatic curve to determine the long-term variations of the astronomic climate index for 400 k.y. BP to 100 k.y. AP (Fig. 10.6). The model predicts the next peak of natural cooling at around 5 k.y. AP, one at 22 k.y. AP and another much stronger cooling at ~60 k.y. AP.

The long-term variations in the magnitude and geographic distribution of

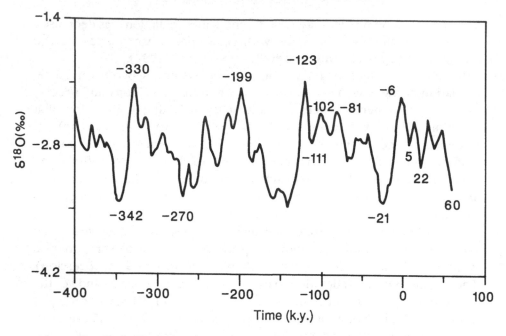

Fig. 10.6. *Variations in the global climate during the last 400 k.y. BP and future 60 k.y., derived from insolation changes due to orbital forcing of the global climate (mainly temperature). The numbers on the curve are the ages of specific peaks. From Berger et al. (1981) with permission of* Geologische Rundschau.

incoming solar radiation, due to variations in eccentricity, precession and obliquity of the Earth, are known to a high degree of accuracy (Berger and Pestiaux, 1984). However, to translate these known variations into forcing functions for the internal variables of the mass of ice sheets, the mass of marine and continental marginal ice and the mean ocean temperature is most difficult (Saltzman, Hansen and Maasch,1984).

The eccentricity variations, with major periods near 100 and 400 k.y., are the only ones that lead to net increases and decreases of total energy incoming to the Earth; these changes have an amplitude of the order of 10^{-1} W m². This quantity is comparable to the net global latent heat release and consumption involved in long-term Quaternary ice variations. However, the greater part of this quantity is made unavailable for water phase changes and heat storage by compensating changes in outgoing longwave radiation that lead to rapid equilibration of the atmosphere and upper layer of the Earth and ocean. Thus, we can expect relatively little direct forcing of the deep cryosphere and oceans by these variations (Saltzman, 1987).

The precessional and obliquity variations do not lead to a net energy input into or out of the climatic system; rather they lead to relatively high amplitude variations in the geographical distribution of incoming solar radiation. However, it is expected that the general circulation of the

atmosphere and oceans can be affected in a way that leads to ice-volume changes, mainly in the form of the more sensitive marginal marine ice masses, and these effects can be an order of magnitude greater than the eccentricity effects.

Both spectral analysis of oxygen isotopic data (Imbrie *et al.*, 1984) and climate modelling (Pollard, 1983; Saltzman, 1987; Hyde and Peltier, 1987) provide evidence that Earth-orbital radiation variations can play the critical role as 'pacemaker' for the ice age variations as was proposed by Hays *et al.* (1976). However, many fundamental questions remain to be answered. For example, how can such small amplitude periodic forcing control the phase in a complex non-linear oscillatory system, and, is there a good physical explanation for this 'phase locking' phenomenon? Combination of the three main orbital variations permits accurate estimation of insolation for any given time and place. The astronomical theory of ice ages states that northern hemisphere glaciers grow when summers receive less insolation and melt when they receive more. However, the climatic state of the Earth is not a simple function of solar insolation. In fact, the Earth modulates insolation forcing by reflection, retention and redistribution of received energy. The terrestrial mechanisms and feedbacks are the major unknowns in predicting climate changes.

Analyses of phase relationships of various climate characteristics during each Milankovitch cycle show that if dated against ice-volume changes, changes in deep-water chemistry and low-latitude sea surface temperature (SST) lead in time. Changes in SST at 50° S latitude occur somewhat later, approximately in phase with changes in $\Delta\delta^{13}C$ and ice core CO_2 (at 41 and 100 k.y. periods), but still precede the ice-volume changes. In the 23 k.y. cycle band, changes in North Atlantic SST and CO_2 lag behind ice-volume changes. SST changes associated with African and Indian monsoons are out of phase (Imbrie, 1988). The variations in Indian monsoon intensity are caused by the seasonal distribution of solar radiation, especially over the 23 k.y. precession band, and by changes in surface boundary conditions, such as the size and shape of ice sheets and the structure of the SST field. A combination of palaeoceanographic data and general circulation model (GCM) climate modelling (Prell and McIntyre, 1988; Prell and Kutzbach, 1987) for the last 150 k.y. reveals that increased summer radiation in the northern hemisphere produced stronger monsoons that were associated with more upwelling and lower SST in the western Indian Ocean (Fig. 10.7). A strong monsoon weakens the trade winds in the eastern equatorial Atlantic, resulting in less divergence and higher SSTs. According to Prell and McIntyre (1988) the monsoon thus provides a physical mechanism that explains the opposite phase of SST records in the Atlantic and Indian Oceans over the 23 k.y. frequency band. Longer (450 k.y.) records of monsoon upwelling reveal 41 as well as 23 k.y. power.

Response of the Southern Ocean in the orbital frequency band was

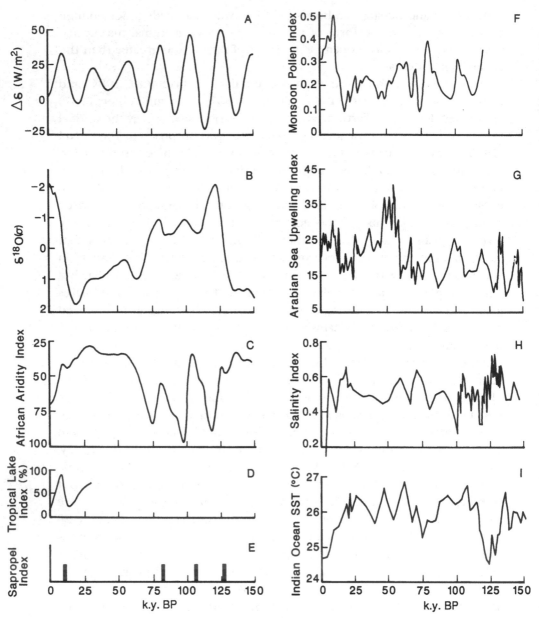

Fig. 10.7. Monsoon related palaeoclimate records for the past 150 k.y. (A) Average northern hemisphere summer radiation. (B) The SPECMAP oxygen isotope record (Imbrie et al., 1984). (C) The record of African aridity indicating wet and dry periods in equatorial Africa (Pokras and Mix, 1985). (D) Tropical lake level record, indicating percentage of lakes in the intertropical zone that are at high or intermediate levels (Street-Perrot and Harrison, 1985). (E) The Mediterranean sapropel index, indicating the presence (vertical bars) or absence of sapropels (Rossignol-Strick, 1983). (F) The record of monsoon-related pollen in the Gulf of Aden (Van Campo et al., 1982). (G) The faunal record of monsoon-related upwelling off Arabia (Prell, 1984b). (H) The faunal record of salinity. (I) The SST record of the western Indian Ocean. From Prell and Kutzbach (1987). Copyright © 1987 by the American Geophysical Union.

investigated by Morley, Prell and Howard (1988). Changes in surface temperature and circulation over the last 450 k.y., as recorded by fossil plankton in piston cores located between the Subtropical Convergence (STC) and the Polar Front (PF), allowed determination of the positions of the STC and the PF. Calculated frontal positions indicate that conditions were colder than present during much of the last 450 k.y. During glacial periods, the latitudinal spread between these two fronts averaged 12°, whereas during interglacials it expanded to a maximum of 16°. Variations in SST associated with these changes in circulation were found to be strongly coherent with orbital variations in the three orbital bands (23, 41, 100 k.y.). In each cycle, the sector-average SST response occurs early and leads the change in ice volume, with southern sites showing longer leads. In all three bands, SST changes in the Southern Ocean lead the corresponding SST changes in the mid-latitude North Atlantic.

Analysis of leads and lags among SST, ice volume and CO_2 within the orbital frequency band and in different parts of the Earth show that, in each frequency band, the SST change in each ocean region has a distinctive amplitude and phase. In the 23 and 41 k.y. bands, SST changes in the Antarctic occur in phase with $\Delta\delta^{13}C$ (proxy for CO_2), but lead changes in ice volume, which in turn lead SST changes in the boreal latitudes of the North Atlantic. In all three bands, changes in $\Delta\delta^{13}C$ lead ice-volume changes. Changes in the eastern Equatorial Atlantic at 23 k.y. dominate the signal and are in phase with changes in high southern latitudes (Imbrie *et al.*, 1986). Since orbital variations mainly affect the hemispheric and latitudinal distribution of solar energy rather than its total, continents exibit far more direct response to these changes than oceans. Therefore the northern hemisphere response is strongest because of the large proportion of land.

Physically based models have been used to model the glacial cycle. These models include prognostic equations for continental ice, its deformable bedrock, thick marine ice shelves and grounded ice, and deep-ocean temperature (see for review Saltzman, 1985). Some of these models are able to simulate a realistic glacial cycle, with slowly growing ice sheets and rapid terminations (e.g. Pollard, 1983; Saltzman and Sutera, 1987). Pollard (1984) found that with orbital forcing the phase of the 100 k.y. response is controlled by eccentricity via modulation by the precession cycle. The same cycle can also be produced by obliquity and precessional forcing acting together without eccentricity, but only over the last 0.6 m.y.

At present we do not have an adequate model for climatic change on any time scale. Glacial cyclicity observed in the climatic record is explained by some plausible scenarios involving orbital forcing applied to a complex non-linear system in which inertial components such as continental ice and its underlying bedrock, thick marine ice shelves, and the thermal state of the deep ocean are likely to play dynamic roles. The evidence seems to indicate

that orbital variations along with the carbon (or geochemical) cycle are the most likely forcing elements in short-term climate change.

Orbital variations in the climate record

As shown above, proxy palaeoclimatic data from the geological record provide verification of the astronomical theory of climatic change, and empirical evidence from ocean, cryosphere and land data supporting orbital forcing has been documented (Hays *et al.*, 1976; Ruddiman and McIntyre, 1981a; Molfino, Heusser and Woillard, 1984, among others). The glacial oscillations which dominate Quaternary global climates were closely related to changes in incoming solar radiation, but details of the palaeoclimate record indicate that other factors, usually referred to as 'feedback', significantly increased the climatic response to extraterrestrial forcing.

Much of the cyclicity seen in the Pleistocene record displays precessional forcing and is from low latitudes, where modelling with general circulation models (GCMs) has already suggested that a significant response does not depend on ice-sheet feedback. On the other hand, information about the pre-Pleistocene Tertiary is from deep-water sediments and high latitudes; these data indicate that obliquity was the control on the signal (de Boer and Shackleton, 1984). However, it is possible that low eccentricity may have contributed by leading to a reduction in poleward heat transport through a decrease in tropical heating and precipitation. Such a reduction could cause substantial ice growth due to non-linear amplification of relatively small forcing (Schneider, 1984).

Ice-rafted debris and planktonic foraminifera from the north-east Atlantic suggest a strong reaction to precessional forcing (Hartmut, 1988). Since 130 k.y. BP, ice-rafted debris has been concentrated in a periodicity equal to half the precessional cycle. Ice-rafting is most effective during times of winter minimum/summer maximum insolation and summer minimum/winter maximum insolation. Both situations result in southward propagation of polar water which then restricts warm NADW from the northern part of the basin and reduces ice growth by lessening moisture transport to the ice sheets.

High-latitude surface waters in the North Atlantic were dominated by the 41 k.y. obliquity signal and low latitudes are dominated by the 23 k.y. precessional signal (Ruddiman and McIntyre, 1984a; Ruddiman, Raymo and McIntyre, 1986). In subpolar latitudes, Pacific radiolarian species show 41 k.y. and 23 k.y. frequencies, while these same species from the equatorial Pacific show the 23 and 19 k.y. frequencies (Morley and Shackleton, 1984; Hays *et al.*, 1976). North-west Pacific marine sediments illustrate the spectra of the orbital parameters in palaeoceanographic data (Janecek and Rea, 1984, 1985) and in aeolian quartz (a proxy indicator of atmospheric activity). The latter shows highest concentrations during glacial periods on

both long period (100 k.y.) and short period (23 k.y.) cycles (Pisias and Leinen, 1984; Morley, Pisias and Leinen, 1987).

Hays *et al.* (1976) showed that, at the 41 k.y. frequency, $\partial^{18}O$ lagged the obliquity by 9 k.y., while at 23 k.y. $\partial^{18}O$ lagged precession by ~ 5 k.y. In the high southern latitudes summer temperatures led the $\partial^{18}O$ record by ~ 2 k.y. Summer temperatures from the North Atlantic are in phase with $\partial^{18}O$ during deglaciation but lag the $\partial^{18}O$ isotope record at times of ice build-up.

Pollen data of a 0.5 m.y.-long record from Lake Biwa (Japan) have been analysed spectrally and show four peaks at the periods of 104, 44, 25 and 12.7 k.y., the first three corresponding to orbital forcing (Kanari, Fuji and Horie, 1984). Pollen power spectra from the Grande Pile (Europe) show major components of the precessional index (23 and 19 k.y.) and also periodicities around 9 and 6 k.y. (Molfino *et al.*, 1984). The periodicities at 9 and 6 k.y. are related to second and third harmonics of the 19 k.y. component and thus are an indication of non-linearity in the response of continental vegetation to precessional forcing.

Significant influence of the Earth's precessional cycle on the monsoonal climate of tropical Africa has also been shown. A 260 k.y.-long record of the freshwater diatom *Melosira* in an equatorial Atlantic core strongly suggests direct modulation of monsoonal climates in tropical Africa by low-latitude solar radiation variations, which are dominated by the 23 k.y. precessional period (Fig. 10.8). The apparent lack of a 41 k.y. cyclicity in the core suggests that changes in high-latitude solar radiation (obliquity) and feedback from global ice-volume changes did not contribute strongly to climate change in this area (Pokras and Mix, 1985, 1987). This is supported by GCM results (Kutzbach and Guetter, 1986) which show that tropical precipitation is more sensitive to changes in seasonal radiation than to changes of glacial boundary conditions. Short and Mengel (1986) used an energy balance model to demonstrate that tropical climatic response reaches a maximum up to 3 k.y. after solstice–perihelion alignment (precessional cycle), in a direct thermal response to astronomical forcing.

Formation of sapropels in the Mediterranean during the Quaternary was driven by run-off from the Ethiopian highlands, which gave rise to the Nile River summer floods. The plume of the flood in the Mediterranean spreads at the surface and contributes to a low salinity layer which causes stratification of the water column and stagnation of the bottom waters. Sapropel deposition during the last 465 k.y. corresponds to insolation maxima during the summer in the northern hemisphere tropics, thus further demonstrating the physical link between orbital forcing and African monsoonal precipitation (Rossignol-Strick, 1984, 1987).

Solar radiation forced changes in the monsoonal circulation in the Arabian Sea region during the late Quaternary (Prell, 1984b). However, related upwelling changes lagged the changes in summer insolation, suggesting

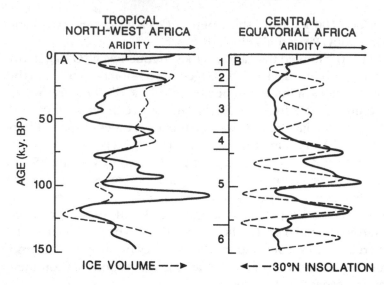

Fig. 10.8. *Summary of climatic changes in tropical Africa as indicated by the aeolian record. (A) Aridity in tropical north-west Africa is indicated by the record of diatom* Melosira *(solid line) superimposed on a smoothed ice-volume curve (dashed line). (B) Aridity in central equatorial Africa indicated by the diatom* Melosira *(solid line) superimposed on the 30° N May to July insolation curve (dashed line). From Pokras and Mix (1985), with permission of Academic Press.*

that other processes, such as seasonal snow cover, may have altered the monsoonal response to precessional forcing. Numerical simulations with a global atmospheric circulation model suggest that large-scale variations in the amount of snowfall over Eurasia in the springtime are linked to the strength of the Indian summer monsoon (Barnett, 1988).

An increase in evaporation in the Arabian Sea during the last glacial maximum (LGM) was caused by north-eastern winds (winter monsoon) blowing from the continent to the ocean (Duplessy, 1982). Enhanced continental aridity during the LGM is indicated by an increase in quartz input to the Arabian Sea and an increase in dune activity in Arabia (Sarnthein, 1978). Stronger north-eastern winds are also indicated by the pollen assemblage in Arabian Sea sediments (Van Campo, Duplessey and Rossignol-Strick, 1982). Studies of planktonic foraminifera (Prell *et al.*, 1980) show warmer SSTs along the Arabian coast during the glacial summer than at present, implying weaker upwelling and a weaker south-west monsoon. Changes in the local climate and circulation regime during the LGM resulted in significant hydrographic changes in surface and bottom waters in the Red Sea and in intermediate waters in the Gulf of Aden (Locke and Thunell, 1988). High aridity and glacioeustatic sea-level lowering were the main causes of these changes.

There is an indication that monsoonal conditions intensified in northern

Africa from 12 to 8 k.y. BP. Pollen and lake level records suggest increased rainfall (Ritchie, 1985; Street-Perrott and Harrison, 1985; Prell and Kutzbach, 1987). Similarly, pollen records from south-west Arabia suggest very wet conditions at the beginning of the Holocene (Van Campo *et al.*, 1982).

Quaternary CO_2 variations and climate

Understanding of palaeoclimatic and palaeoceanographic changes in the Quaternary has become increasingly important because of the links between atmospheric CO_2 and oceanic chemistry and circulation. Changes in Pleistocene climate are now known to have been accompanied by large changes in atmospheric and ocean chemistry. In particular, ice-core analyses of trapped air bubbles show that a large increase in pCO_2 of the atmosphere accompanied the last deglaciation. This change in atmospheric pCO_2 amplified the changes in climate caused by variations in the Earth's orbit. In fact, it has been shown that the changes in pCO_2 preceded the changes in climate and may have contributed to the deglaciation. Shackleton (1977) demonstrated that the $\delta^{13}C$ of benthic foraminifera is lower in glacial than in interglacial sediments. He also suggested that the carbon storage in terrestrial forests and soils during interglacials and the transfer of this carbon to the ocean during glaciations is the likely mechanism for carbon depletion in glacial age sediments. However, detailed study shows that the relationship between carbon isotope variability and glacial–interglacial cycles is complex.

Carbon isotope fluctuations are significantly coherent with the Earth's precession parameter (Keigwin and Boyle, 1985). Low-latitude summer insolation in the northern hemisphere leads maximum oceanic carbon isotope values (maximum continental biomass) by 6 to 4 k.y. Keigwin and Boyle propose that this relationship is due to tropical humidity variations driven by insolation distribution changes, as is evident from the GCM study of Kutzbach and Otto-Bliesner (1982).

Measurements of the CO_2 content of air bubbles trapped in ice cores from Greenland and Antarctica (Berner, Stauffer and Oeschger, 1979; Oeschger *et al.*, 1984; Barnola *et al.*, 1987) show that glacial age ice has low CO_2 concentration (200–180 p.p.m.), whereas interglacial age ice has values close to 300 p.p.m. Broecker (1982) suggested that these changes were related to the nutrient chemistry of seawater. Reduction of the high nutrient content of high-latitude surface waters by more efficient biological removal or increased residence times during glacial periods (Broecker and Peng, 1987) could also account for CO_2 changes. Recent work on the changes in ocean structure during the last glacial maximum (Keigwin, 1987c) supports the hypothesis of vertical oceanic nutrient fractionation and the role of oceanic alkalinity in controlling atmospheric carbon dioxide (Boyle, 1988a). In particular, an increased flux of North Atlantic Interme-

diate Water may have contributed to global intermediate-water nutrient depletion, thus eliminating the problem of glacial anoxia implied by models for atmospheric carbon dioxide.

Processes internal to the oceans can lead to rapid changes in pCO_2 (Sarmiento, Herbert and Toggweiler, 1988). Exchange between the atmosphere and deep ocean (through the high-latitude outcrop regions of the deep waters) is regulated by the rate of exchange between the surface and deep ocean as against the rate of biological uptake of the excess carbon brought up from the deep ocean by this exchange. In a multiple-box three-dimensional model, changes in the relative magnitude of these two processes can lead to atmospheric pCO_2 values ranging from 425 p.p.m. to 165 p.p.m. Another such mechanism is related to the fact that organic carbon is regenerated primarily in the upper ocean whereas $CaCO_3$ is dissolved primarily in the deep ocean. The resulting chemical balance depends on the rate of calcium carbonate formation at the surface against the rate of upward mixing of deep waters. This mechanism can theoretically lead to atmospheric CO_2 concentrations in excess of 20 000 p.p.m., although oceanic values greater than 1100 p.p.m. are unlikely because calcareous organisms would have difficulty surviving in such environments.

Detailed measurements on ice cores from Greenland (Dansgaard *et al.*, 1984) indicate that the deglacial climatic changes occured rapidly, on time scales of hundreds and not thousands of years. Ice core pCO_2 values cannot be explained by carbon transfer and nutrient chemistry hypotheses, since such chemical processes require time scales longer than 10^2 years (Broecker, Peteet and Rind, 1985).

A possible model for the change in atmospheric CO_2 concentration during the transition from glacial to interglacial time involves increases in deep-ocean nutrients during glacial time due to stripping of the continental shelves of nutrient-rich organic material (Broecker, 1982). This would increase the storage of carbon in the deep sea, decrease the amount of CO_2 in ocean surface waters and atmosphere and thereby provide positive feedback to cold climates. However, the ice-core data for significant carbon dioxide changes on time scales of a few hundred years (Stauffer *et al.*, 1984) similarly led to rejection of this hypothesis. Faster changes might be explained by hypotheses involving changes of circulation or biological productivity in the high-latitude ocean (Broecker, 1984; Broecker and Peng, 1987).

The CO_2 content of the deep ocean is increased by polar ocean cooling because the solubility of carbon dioxide is thereby increased. This process and others, including reduction of deep ocean ventilation rates, would reduce atmospheric pCO_2. Boyle (1986) suggests that CO_2 changes are caused by polar temperature changes together with variability in low-latitude upwelling.

Atmospheric pCO_2 is a function of the efficiency with which carbon is removed from the surface waters of the oceans, whose carbon chemistry is in equilibrium with the atmosphere. The efficiency of extraction is dependent on the amount of nutrients brought up by the upwelling deep water (Berger and Mayer, 1987). Numerous hypotheses with various dominant mechanisms have been proposed to account for changes in nutrient supply and upwelling, and several authors have explored mechanisms which have a high-latitude origin and also significantly influence ocean productivity (e.g. Wenk and Siegenthaler, 1985; Toggweiler and Sarmiento, 1985; Ennever and McElroy, 1985; Shackleton and Pisias, 1985; but see Boyle, 1986). The importance of changes in productivity and the efficiency of nutrient utilization in Antarctic surface waters is usually emphasized. Geological evidence of productivity variations in the Antarctic waters provides some support for such hypotheses (Toggweiler and Sarmiento, 1985), and the phase relationships between $\Delta\delta^{13}C$ and insolation are consistent with high-latitude forcing (Shackleton and Pisias, 1985). However, a strong direct correlation between temporal variations in high-latitude productivity and $\Delta\delta^{13}C$ has yet to be established.

Several explanations for the 200–280 p.p.m. glacial to interglacial change in atmospheric CO_2 concentration invoke variations in high-latitude phytoplankton productivity and nutrient supply. Martin (1990) proposed a scenario where productivity is limited by iron supply, despite nutrient availability, and thus CO_2 concentration is high. During the last glacial maximum, the supply of iron in atmospheric dust was some 50 times higher than now, which led to greatly enhanced phytoplankton growth and thus drawdown of atmospheric carbon dioxide to a level of 200 p.p.m. Sarnthein *et al.* (1988) proposed a mechanism for rapid glacial to interglacial changes in atmospheric pCO_2. Variations in ocean productivity are highly sensitive to external forcing of the Earth's orbital parameters, because ocean productivity changes are driven by meridional surface winds, which in turn are linked to the extent of high-latitude sea ice and insolation.

In these models the main controls on atmospheric CO_2 are considered to be (1) sea surface temperature, because of dissolution properties, and (2) the strength of the biological pump, which is strongly affected by sea ice extent. Sea ice extent determines both the accessibility of cold nutrient-rich, high-latitude waters to solar radiation, and the extent of surface water mixing between high-latitude cold water and lower latitude warm water. When sea ice extent is large, exposed cold nutrient-rich waters are displaced equatorward where more radiation is available for the biota; the resulting strong baroclinic zone promotes greater transport of nutrient-rich but carbon-depleted waters to high latitudes. Both conditions enhance the uptake of CO_2 from the atmosphere. Saltzman (1987) assumes that the uptake of CO_2 is also enhanced by a shallower thermocline or by colder bulk

deep-ocean temperature (probably related to the surface nutrient supply) and by a diminished grand thermohaline circulation, the strength of which is controlled by total snow-derived ice mass.

There are three types of ocean carbon pumps, a process that depletes surface waters of CO_2 relative to deep-water CO_2 (Volk and Hoffert, 1985). Carbonate and soft-tissue pumps result from the flow of $CaCO_3$ and organic matter detritus respectively from the surface, and the third, the solubility pump, results from increased CO_2 solubility in downwelling cold water.

Keir (1988) devised a 14-box ocean and atmosphere model for geochemical changes during the Pleistocene. Regional distributions of $CaCO_3$ dissolution, CO_2, and $\delta^{13}C$ were included. When the downward particulate carbon flux in the Antarctic region was increased by a factor of 2 to 3 and Atlantic section (Antarctic bottom water) was increased, the model predicted several changes recorded in the sedimentary record, in particular CO_2 decreases of up to 90–110 p.p.m. Lindzen (1988) used a simple climate model with orbital forcing centred at 20 k.y. and included a small 100 k.y. component to reproduce the 100 k.y. cycle of glaciation to investigate dependence of CO_2 on snow/sea ice position and glaciation. The results show strong amplification of the 100 k.y. glacial cycle. From this simple experiment we should expect to see in the CO_2 record both high frequency fluctuations related to snow/sea ice cover and lower frequency fluctuations related to the glacial cycle.

The difference between the carbon isotope record of planktonic and benthic foraminifera was used by Shackleton *et al.* (1983) as a proxy indicator for atmospheric CO_2 partial pressure in surface waters. This approach demonstrated that atmospheric CO_2 content varied in phase with global climate. Furthermore, Shackleton and Pisias (1985), from analysis of a 340 k.y. record of benthic and planktonic oxygen and carbon isotope records from an equatorial Pacific core, estimated both global ice volume and atmospheric carbon dioxide over this period. This record is dominated by orbital frequencies. Examination of phase relationships indicates that atmospheric CO_2 concentration leads ice volume over the orbital bandwidth, and lags changes in orbital geometry. However, data from the Vostok ice core (Barnola *et al.*, 1987) show that for the abrupt transition from glacial to interglacial conditions at ~130 k.y., no lag exists between carbon dioxide and temperature change. This creates a serious problem for any nutrient-based scenario, since there is no time lag between the temperature-forced increase in productivity and the carbon dioxide increase.

Using a model describing ice-volume changes as a function of insolation changes at 65° N, Pisias and Shackleton (1984) tested the role of carbon dioxide variations in cyclic climate changes. Fig. 10.9 shows the ice-volume curve simulated by orbital forcing compared with ice volume derived from oxygen isotope ratios of benthic foraminifera (top), and with CO_2 changes

Fig. 10.9. The ice-volume record ($\delta^{18}O$) from deep-sea sediments (middle graph) compared with the model-produced ice volume with orbital forcing only (upper graph) and orbital forcing and carbon dioxide changes (bottom graph). The fit of the model to the observed data is improved considerably when carbon dioxide variation is added to the model. Adapted from Pisias and Shackleton (1984). Copyright© 1984 by Macmillan Magazines Ltd.

(bottom). The fit of the model to the observed data is significant, thus supporting the idea that changes in atmospheric carbon dioxide play a role in causing climate change.

Variations in atmospheric CO_2 thus appear to be part of the forcing of ice-volume changes. Changes in obliquity have had the major effect on atmospheric CO_2 changes, implying a high-latitude control. The mechanism probably involves changes in the internal cycling of nutrients within the ocean, perhaps as a result of changes in deep-water circulation, and not transfer of nutrients into or out of the oceanic reservoir. These atmospheric carbon dioxide variations contributed, together with direct insolation changes, to controlling the growth and decay of continental ice volume (Shackleton, 1986b, 1987b). Development of the concept of shallow to deep $\delta^{13}C$ differences was greatly stimulated by the model of productivity-controlled pCO_2 variations developed by Broecker (1982). Shackleton *et al.* (1983) related the record of $\delta^{13}C$ to the productivity pump in order to

propose a constraint on atmospheric pCO_2 as outlined above. More recently, Curry and Crowley (1987) re-examined this concept and concluded that nutrient-driven pCO_2 changes account for only about one-third of the total decrease observed in ice cores, and consequently, that the difference concept should not be used as a proxy pCO_2 index. CO_2 data from the Vostok core confirms such a conclusion (Barnola *et al.*, 1987). Curry and Crowley point out that as much as 40 p.p.m. pCO_2 change still has not been accounted for by models of past chemistry and physics of the ocean (see Table 6 in Curry and Crowley, 1987). Possible 'missing sources' remain to be determined, but several possibilities can be identified. For example, the magnitude of the 'biological pump' could be greater (see Curry and Crowley for discussion), or glacial SST estimates by CLIMAP could be lower by 2°C, as concluded by Rind and Peteet (1985). SST changes alone could contribute another 28 p.p.m. to atmospheric pCO_2 change. A third possibility involves changes in the organic carbon/carbonate ratio (Broecker and Peng, 1987; Curry and Crowley, 1987). Boyle (1986) has suggested that stronger ice age winds and increased upwelling may have caused changes in nutrient cycling and in the rain ratio of carbonate versus organic carbon by causing a change in surface-water ecology from one suited to calcium carbonate organisms to one best for siliceous organisms. Since upwelling may be related to precessional forcing of the wind field (Kutzbach and Guetter, 1986; Pokras and Mix, 1987), one would expect fluctuations in pCO_2 caused by these mechanisms to occur in the 23 k.y. orbital period. All the above mechanisms may contribute to a pCO_2 decrease.

Broecker and Peng (1989) proposed another possible mechanism of carbon dioxide depletion during the glacial cycle. The idea is that sea ice formation lowers the salinity of polar surface waters. They suggest that this reduction in turn lowers the CO_2 partial pressure of polar surface waters and hence also of the atmosphere. During glacial times when more sea ice existed this pumping action may have been stronger, so that atmospheric CO_2 was depleted.

Another new mechanism to control atmospheric carbon dioxide and climatic cycles was proposed by Faure (1987). This is based on the assumption that cold high-latitude ocean waters are a sink and low-latitude waters a source for atmospheric carbon dioxide and that the main controlling mechanism on carbon dioxide fluxes during the glacial–interglacial cycle is sea ice cover. Sea ice extension near the end of a glacial cycle constitutes a CO_2-tight lid on the cold oceanic surface waters of both hemispheres. Thus, the global ocean is no longer a sink for atmospheric CO_2. The resulting rapid increase of CO_2 in the atmosphere, together with a rise in sea level, is responsible for accelerated warming and melting of the sea ice. The cold oceans, freed from their ice lid, can again become a CO_2 sink and a new glacial cycle may be triggered.

The role of changes in atmospheric CO_2 in the complex system of feedback mechanisms that modulate global climate on orbital time scales has been addressed by Pisias *et al.* (1988). Biochemical processes in the ocean carbon system (nutrient cycling, photosynthesis, respiration, carbonate production and dissolution) are controlled by physical processes (thermohaline circulation, open ocean convection, run-off, burial of organic and carbonate carbon). Each of the above process has a particular time constant and spatial distribution, and each should have its characteristic orbital response. The marine nutrient pump is a process in which nutrient-limited biological productivity moves ^{13}C depleted organic carbon from surface to deep waters, leaving surface waters and the atmosphere depleted in CO_2 but enriched in ^{13}C. Comparison of deep versus shallow $\delta^{13}C$ values with ice-core carbon dioxide records suggests that this mechanism was responsible for substantial changes in CO_2 over the last 160 k.y. Phase analysis indicates that in the 100 k.y. and 41 k.y. cycles, changes in thermohaline circulation lead the changes in the nutrient pump and CO_2 and are followed by changes in ice volume. In the 23 k.y. cycle, atmospheric CO_2 appears to lag ice volume and thus does not play an early role in the feedback chain.

Late Quaternary climatic history

The single most important feature of Pleistocene climates was the existence of ice sheets and polar land masses. While these contributed to refrigeration of the high latitudes, they also exported their effects to middle and low-latitude zones, where temperature, humidity and wind strength were all affected in concert with orbitally controlled fluctuations of the ice masses in both hemispheres.

During the last 20 years considerable progress has been achieved in reconstruction of the past climates, establishment of reliable and detailed chronologies and stratigraphic correlation of climatic events, and in modelling experimentation in space, time and frequency domains (see for discussion Hecht, 1985; Shackleton, Van Andel *et al.*, 1990). Correlation of Quaternary glaciations in the northern hemisphere (Sribrava *et al.*, 1986) contributed to land–sea stratigraphic correlation and comparison of the Quaternary palaeoclimatic changes in marine and continental realms. However, estimated astronomical ages of the oxygen isotope stage and substage boundaries do not necessarily provide bracketing ages for the terrestrial glaciations since oxygen isotope variations in the marine record appear to lag by 500–3000 years behind the corresponding changes in ice volume on the continents (Mix and Ruddiman, 1984). If so, then glaciations may have been initiated during the late parts of the warm isotope stages (see Richmond and Fullerton, 1986).

However, the equation of oxygen isotope maxima in deep-sea cores with maximum volumes of ice on the continents is generally accepted (Mix and Ruddiman, 1984; Shackleton, 1987b), and the relationship has been extended to include other proxy records of climate such as stratigraphic sequences of loess and palaeosol deposits (Kukla, 1987) and terrestrial pollen records (Kukla, 1977; Bradley, 1985). However, pollen, loess, and oxygen isotope records are all proxy records of glaciation; none is necessarily a direct record of glaciation on continents. Confirmation of the marine oxygen isotope record as directly reflecting glaciation is possible only through precise dating of glaciation on the continents (see for discussion Fullerton and Richmond, 1986).

Two basic models for the mechanism of glaciation have been advanced. The first model (Flint, 1971) suggests that ice sheets originated as valley glaciers, probably in coastal mountains of Baffin Island, Labrador and Scandinavia. As these valley glaciers grew, piedmont glaciers formed, which coalesced and thickened. Flow away from the original source areas built large ice domes. The second model (Ives, Andrews and Barry, 1975), assumes that snowfields accumulated rapidly over large upland areas of Baffin–Labrador–Ungava and possibly Scandinavia. Positive albedo feedback then resulted in rapid accumulation of snow with very little ablation. A variation of this model (Denton and Hughes, 1981) expands the area available for instantaneous glaciation to include interior seas (Hudson Bay, Baltic Sea), by adding ice shelves that trapped virtually all the precipitation and built vertically into ice domes.

Geological data indicate that high-latitude and mid-latitude ice sheets may respond to forcing in different ways (see Hughes, 1987). Boulton (1979) invoked moisture-flux arguments to explain geological evidence for asynchronous high-latitude and mid-latitude glacier advances and indicated that the ice-volume records in these different areas should generally be out of phase. Since solar insolation varies dramatically in spectral character with latitude (from precession dominated over the high latitudes to obliquity dominated over the middle latitude ice sheets), ice sheets may respond volumetrically to orbital forcing with different frequencies depending on their latitude. Ruddiman and McIntyre (1981a; 1984b) show that down-core foraminiferal temperature estimates, which should respond to local ice-volume change, contain mostly 41 k.y. and 100 k.y. power north of 50° N in the Atlantic. South of this latitude records contain mostly 23 k.y. power.

The detailed record of sea-level over the past 240 k.y. from uplifted terraces in New Guinea (Chappell and Shackleton, 1986) was compared by Shackleton (1987b) with the benthic oxygen isotope record (Aharon and Chappell, 1986) (Fig. 10.10). A major discrepancy between these records spans isotopic stages 5 to 3. Shackleton explains this discrepancy by proposing that the temperature of Pacific deep water was not constant, as

Fig. 10.10. Sea level variations over the last 160 k.y., determined from raised reefs in New Guinea (Chappell and Shackleton, 1986) and from deep sea $\delta^{18}O$ records (Shackleton, 1987b). Copyright© 1987 by Pergamon Press plc.

previously assumed, but was continuously lower by ∼ 1.5°C from substage 5c to the beginning of the Holocene. Labeyrie, Duplessey and Blanc (1987) also proposed a widespread deep-ocean cooling of 2–3°C during glacial times. Improved dating of sea-level changes during the past 130 k.y. (Bard *et al.*, 1990) from Barbados shows that the last deglaciation started some 3 k.y. earlier than previously measured and confirms a two-phase deglaciation with meltwater surges at ∼ 14 k.y. and 11 k.y. BP.

If large floating ice shelves existed at the LGM, they would contribute to oceanic $\delta^{18}O$ change but not affect sea level (Johnson and Andrews, 1986). Continuous sedimentary sequences in the Arctic (Clark, Byers and Pratt, 1986) weaken the hypothesis of extensive ice shelves. It appears likely that shelves might have existed in Antarctica and possibly the North Pacific.

There is substantial evidence that during the late Quaternary there was a transition (between ∼ 0.9 and 0.4 Ma) from relatively small ice volume and low-amplitude glacial oscillations (mainly with a period of 40 k.y.) to large ice-volume and high-amplitude (near 100 k.y.) glacial cycles (e.g. Prell, 1982; Ruddiman, McIntyre and Raymo, 1986). Since orbital forcing was of the same character over the last 2 m.y., it has been suggested that another unknown long-term forcing may be at work that had a direct external effect or could change some internal parameter controlling the nature of the free response to orbital forcing (e.g. Saltzman *et al.*, 1984; Oerlemans, 1984; Ghil, 1984).

Saltzman and Sutera (1987) used a simple dynamical model governing global ice mass, atmospheric and mean ocean temperatures to simulate climatic transition during the last 2 m.y. No external Earth-orbital forcing was prescribed; the model represents only internal dynamics. The model successfully reproduces the climatic transition at ~0.9 Ma and in essence accounts internally for some of the variability of the $\delta^{18}O$ measure of ice mass and of carbon dioxide during the Quaternary. Saltzman and Sutera conclude that if the physical aspects of the model are correct, we can say that ice ages are the consequence of a sensitive balance between water in its liquid and solid phases. This balance depends critically on the controlling influence of liquid water (the ocean has a large thermal and chemical capacity) on atmospheric carbon dioxide (via temperature structure related to chemical or nutrient variations). Positive feedback involving CO_2 introduces a basic instability that drives the ice volume variations in the system. The conclusion from this exercise is that even if orbital forcing were absent, cyclic oscillations of climate should still occur.

North American glaciations

The Laurentide ice sheet was a glacier complex that covered large parts of eastern, central and northern North America during the last glaciation and was made up of at least three major coalescent ice masses. At its maximum the Laurentide ice sheet covered an area comparable to that of the present Antarctic ice sheet (12.6 million km²) and its thickness was up to 4 km. Models constructed by Hughes *et al.* (1981) suggest a maximum ice volume of 34.8 km³. It may have made up ~35% of the world's ice volume at the late Wisconsinian maximum and accounted for 60 to 70% of the ice which melted at the end of the last glacial cycle (Fullerton and Prest, 1987). Thus, the ice sheet presented a substantial barrier for atmospheric circulation which, combined with high surface albedo, significantly influenced climate by its mere presence (Kutzbach and Wright, 1985; Manabe and Broccoli, 1985).

The $\delta^{18}O$ isotope record of foraminifera indicates that inception of the Laurentide ice sheet occurred in the interval between 120 and 80 k.y. at high latitudes (Andrews and Fulton, 1987). There are two basic ideas about where and how the ice sheet developed. The first hypothesis is that the ice sheet originated over the high plateaux of eastern North America and, possibly Keewatin (Ives *et al.*, 1975) through lowering of the snowline. A second hypothesis is that of Denton, Hughes and Karlen, (1986), suggesting that the ice sheet developed *in situ* over Hudson Bay through a mechanism of basal freezing. Andrews and Fulton (1987) pointed out that the second hypothesis requires an extremely cold and arid climate, whereas the alternative hypothesis proposes only increased snowfall and mild climatic conditions. Atmospheric general circulation modelling of climate for orbital

forcing specified for cycles of 115 k.y. and 125 k.y. indicates that at 115 k.y., surface cooling over the Labrador area was sufficient to allow permanent snow cover, and thus conditions favourable for formation of the Laurentide ice sheet were created merely by an insolation anomaly (Royer, Deque and Pestiaux, 1983).

Aksu (1985) and Aksu and Mudie (1985) documented the presence of open marine conditions in the Labrador Sea during the initial phase of the last glacial cycle. Palynological and isotopic analyses of deep-sea cores from the Labrador Sea show that during the late Quaternary the Labrador Sea was characterized by strong exchanges between North Atlantic water masses, Arctic outflows, and meltwater discharges from the Laurentide, Greenland and Inuitian ice sheets. The penetration of temperate Atlantic waters persisted throughout most of the late Quaternary, with a brief interruption during the late Wisconsinian. The advection of southern air masses along the Laurentide ice margins, as evidenced by pollen assemblages, caused high precipitation and thus high snow accumulation (Aksu and Mudie, 1985; Andrews *et al.*, 1985; de Vernal and Hillaire-Marcel, 1987).

Greatest accretion of the Laurentide ice sheet may have occurred when warm water and moist air extended northwards, creating a steep tempera-ture gradient and a moisture source (Ruddiman and McIntyre,1979; Ruddi-man, Molfino *et al.*, 1980). Sea-level curves and oxygen isotope records indicate high ice volume during the early and late Wisconsinian and a reduction of up to 55% in ice volume during the middle Wisconsinian. The cause of this ice-volume reduction remains uncertain. Ruddiman and McIn-tyre (1981a) argue that ice volume is driven by planetary phenomena, based on the 23 k.y. precessional cycle. Johnson (1982), on the other hand, correlates glacial minima to precessional insolation peaks.

The Laurentide ice sheet reached its maximum southern extent between 22 and 14 k.y. but the northern margin apparently did not begin significant retreat until ~ 12 k.y. (Andrews and Fulton, 1987). At 18 k.y. the ice sheet consisted of sectors with an interlocked system of ice divides joined at intersector saddles (Dyke and Prest, 1987). The ice sheet retreated slowly between 18 and 13 k.y. BP. Retreat between 13 and 8 k.y. was much more rapid in the west than in the east. The present existence of the Penny and Barnes ice cap (remains of the Laurentide ice sheet) indicates how strong was the east–west asymmetry of deglaciation; the eastern margin of the Barnes ice cap has retreated only 35–60 km since the late Wisconsinian maximum.

Recent correlation of the late Quaternary glacial deposits in Arctic Canada (Andrews *et al.*, 1984) reveals that retreat of Arctic glaciers from the late Wisconsinian maximum lags that along the southern Laurentide ice sheet margin by several thousand years, but appears to be correlative with glacial advances in other Arctic regions such as Greenland and Spitsbergen.

This marine-based section of the Laurentide ice sheet had two intervals of rapid calving and massive ice discharge, the first at ~ 10 k.y. and the second at 8 k.y. BP. In both instances the rate of melting was almost catastrophic (Andrews and Fulton, 1987). On the other hand, deglaciation of northern Labrador is first indicated at 17 k.y. (de Vernal and Hillaire-Marcell, 1987) and again at 11 k.y. BP, when the main ice retreat resumed.

Eurasian glaciations

A considerable portion of northern Eurasia, especially the Arctic continental shelves, was glaciated during the late Quaternary (Grosvald, 1979). Reconstruction of ice-front positions reveals that the last ice sheet in northern Eurasia extended over a distance of 6000 km and covered an area of 8 370 000 km^2. One half of this area was located on present-day continental shelves. The Eurasian ice sheet impounded the waters of the Northern Dvina, Mezen, Pechora, Ob and Yenisey rivers, resulting in the formation of vast ice-dammed lakes in the north-east of the Russian plain and in the West Siberian plain. These lakes existed over the period 12.4 to 10.5 k.y. BP and drained southward. For physical reasons, the ice cover could exist only if the outer edges did not hang freely over deep-ocean water, but were supported by floating ice shelves, which in turn were stabilized by islands, submarine ridges, sea coasts, or by other glaciers. According to Grosvald (1979) this forces the conclusion that at the time of glaciation the Arctic and North European basins were covered by floating ice shelves. In this interpretation, ice shelves linked the Eurasian ice sheet with the Laurentian, Inuitian and Greenland ice sheets, resulting in the formation of a gigantic, complex northern hemisphere ice system.

The development of Pleistocene marine ice sheets was linked to hydrological subarctic fronts, in both the Atlantic and Pacific, where mass- and energy-exchange was most active in the ocean–atmosphere system. The growing isolation of the Sea of Okhotsk and the Bering Sea, caused by glacioeustatic regression, interrupted the inflow of warm ocean water via the Tsushima and Aleutian currents respectively. These conditions in both oceans provided the basis for the development of a stable ice cover. Vozovik (1985) presents arguments supporting more massive ice shelves in the Bering and Okhotsk Seas than in the Atlantic during the late Quaternary. Large-scale denudational relief in the deep basin of the Bering Sea and very thick glaciomarine sediments in the North Pacific are taken as the main evidence for this.

Glaciations of continental shelves

During glacioeustatic lowering of sea level, the shallow north-eastern portion of the Bering Sea became a part of the Bering Land Bridge. Grosvald and Vozovik (1984) see an analogy between the Bering Sea and the Norwe-

gian–Greenland Sea, which strongly suggests that conditions conducive to ice-sheet growth existed over both seas (Lindström and MacAyeal, 1986).

The intensified polar front in the glacial North Pacific acted as a barrier to northward flow of subtropical water. This resulted in a negative heat balance in the sea, which forced the inception of a marine ice sheet through an ice-shelf mechanism. Grosvald and Vozovik (1984) propose the existence of a marine ice cap in South Beringa with a total area of 2.6–2.7 million km², on the basis of a number of geological and geomorphological features in the Bering Sea. Consequently they propose that in SSTs reconstructed by CLIMAP, the 0°C isotherm should be plotted south of the Aleutian–Commander Ridge, and that the error in CLIMAP SSTs in these areas is 8°C.

Geological investigations in the shallow areas of the Bering Sea and on the adjacent coasts show an abundance of glacial sediments. On the basis of these, mountain-type ice sheets are proposed to have existed in Kamchatka, the Koryak Upland and Chukotka. The total area of glaciation was estimated to be around 550 000 km². Existence of fiord coasts along the north-eastern part of the USSR was used as evidence of ice cap glaciation in coastal areas with ice shelves in the sea. The complex of exogenic forms in the submarine relief of South Beringia, including the system of spectacular submarine canyons on the continental slope, may be explained by the process of glacial erosion, while the young sedimentary facies on the continental slopes and the bottoms of the deep basins are interpreted as glaciomarine turbidites.

Velichko (1983) proposed a model of atmospheric circulation explaining latitudinal asymmetry of late Quaternary glaciation. This model of the evolution of the circulation system over glaciated areas proposes that, during the early stages of the glaciation cycle, Arctic air masses expand and thus establish more favourable conditions for cyclogenesis in the boundary zone between air masses which occur over the Gulf Stream. The role of the Icelandic Low as an area of cyclogenesis is increased. Frontal precipitation is transported to northern Eurasia and partly westward, towards North America. The presence of cold air masses over warm ocean surface waters causes more precipitation near developing ice sheets (Ruddiman *et al.*, 1980). Due to the influence of the cold Labrador current, the North American east coast had potentially more favourable conditions for the accumulation of solid precipitation brought by the cyclonic activity from the cyclogenetic zone above the Gulf Stream. As a result, one could expect earlier formation of ice sheets in this area than in Europe. North-east Asia, however, lacked climatic conditions favourable for the development of glaciation. This area was under the influence of stable winter Siberian and Central Asian high-pressure zones; consequently winter precipitation was very limited. As the cooling increased, the anticyclone extended eastward, thus preventing snow accumulation in western locations. This resulted in a pronounced decrease in glaciation from west to east.

After the maximum glacial advance around 18 to 20 k.y. BP, atmospheric circulation underwent major changes. The North Atlantic was covered by sea ice and the warming influence of the Gulf Stream was reduced. Consequently, the Icelandic Low was weakened and the Eurasian Arctic underwent strong cooling. Meanwhile, a strong, cold, circumpolar high-pressure belt developed over the ocean and the continents. The Polar High, in eastern Eurasia, joined with the Central Asian and Siberian Highs which extended far to the west. In Europe a high-pressure system had formed over ice-covered areas. Velichko (1983) argues that during the last glacial maximum, high pressure dominated over most of the area outside the tropics, at least in the winter. The emergence of high-pressure areas in polar latitudes would have reduced the magnitude of the atmospheric pressure gradient between polar regions and the tropics, and as this gradient determines the strength of westerly air flow, an easterly flow thus prevailed near the surface. The disappearance of the North Atlantic current was the cause of weakening of the westerly air flow. As a result of these pressure changes, the Icelandic Low was replaced by a high-pressure system and only the Aleutian Low was present in high and middle latitudes.

The conceptual model of Velichko is supported by modelling carried out by Gates (1976). During the last glacial maximum, the climate in Eurasia was characterized by westward and southward flow of very dry and extremely cold air masses originating in polar and easterly regions. As a result, perennial permafrost and periglacial tundra–steppe vegetation extended as far as western Europe. Dry conditions hindered further glacial advance and growth in Eurasia, with the exception of Kamchatka, where ice growth was sustained by precipitation from the Aleutian Low. In the North Pacific, in contrast to the Atlantic, warm water currents were flowing northward, maintaining precipitation in high latitudes. In the Atlantic, the Gulf Stream had not yet extended towards Europe; instead the North Equatorial and the Antilles currents formed a closed loop. Therefore, the region of cyclogenesis was preserved and North America had abundant precipitation. An increasing contrast between air masses was conducive to formation of cyclones since the dominance of an easterly air flow supported the transfer of precipitation into the interior of North America (moisture-laden air masses from the west Atlantic were moving eastward). Precipitation occurred at the fringes of the ice sheet adjacent to the open ocean. In polar latitudes, which correspond to the central areas of glaciation, the ocean was frozen and it could not serve as a source of major precipitation. The growth of the North American glaciers probably resembled that of the present Antarctic ice sheet, where glaciers grow mostly peripherially. Variable precipitation regimes determined the different development of these glacial systems during the second phase of the late Quaternary.

The proposed model of atmospheric circulation during the late Quatern-

ary can also explain lake levels in the northern hemisphere during the glacial maximum. The asymmetrical occurrence of dry and pluvial periods, in the eastern and western hemispheres respectively, can be explained by suppression of the sources of cyclogenesis in the Atlantic. This occurred because activity along the southern storm tracks, which originate in the Atlantic, was also sharply reduced. In North America, the conditions were quite different; because of the presence of the Aleutian Low, zones of cyclogenesis existed along the west coast of North America. Consequently, air flow to the continent was from the southern part of the low-pressure system. Upon reaching the continent, this flow was deflected southward because an extensive high-pressure system dominated the glaciated area of northern North America. The proximity of the Aleutian Low to a high-pressure system in the eastern Pacific could have increased frontal precipitation on the continent. Precipitation in western North America could have occurred along both the northern and southern fringes of the high-pressure system (Velichko, 1983). The proposed model allows the existence of dry conditions in the eastern hemisphere and pluvial conditions in the western hemisphere during the glacial interval. However, on the global scale, periods of glaciations were accompanied by increased aridity.

The initial disintegration of ice sheets probably resulted from the ice mass reaching its size limit. This caused climate warming, which in turn resulted in the breakdown of the circum-polar high-pressure system. The glacial retreat was further intensified by warming of northern Europe by the Gulf Stream, which by then extended northward. At the same time, the retreat of the Laurentide ice sheets had been delayed due to the presence of the cold Labrador current. Thus, ocean currents with contrasting thermal properties determined the different and non-synchronous nature of glacial retreat in the North European and North American ice sheets.

Antarctic glaciations

The history of the melting of Antarctic ice during the late Quaternary had an important bearing on the formation of bottom and intermediate water, and consequently on CO_2 concentration in the atmosphere. It is possible that large surges of part of the West Antarctic ice sheet influenced global climate in the Quaternary (Hollin, 1962, 1980; Bowen, 1980). Labeyrie *et al.* (1986) make comparisons of the oxygen isotopic composition of planktonic and benthic foraminifera of the Southern Ocean with SSTs estimated by a transfer function derived from the diatoms in the same sediments. They propose that isotopic anomalies over the last 60 k.y. were caused by a succession of pulsed ice surges. Floating shelf ice is very sensitive to destabilization by sea-level variations (Denton and Hughes, 1986), and periodic surges of Antarctic ice sheets thus would raise sea level at the time when the northern hemisphere ice sheets were near maximum volume.

This may have led to destabilization of marine-based ice sheets of the northern hemisphere and to a sea-level rise at around 25–30 k.y. (Chappell, 1983).

There is good evidence that the Southern Ocean led by 2 k.y. the North Atlantic warming and northern hemisphere deglaciation. There were several large isotopic anomalies during the LGM (particularly between 35 and 17 k.y. BP), probably due to large inputs of meltwater from the rapidly eroding periphery of the Antarctic ice sheet (Labeyrie *et al.*, 1986). Rapid erosion during maximum extension of the marine-based ice sheet was due to ice streams flowing towards the ice shelves. When most of the ice available for feeding the coastal ice streams had been eroded, the ice shelves were destroyed, and iceberg calving decreased until a new equilibrium ice sheet was established.

The last interglacial

The last interglacial (~ 122 k.y. BP) was a time of small ice volume, similar to the present day. This period can be detected from benthic foraminifera oxygen isotopes (Fig. 10.11) indicating low $\delta^{18}O$. It began at ~ 127 k.y. and finished at 116 k.y. BP and thus coincides with isotopic substage 5e. CLIMAP investigated the last interglacial by means of oxygen isotope analyses and biotic census counts of 52 cores across the global ocean (CLIMAP, 1984). The reconstruction shows that SSTs were very similar to the present. In particular, the North Atlantic and North Pacific were slightly warmer (1–2°C) and the Gulf of Mexico slightly cooler. Non-marine and shallow-marine sequences provide strong evidence of conditions significantly warmer than present, especially during the summer. This difference may be ascribed to the different climatic responses of land and water. Land has low thermal inertia and responds directly to solar forcing. The higher summer insolation at 122 k.y. BP (Berger, 1978) would have caused increased summer heating of the continents (e.g. Kutzbach, 1981). During the winter, insolation was reduced, but the varied mid-latitude vegetation ensured a response to growing season conditions. In contrast, the ocean can largely integrate seasonal changes; the similar conditions of the last interglacial ocean may thus reflect that mean annual insolation was similar to that of the present.

The CLIMAP study also focused on the time relations between ice growth and oceanic cooling during the interglacial to glacial transition. In the high-latitude southern hemisphere, oceanic response (as indicated by palaeobiogeography) preceded the global ice-volume response (indicated by the oxygen-isotope signal) by several thousand years. In the subpolar North Atlantic from 40° to 60° N the reverse pattern prevailed. There, warm SSTs were maintained during the first half of the ice-growth phase (Ruddiman and McIntyre, 1979; Pujol and Duplessy, 1983).

Howard and Prell (1984) compared radiolarian and foraminiferal records

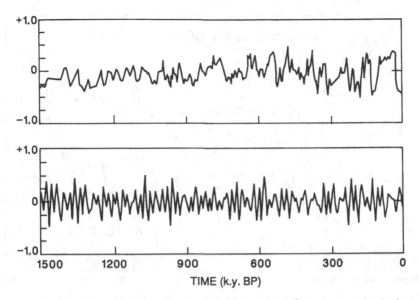

Fig. 10.11. Comparison of variations in the δ¹⁸O record over the last 1.5 m.y. (upper curve) with the index of effective radiation changes in high latitudes due to orbital forcing (lower curve). From Saltzman and Sutera (1987).

from the southern Indian Ocean in order to determine why warming occurred before ice-volume reduction (Fig. 10.12). They deduced that the warming shown by SSTs from radiolaria occurred before the warming recorded by foraminifera and before the isotopic depletion, as a result of two-phase warming in the mid-latitude southern Indian Ocean. The first phase involved a poleward expansion of subtropical lower water (STLW), and the second phase a surface shift of the Subtropical Convergence (STC) as northward Ekman drift was reduced by a relaxation of westerly winds. This study also showed that southerly migration of the STC is delayed with respect to the southerly migration of the STLW front. Further, the cause of the earlier cooling in the southern versus the northern hemisphere (Hays *et al.*, 1976) may lie in the large thermal inertia of northern hemisphere ice sheets and to the impact of icebergs and meltwater in the North Atlantic (Ruddiman and McIntyre, 1979, 1981a).

Non-marine diatoms from the Camp Century ice core indicate major variations in the size of the Greenland ice sheet and suggest reduction in size of the ice sheet by half during an (undated) Pleistocene interglacial. This reduction implies that previous interglacials were warmer and/or of longer duration than the Holocene interglacial (Harwood, 1986). The CLIMAP (1984) study concluded that such a decrease in ice volume relative to today is equivalent to the volume of water stored in the Greenland and the West Antarctic ice sheets and to a sea-level rise of between 2 and 7 m. However,

Fig. 10.12. Oxygen isotope δ¹⁸O record of planktonic foraminifera (upper curve) and benthic foraminifera (lower curve) from a southern Indian Ocean deep-sea core. Both records show strong warming and ice-volume depletion at around 122 k.y. BP, the time of the last interglacial (see Pujol and Duplessy 1983, for details). From Pujol and Duplessy (1983). With permission of Kluwer Academic Publishers.

the question of which ice sheet melted and to what degree is still open (see discussion in CLIMAP, 1984).

Reconstructions and modelling of the last glacial maximum (LGM)

During the last 18 k.y., the climate changed from full glacial to interglacial conditions. Climatic changes were large, and geologic evidence for these changes is well documented and dated. At the LGM large ice sheets covered parts of North America and Eurasia, sea ice in both hemispheres extended 10° latitude or further towards the equator than it does now, ocean temperatures were generally lower, boreal vegetation covered much of the continents in mid-latitudes, and the tropical belt was more arid and extensive than now (CLIMAP, 1976, 1981; Kukla, 1977; Petersen *et al.*, 1979). The peak of glaciation at 18 k.y. was followed by a rapid deglaciation between 15 and 12 k.y. BP, and by ~9 k.y. the North American ice sheets had disappeared, and sea level and SSTs were similar to present conditions.

Reconstruction of conditions prevailing over the Earth at the culmination of the last ice age (18 k.y. BP) is a major scientific accomplishment of the CLIMAP program. Uncertainties in the geologic data include how extensive marine-based ice sheets were and how widespread were ice sheets in large areas of northern Canada and Asia. The major conclusions of the CLIMAP program after Hecht (1985) are as follows:

1 During the LGM several large land-based ice sheets as much as 3 km thick existed in the northern hemisphere. In the southern hemisphere the most striking contrast with today was a greater extent of sea ice. Global sea level was lower by at least 120 m.

2 Global average sea surface temperatures associated with the ice age maximum were $\sim 1.7°C$ cooler in August and $1.4°C$ cooler in February than now.

3 The average surface albedo at the last glacial maximum was higher than today mainly due to expanded glaciers and sea-ice fields. The albedo of non-glaciated areas was also higher than today due to lower density of vegetative cover and to contrasting characteristics of glacial soils.

4 The ice age ocean was characterized by steepening of thermal gradients along frontal systems; an equatorward displacement of the Subtropical Convergence in the southern hemisphere; a large shift in the path of the Gulf Stream (which flowed due east at $40°$ N latitude during the glacial maximum); nearly stable positions and temperatures of the central gyres; and increased upwelling along the equatorial divergences.

However, as pointed out by Rind and Peteet (1985) sea surface temperatures reconstructed by CLIMAP are inconsistent with terrestrial conditions during the LGM, especially temperature estimates from high-altitude equatorial regions. Analysis of available data and climate modelling with GCMs indicates that SSTs lower by $2°C$ would be a better fit to the data, particularly since lowered carbon dioxide levels, which could contribute substantially towards lower SSTs, were not considered (Hansen *et al.*, 1984). Reconstructed continental conditions at 18 k.y. BP indicate temperature decreases from 1 to $12°C$ or more, depending on the location. Large portions of the continents were drier during the LGM, especially the Mediterranean, the Middle East and parts of Australia and Africa (Petersen *et al.*, 1979). A cold and dry climate in high northern latitudes led to the development of polar deserts and to steppe conditions further south. Development of the Scandinavian ice sheet commenced after cold waters had invaded the northern Atlantic.

Investigation of the influence of large ice sheets on global temperature during the LGM (18 k.y. BP), using an atmospheric general circulation model (AGCM) coupled to a static mixed-layer ocean leads to the conclusion that the presence of large ice sheets alone is insufficient to explain the glacial climate of the southern hemisphere (Manabe and Broccoli, 1984). Orbital forcing may be responsible for large fluctuations in the extent of the northern hemisphere ice sheets during the Quaternary, but we need an additional mechanism plus lowered interhemispheric heat transport by the atmosphere to account for decreased temperature of the southern hemis-

phere. The favourite agent involved is decreased carbon dioxide concentration in the glacial atmosphere. AGCM modelling of the LGM with boundary conditions including expanded ice sheets and reduced CO_2, yields significant cooling of the southern hemisphere. The cooling in high southern latitudes is due to a lower level of carbon dioxide (Broccoli and Manabe, 1987). These boundary conditions for the LGM substantially modify the tropospheric circulation, particularly in the northern hemisphere during winter. The North American ice sheet splits the jet stream, sending a southern branch, stronger than today, over the southern part of the United States. The jet stream is the boundary between cold, dry high-latitude and warm, wet low-latitude air masses, and southward displacement of the jet led to the condition where cold air dominated over most of the United States during the LGM (Kutzbach, 1985).

Rind (1987) also investigated with an AGCM the role of prescribed ice sheets in the LGM climate. The results show that realistic topography (avoiding, for example, a standard 10-m thick representation of ice sheets) cools the land in winter, further increases storminess off north-eastern North America, includes subsidence-warming downstream of the European ice sheet in summer, and increases stationary and eddy energy through increased baroclinicity. Other results show that the broad, raised ice sheets have both a topographic and a thermal effect that occur simultaneously and lead to a greater stationary–eddy forcing of the zonal mean flow.

An amplified ridge–trough pattern at the 500 mb level over North America and the North Atlantic is caused by anomalous northerly flow over the eastern part of the Laurentide ice sheet. The middle latitude westerlies are also strengthened from the western Atlantic across to Eurasia. These enhanced westerlies are connected with a belt of reduced sea-level pressure and increased storminess (Manabe and Broccoli, 1985; Kutzbach and Guetter, 1986; Broccoli and Manabe, 1987). Kutzbach and Wright (1985) compared AGCM-produced atmospheric circulation during the LGM over North America, the North Atlantic and Europe with extensive proxy climate indicators (i.e. sand dunes and loess records, estimates of snowline depression, pollen records and lake levels). The geologic evidence is generally in agreement with the model simulation.

Ice-sheet growth in North America was favoured when northern hemisphere summer radiation was low and SSTs in high latitudes were still high (temperature fall lagged behind ice growth by 1.5 k.y.; Ruddiman and McIntyre, 1981a). Such conditions provided moisture for ice build-up on the continent. Once ice build-up proceeded beyond a certain point, isostatic adjustment and positive feedback mechanisms stimulated ice growth, even though adjacent ocean surface temperatures were low (Fig. 10.13). SSTs may also have played an important role in deglaciation, as the last two deglaciations have coincided with high summer and low winter insolation

(A)

ICE-GROWTH SEQUENCE

(B)

ICE-DECAY SEQUENCE

Fig. 10.13. Schematic diagrams representing changes in surface circulation in the North Atlantic associated with (A) ice growth and (B) ice decay. Generalized curves showing lag in ice-sheet growth after a decrease in summer insolation, and a further lag in sea-surface temperatures. For the deglacial phase increased summer insolation causes a delayed reduction in ice volume, but ocean temperatures increase even later because of the cooling effects of glacial meltwater. From Ruddiman and McIntyre (1981a). Copyright © 1981 by AAAS.

(principally due to precession). As ice began to melt as a result of increased summer radiation, meltwater entering the Atlantic during the winter season led to development of extensive sea-ice cover, which prohibited evaporation and therefore drastically restricted growth of the ice sheet. In addition, calving icebergs reduced SSTs in the North Atlantic, further decreasing the moisture flux from the North Atlantic to the decaying ice sheets. As the ice sheet melted, sea level rose, thus increasing iceberg calving. The model describes a mechanism of self-regulation of ice sheet

growth and decay and has found substantial support from land and ocean climatic data (Ruddiman, Molfino *et al.*, 1980; Ruddiman and McIntyre, 1981b). However, modelling of this hypothesis with a coupled climate–ice sheet model (Pollard, 1983) showed that the meltwater layer mechanism was able to produce the glacial cycle (including rapid terminations) only after a substantial increase in sensitivity between snowfall and the meltwater discharge rate. Pollard considered the calving mechanism to be a better candidate for producing 100 k.y. cycles.

Aharon (1984) speculated that cooling of the surface ocean at low latitudes lagged significantly behind glaciation at high northern latitudes during the early stages of the LGM. This lag factor could explain the discrepancy between the sea-level record from raised coral-reef terraces and that estimated from oxygen isotope data.

Proxy data of importance in marine palaeoclimatology include aeolian material blown into the oceans. Quaternary aeolian activity in the northwest Pacific was lower during interglacials, reflecting more humid conditions in the central Asian source region (Hovan, Rea and Pisias, 1988). The grain size of the aeolian sediment, which is an indicator of wind intensity, exhibits fluctuations at periods much longer than 100 k.y. At around 300 k.y. (mid-Brunhes), fluxes of aeolian material and $CaCO_3$ show a change in variability (interpreted as a change in atmospheric circulation) from high-amplitude and possibly lower frequency variations to low-amplitude variations (Pisias and Rea, 1988). Unlike the isotope record from the same cores, the dominant period in aeolian grain size is 31 k.y., as it is in indicators of equatorial divergence. This spectral component is coherent and in phase with the aeolian and divergence ($CaCO_3$) records as indicated by Pisias and Rea, confirming the link between atmospheric and oceanic surface circulation for the first time in the palaeoclimatic record. Presence of the 31 k.y. periodicity in sediments from the equatorial Pacific could be explained by non-linear responses of the climate system to orbital forcing (e.g. a combination of obliquity and eccentricity).

The mid-Brunhes change observed in the climate record gives us information about the nature of the climate system; however, the cause for this transition is not known. We know from the oxygen isotope record that from ~ 0.9 until ~ 0.4 Ma there was a gradual transition in the amplitude and frequency distribution in global ice volume (Williams *et al.*, 1988). This gradual increase in the importance of the 100 k.y. rhythm has been ascribed by Pisias and Moore (1981) to the response of ice sheets to marine calving, and by Ruddiman, Molfino and Raymo (1986) to long-term changes in global topographic elevation, which in turn cause changes in the response of the global atmospheric circulation to orbital forcing. Therefore, the mid-Brunhes change in climate response may be part of a long-term climate evolution in the late Pleistocene.

The pelagic sediments deposited in much of the north-west Pacific are dominated by terrigenous material transported to the region through the atmosphere. The composition of the clay and silt-size sediment reflects climatically induced changes in weathering and source of the terrigenous material. Hay, McClain and Sloan (1988) calculated sediment fluxes to the ocean during the Holocene and the LGM and concluded that accumulation rates during the last glacial episode were about three times higher than during the Holocene. They propose that rapid climatic changes during the Pleistocene prevented soil profiles from reaching equilibrium and thus forced higher weathering rates. Time series analysis of the abundance of certain minerals, like quartz, allowed an evaluation of the periodicity of the variation (Leinen, 1988). These time series indicate latitudinal variation in the periodicity of mineral abundances. At 40° N latitude in the Pacific, quartz variation is dominated by 100 k.y. and 23 k.y. periods, while at 47° N, 100 k.y. and 50 k.y. periods are dominant.

Atmospheric circulation changes during the Quaternary can be deduced from mass accumulation rates and grain size of the total aeolian component isolated from pelagic sediment (Janecek and Rea, 1985). A piston core located under the prevailing westerlies and material from DSDP site 503 beneath the trade winds in the north Pacific were used to assess changes in the intensity of atmospheric circulation and of source-area aridity during the last 700 k.y. Both sites display greater variability in wind intensity prior to 250 k.y. Aeolian accumulation rate, an indicator of source-area aridity, fluctuates at periodicities similar to those of the glacial–interglacial cycles. Lower aeolian accumulation rates during colder (glacial) periods most probably reflect increased glacial humidity in central Asia and central America (Rea, 1986). Together, these proxy indicators of atmospheric activity emphasize the importance of regional variability in atmospheric and oceanic circulation.

Orbitally controlled changes in insolation at any point in space are important to long-term climatic change, but equally important may have been changes in the insolation gradient between low and high latitudes. Young and Bradley (1984) argued that gradient changes would directly alter the intensity of the atmospheric circulation north of 30° N latitude, with stronger gradients leading to more vigorous (winter-like) circulation. Ruddiman and McIntyre (1984b) suggested that insolation-gradient changes will alter atmospheric circulation cells and hence the net northward advection of oceanic heat from the South Atlantic to the North Atlantic. They showed a record of the equatorial Atlantic SST for the last 250 k.y., with a strong 23 k.y. (precession) signal and coherence to insolation gradients supporting this point.

Climates on land

The record of Pleistocene climates on land is pieced together from continuous sedimentary sequences in periglacial areas. The record correlates well with the climatic history reconstructed from deep-sea sediments and shows ~ 20 major cycles within the last 2 m.y. The last glacial cycle, which started ~ 127 k.y. BP and ended 10 k.y. BP, consisted of several mild interstadials separated by cold stadials and was the subject of detailed studies within CLIMAP (1976, 1981, 1984).

Few of the land records that are well dated, continuous, and long enough to cover several periods of orbital perturbations, are found to contain clear climatic information. On the other hand, the land record exhibits a wide variety of proxy indicators that, in conjunction, offer valuable information about Pleistocene climate changes. Information is provided which bears on sea level and temperature of the marginal oceans, surface temperature on land, moisture conditions on land (evaporation/precipitation ratio), patterns of vegetational changes, variations in wind direction and intensity, and on ice conditions (advance–retreat cycles of continental ice caps and aerial extent of ice sheets and mountain glaciers; Kukla and Aharon, 1984). The most useful land-based proxy climate records are lake levels, pollen and loess–soil sequences, speleothems, terminal moraines on tropical highlands, valley-fills related to glacial deposits, and periglacial phenomena (e.g. ice wedges, pingos, etc.).

Lake level data for the period 24 k.y. BP to the present reveal large fluctuations in water balance. The maximum extent of aridity lagged behind the global ice volume maximum by ~ 5 k.y. according to Street-Perrott and Roberts (1983). At 24 k.y. BP the present northern subtropical arid zone was occupied by a belt of enlarged lakes, suggesting, according to Street-Perrott and Harrison (1985), an equatorward shift of the westerly storm tracks of ~ 5° of latitude. The sparse data indicate a narrow dry zone located on the equator, suggesting a displacement of the equatorial rainy belt. At 18 k.y. BP the mean westerly storm track in the northern hemisphere was still situated at ~ 35° N, but a dry zone opened up in the tropics. Wetter conditions are indicated in Australia at 30–38° S.

The maximum extent of aridity was at 13 k.y. BP in the late Quaternary throughout Africa and western Eurasia. Only a few data points indicate wet conditions. The minimum of atmospheric moisture, convergence and run-off over Africa, western Eurasia, and tropical Australia implies a strong suppression of summer monsoon precipitation and thus a different mode of operation for the Asian–Indian Ocean monsoon system from the present day. This arid period followed a peak of explosive volcanism at ~ 14.5 k.y. BP and may also have been linked to the spread of a meltwater layer across the mid-latitude oceans (Street-Perrott and Roberts, 1983). The late-glacial to

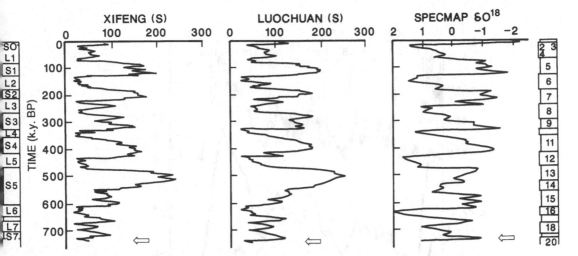

Fig. 10.14. SPECMAP oxygen isotope record tuned to astronomical chrono-logy compared to magnetic susceptibility in Xifeng and Luochuan (central China) loess sequences. From Kukla (1987). Copyright© 1987 by Perga-mon Press PLC.

early post-glacial phase of high lake levels was interrupted by two dramatic falls which culminated around 10.2 and 7.4 k.y. BP. These falls in level also were apparently preceded by volcanic eruptions and by meltwater discharge from the ice sheets.

At ~6 k.y. BP a wide belt of enlarged lakes occupied the tropics (32° N–18° S). This implies a northward shift of the equatorial rain belt as well as greatly enhanced monsoonal transport of moisture into the tropical continents. At the same time, in North America, a well-developed arid zone appeared between 32 and 51° N, suggesting displacement of the northern hemisphere westerly storm tracks. A less pronounced dry zone developed in the Near East north of 26° N, and east of 25°E. Kutzbach (1983) suggests that this represented a zone of enhanced north-easterly flow on the western side of an intensified Asian summer-monsoon low. Southern hemisphere westerlies were displaced ~4° equatorward of their present position, as suggested by widespread signs of increased moisture in temperate latitudes of the Australian continent. A broad envelope of wet conditions in the northern hemisphere tropics between 12.5 and 5 k.y. BP is attributed by Kutzbach and Street-Perrott (1985) to increased solar radiation due to orbital changes.

The loess deposits in central China and central Europe record world-wide climatic changes of the last 700 k.y. (Kukla, 1987) (Fig. 10.14). Climatic oscillations are marked by alternating loess and soil units and these are correlated with the oxygen isotope record of deep-sea sediments. Loess records show prolonged warm and humid climates between 560 and 460

Fig. 10.15. *Vostok ice core record of temperature and CO_2 changes during the last 160 k.y. (a) shows deuterium content versus SMOW. (b) smoothed Vostok isotope temperature record expressed as the difference from current surface temperatures. (c) marine $\delta^{18}O$ record from Martinson et al. (1987). (d) CO_2 concentrations (ppmv). From Lorius et al. (1988). Copyright © 1988 by the American Geophysical Union.*

k.y. BP and several glacial episodes in agreement with the oxygen isotope record (Kukla *et al.*, 1988). The earliest known deposition of loess at 2.5–2.3 Ma marks the transition from warm and humid climates to arid and cold, continental steppe-type environments and is synchronous with the development of northern hemisphere glaciation.

Distribution patterns of windborne pollen from southern Europe and

north-west Africa at 18 k.y. were reconstructed by Hooghiemstra, Bechler and Beug (1987). On this basis there was little change in atmospheric circulation patterns. In particular the belt with maximum easterly jet transport in Africa did not shift latitudinally during the last glacial–interglacial transition and as a consequence, the northernmost position of the ITCZ was also stationary. The north-east trade winds did not shift latitudinally although they varied in intensity and compared with the present, the glacial trade winds increased in intensity.

The accumulation of past snowfall in the polar ice sheets provides a valuable record of palaeoclimatic and palaeoenvironmental conditions. Ice cores yield information on isotopic composition of the water precipitated, concentration of dissolved and particulate matter, and chemical composition of air bubbles trapped in ice (Lorius *et al.*, 1984, 1985). The best climate record produced to date comes from the Antarctic Vostok ice core, with a record extending back to 160 k.y. BP. The Holocene was preceded by a long glacial interval marked by two relatively warm interstadials (Fig. 10.15) (Lorius *et al.*, 1988). The well-marked last interglacial at Vostok was significantly warmer than the Holocene, and the last stages of the previous glacial event, at ~ 150 k.y., were similar in conditions to the LGM around 20 k.y.

The last deglaciation is seen (from deuterium data) to take place in two steps, with two warming-trend periods interrupted by a 2°C temperature reversal lasting ~ 1 k.y. The Vostok spectrum shows concentration of variance in the order of 100, 40 and 20 k.y. and is dominated by a strong 40 k.y. component, indicating the major influence of the obliquity cycle. There are indications that during the LGM snow accumulation was slower and relative humidity was greater over the ocean than at present. Increases in continental dust (up to 20 times) and marine aerosols (up to 5 times) suggest that atmospheric circulation was more intensified and there was also an increase in the extent of desert area and of sea ice (Lorius *et al.*, 1984).

The carbon dioxide data from the Vostok core (Fig. 10.15) confirm the deglacial CO_2 increase recorded in previous ice-core studies and also show dramatic shifts, from ~ 190 to 280 ppmv, during the time of Terminations II and I (140 and 15 k.y. BP) (Barnola *et al.*, 1987). This is strong evidence for a fundamental link between the climate system and the carbon cycle (Sundquist, 1987).

Genthon *et al.* (1987) considered the role of CO_2 changes in amplifying relatively weak orbital forcing and concluded that CO_2 is the dominant forcing factor which mediates between two asymmetrically forced hemispheres. It is clear that other mechanisms also must be considered to explain interactions between the hemispheres, such as a decrease in production of warm NADW during glacial conditions, thereby reducing southern hemis-

Fig. 10.16. Comparison of Vostok ice core climate record with insolation parameters. (a) shows time series of CO_2 record. (b') (heavy line) is the smoothed record of Vostok isotope temperatures and (b) (light line) is the calculated Vostok temperature change from 78°S annual insolation and atmospheric CO_2 inputs. (c) shows annual insolation at 78°S expressed as the percentage deviation from the current value. (d) is the July insolation at 65°N, expressed as the percentage deviation from the current value. From Berger (1978). From Lorius et al. (1988). Copyright © 1988 by the American Geophysical Union.

phere high-latitude ocean temperatures (Corliss, 1982; Labeyrie *et al.,* 1987).

The 100 k.y. cycle may have its origin in large associated CO_2 variations (Genthon *et al.,* 1987) rather than in postulated non-linear interactions associated with ice-sheet growth and decay (Imbrie and Imbrie, 1980; Pollard, 1983). Development of the theory of Pleistocene climate changes

needs to include both the non-linear response of ice sheets to orbital forcing (Berger, Gall, Fichefet *et al.*, 1990; Le Treut *et al.*, 1988; Saltzman and Maasch, 1988; Ghil, 1988) and the likely contribution from carbon dioxide and aerosols (Shackleton and Pisias, 1985; Broccoli and Manabe, 1987; Saltzman, 1987). Together these can explain more than 90% of the temperature variance recorded in the Vostok ice core (Fig. 10.16). The relative contribution of CO_2 to temperature decrease is ~50%; carbon dioxide should therefore be considered as a major forcing agent in cyclic climate changes (Lorius *et al.*, 1988).

Measurements on ice core samples from Byrd Station (Antarctica) and Dye 3 (Greenland) show that the atmospheric methane concentration was only ~350 ppbv during the last glacial maximum, compared with ~650 ppbv at pre-industrial level (Holocene) and a present value of 1650 ppbv. (Stauffer *et al.*, 1988). Decreased production of methane during the LGM may have been due to decreased areas of wetlands, a major source of methane. The Vostok ice core provides a long record of CH_4 spanning the last 160 k.y., which indicates that, from the end of the penultimate ice age to the last interglacial (160 to 120 k.y. BP), CH_4 concentration increased from 340 to 620 ppbv (Raynaud *et al.*, 1988). The methane contribution to this global warming is estimated to be ~25% of that due to carbon dioxide. Total warming due to an increase in carbon dioxide and methane during a glacial–interglacial cycle is estimated at ~2.2°C (Hansen *et al.*, 1984).

Data from Antarctic ice cores indicate a large increase in the atmospheric aerosol loading during the LGM (De Angelis, Barkov and Petrov, 1987). Harvey (1988) has examined the role of increased aerosol loading in glacial cooling using an energy balance model; an appropriate increase in the atmospheric aerosol optical depth could lead to a 2–3°C global cooling. However, in contrast to carbon dioxide, which appears to have led ice-volume variations, increases in aerosols lag the cooling and ice build-up. Thus, increased aerosols probably served as a positive feedback mechanism, amplifying the initial temperature decrease (see also Legrand, Delmas and Charlson, 1988). There is thus growing evidence that large chemical atmospheric modifications have contributed to glacial–interglacial climatic changes.

Oceanic circulation in the Quaternary

It is widely accepted that changes in orbital forcing, carbon dioxide and deep-water production are the main contributors to ice age climate variations. The areas of deep-water production are associated with high-latitude glaciated areas and are sensitive to changes in ice volume on land and to ice cover extent. The total mass of deep water formed in high latitudes has an effect on global surface temperature through upwelling in equatorial areas. Changes in global water-mass structure may have effects on the carbon cycle, as indicated by Broecker and Denton (1985).

Among the factors contributing to the lower temperature (2–3°C) in glacial times were colder winters in the source region for the Mediterranean deep water, reduced entrainment of warm water during North Atlantic deep-water formation, and a source of cold deep water in the North Pacific (Shackleton and Duplessy, 1986).

There is evidence that continental glaciation preceded changes in North Atlantic SSTs as far north as 60° N latitude at 120 k.y. BP. Keffer *et al.* (1988), Miller, Mountain and Christie-Blick (1988) and Broecker (1987) offer suggestions involving orographic effects of the ice sheet, sea-level fall and loss of the North Atlantic convecting system, respectively, to explain this relationship.

It has long been known that in the eastern equatorial Pacific relatively high productivity occurred during glacial stages. A general ocean-wide hypothesis to explain this, first proposed by Arhenius (1952), is that increased temperature gradients between equatorial and polar areas during glacial periods leads to strengthening of the trade winds (Chuey, Rea and Pisias, 1987). This in turn leads to more vigorous oceanic circulation (stronger mixing), enhanced coastal and equatorial upwelling, and higher primary productivity. Productivity increase in the glacial ocean is connected with CO_2 depleted atmosphere. In addition, shifting climatic zones may have resulted in a high global nutrient discharge to the oceans, for example by erosion of high-latitude interglacial soils by advancing ice sheets or shelves. All these processes are ultimately controlled by cyclic climate changes during the late Quaternary, particularly at the 100 k.y. periodicity (Finney, Lyle and Heath, 1988).

A wide range of palaeontological and geochemical evidence from the deep Atlantic suggests that NADW production during the LGM was reduced relative to the present (Boyle and Keigwin, 1982; Corliss, Martinson and Keffer, 1986; Oppo and Fairbanks, 1987; Boyle and Rosener, 1990). Variations in the Cd/Ca ratio of benthic foraminifera have been used as an index of NADW production rates (Boyle, 1984); the ratio shows high-frequency fluctuations (41 k.y.) over the past 300 k.y. (Ruddiman and McIntyre, 1981b; Johnson, 1985; Ruddiman, Shackleton and McIntyre, 1986; Ruddiman, McIntyre and Raymo, 1986). Labeyrie *et al.* (1986) indicated that NADW, the major source of relatively warm deep water in the world ocean, was actively forming only for a short time during the last two interglacial periods. Pleistocene cores suggest that a strong 41 k.y. signal was present in geologic indices of NADW production, whereas indices of AABW indicate major pulses at longer periods. While NADW production is linked to obliquity-induced variations in high-latitude insolation, NADW lags ~8 k.y. behind insolation (Boyle and Keigwin, 1987). Deep-water temperatures during most of the last 125 k.y. (particularly during isotope stages 4 to 2) were close to freezing, around − 1°C. This cooling of the deep world ocean during the last glaciation increased the solubility of carbon dioxide and oxygen and also

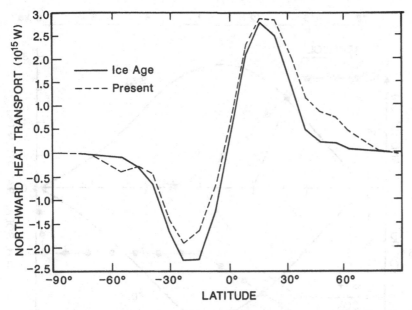

Fig. 10.17. Annual meridional oceanic heat transport for the present climate and for the last glacial maximum. From Miller and Russell (1988). Copyright © 1988 by the American Geophysical Union.

led to changes in vertical density distribution and thermohaline circulation.

The annual mean poleward heat transport by the oceans during the LGM was calculated using surface heat fluxes from GCM modelling experiments with CLIMAP data (seasonally varying SSTs; Miller and Russell, 1988). Comparison of calculated values with present-day values indicates that the general pattern of global transport did not vary significantly (Fig. 10.17). In the Atlantic during the last glacial maximum, however, there was weaker poleward transport in both hemispheres, and a net transport of heat out of the Atlantic. Studies by Mix, Ruddiman and McIntyre, (1986a, 1986b) and Showers and Genna (1988) found that cross-equatorial surface heat transport was reduced during the LGM.

The water-mass structure of the Indian Ocean during the LGM included a deep front separating intermediate and deep-water masses with very different characteristics (Kallel *et al.*, 1988). While intermediate waters show similar characteristics to those of today, deep water was actually cooler than now by 1.5°C, depleted in ^{13}C and poorly oxygenated. The stronger stratification between intermediate and deep waters during the LGM may have controlled the level of atmospheric carbon dioxide (Boyle, 1988b).

Deglaciation

Rapid melting of the northern hemisphere ice sheets began ~17 k.y. BP (when summer insolation in the northern hemisphere increased to its present-day level), and ended by ~6 k.y. BP (Fig. 10.18: ice curve). However,

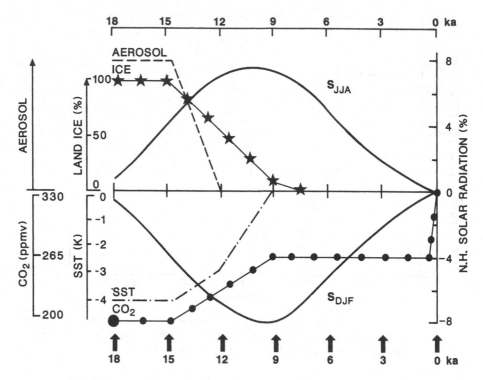

Fig. 10.18. Schematic representation of major changes during the last 18 k.y. in external forcing. Northern hemisphere solar radiation in summer (JJA) and in winter (DJF), extent of land ice as a percentage of the 18 k.y. BP ice volume (stars) and global mean annual SST as a departure from present values (dotted and dashed line). Variations in the concentrations of CO_2 and aerosol is also shown. The global SSTs were calculated in a sequence of GCM experiments for the last 18 k.y. in 3 k.y. increments. The arrows correspond to the seven sets of simulation experiments with CCM (Community Climate Model) by Kutzbach. From Kutzbach (1987).

the pace of deglaciation was uneven (Ruddiman and Duplessy, 1985; Mix and Ruddiman, 1985). The major steps of the isotopic transitions (at 14–12 k.y., 10–9 k.y. and possibly 8–6 k.y.) were separated by pauses in deglaciation at 11–10 k.y. and 8 k.y. BP. The former pause was related to meltwater discharge to the North Atlantic from the Laurentide Ice sheet through Hudson Bay, and the latter is documented by meltwater discharge to the Gulf of Mexico. In general, the mechanisms for discrete steps of deglaciation are taken to be episodes of calving of ice sheets. Also, rising atmospheric carbon dioxide concentration contributed to rapid melting during the last 18 k.y. (Fig. 10.18) (Shackleton and Pisias, 1985; Lorius *et al.*, 1988).

Disintegration of ice sheets can accelerate beyond that expected from direct heating because meltwater would tend to cut off the supply of moisture to the continents, according to Ruddiman and McIntyre (1981b).

Negative feedbacks also include the destabilization and removal of ice shelves by rising sea level. Such effects would accelerate the calving of icebergs, which in turn would further cool and dilute the surface waters of the Atlantic. Denton and Hughes (1981) propose that rising sea level would lead to the creation of relatively high-velocity streams of ice flowing from the central ice sheets to the sea. This would rapidly thin the ice sheets without causing rapid retreat of their edges and would lead to collapse of ice domes. Ice streams would also carry ice to the sea, where the ice could melt more rapidly. Taken together, these mechanisms could cause rapid termination of ice ages every 100 k.y. despite modest changes in the distribution of insolation.

The first Termination IA was an abrupt phase of the deglaciation. It started at the end of the LGM and ended with the beginning of the Bolling warm phase. The second step of the deglaciation, which started at the end of the Younger Dryas cold event and led to the Holocene, is called Termination IB. The major oceanic cooling at around 10.4 k.y. BP coincides with the continental Younger Dryas cold event (Duplessy and Pujol, 1983).

Terminations IA and IB are dated by radiocarbon and oxygen stratigraphy at > 14–13.5 k.y. and 10.3–9.7 k.y., respectively (northern Norwegian shelf). Temperature rose by 4–5°C during Termination IA (Hald and Vorren, 1987) and since 9.7 k.y. BP temperatures were close to those of the present. High-latitude deglaciation dating from the Norwegian–Greenland Sea reveals two-step deglaciation with Termination 1A at 15 k.y. and Termination 1B at 10.1 k.y. BP (Jones, 1987), in agreement with results from the North Atlantic (Duplessy *et al.*, 1986).

Keigwin and Jones (1988) presented evidence for abrupt climatic events in the marine isotope record from the Atlantic and Pacific Oceans during the deglaciation. Brief events with $\delta^{18}O$ as low as Holocene values are seen in planktonic foraminifera between 15 and 11 k.y. BP in both temperate and subtropical regions. Keigwin and Jones suggest that these isotopically light events reflect local lowering of sea surface salinity rather than warm temperatures, and they may correlate in time with meltwater discharge to the Gulf of Mexico. A similar isotope record in the Gulf of California suggests that there may have been local salinity anomalies in the Pacific resulting from pluvial events during climatic transition on the continent. Onset of degaciation in the tropical Atlantic is dated at 15 k.y. BP, and a pause between 12 and 11 k.y. was followed by a second peak centred near 10 k.y. BP. The latter event was followed by a brief excursion towards light oxygen isotopes and may be correlated with the Gulf of Mexico meltwater spike (Berger *et al.*, 1985).

A 12°C deglacial SST rise was recorded west of Portugal at 12.5 to 12 k.y. BP. Bard and Fairbanks (1988) interpreted this change as due to a local decrease in coastal upwelling and not to a shift of the Polar Front. Duplessy

et al. (1981) found that deglacial warmings of the north-eastern Atlantic are closely correlated with climatic events on the adjacent European continent. Also, Alpine deglaciation has been recorded in sediment cores from Lake Zurich (Lister, 1988). These data indicate that release of meltwaters commenced prior to 15 k.y. and was completed by about 12.4 k.y. BP, which is well before an increase in incoming radiation due to changes in orbital configuration or an increase in carbon dioxide concentration in the global atmosphere. The $\delta^{18}O$ temperature calculation indicates a 2°C warming during the first half of the Holocene.

The first direct dating of the last deglaciation in the Southern Ocean indicates that the first deglacial $\delta^{18}O$ decrease took place before 13 k.y. and that Holocene values were reached at ~ 12 k.y. BP (Labracherie *et al.*, 1988). In addition, two $\delta^{18}O$ curves show an increase at ~ 10.5 k.y., which is synchronous with the North Atlantic Younger Dryas. However, it is not certain that this variation is an expression of the Younger Dryas since the diatom-based SSTs from the same cores do not exhibit a concomitant oscillation and the SST record is complicated by the sharp polar-front gradient near this location. Labracherie *et al.* propose alternatively that a pulse-like injection of meltwater from the Antarctic ice sheet at ~ 12–13 k.y. BP depleted the $\delta^{18}O$ of surface water.

The Polar Front shifted north and south in the North Atlantic during the period of glacial retreat (15 to 8 k.y. BP). Analyses by Bard *et al.* (1987) show that the front moved between 35 and 55° N latitude in less than 1000 years during the earliest retreat (12.5 and 12 k.y. BP) and moved almost instantaneously (in less than 400 years) during two abrupt climatic changes associated with the Younger Dryas event. After this, the second phase of deglaciation took place between ~ 10.4 and 9.4 k.y. BP.

As seen above, the warming of surface waters associated with the end of the last glacial (Terminations I and II) occurred very rapidly. This has important implications for theories of cyclic glaciation. Changes in seasonality induced by cyclic changes in the orbital parameters are the most common cause for glacial–interglacial cyclicity (Imbrie *et al.*, 1984). The rapidity of deglaciation has commonly been related to non-linear response of ice cover to the seasonal changes of incoming radiation; that is, during times of strong seasonality excess summer melting more than compensated for the effects of the more severe winters (Weertman, 1976; Imbrie, 1985). However, data from ice cores suggest that the end of the glacial period was abrupt and that the ice then melted in response to this change. Broecker *et al.* (1985) proposed an alternative linkage that involves reorganization of the mode in which the ocean–atmosphere operates, especially in the North Atlantic (Fig. 10.19). Warm surface waters in the North Atlantic may have provided a net water vapour transport through the atmosphere to the Pacific and this leads to a build-up of the salt content of waters in the Atlantic

Fig. 10.19. *Schematic illustration of two possible extreme modes of ocean circulation, corresponding to warm (A) and cold (B) intervals seen in the ice-core record, according to Broecker et al. (1985). (C) shows corresponding variations in ice volume and summer insolation at 65° N. From Broecker et al. (1985). Copyright © 1985 by Macmillan Magazines Ltd.*

(Fig. 10.19). Salty water tends to sink and would have been transported as deep water out of the Atlantic to the Pacific; less salty surface waters were imported to the Atlantic from the Pacific through the Indonesian straits.

During the glacial stage the low seasonal contrast caused the ocean to operate in a different mode from today (Fig. 10.19B). When the seasonality contrast increased towards its maximum in the northern hemisphere during the last 16 k.y., the system flipped back into its current mode of operation (Fig. 10.19A). The Younger Dryas event marked a short return to the glacial mode of operation (Fig. 10.19C).

Abrupt terminations of the late Pleistocene glaciations may be a normal

response of the climate system to changes in incoming solar radiation. Fig. 10.20 shows the sum of responses to orbital forcing at a number of frequencies and their harmonics (upper four curves), and the $\delta^{18}O$ data (middle curve). Comparison of these data sets shows a remarkable agreement; however, differences are seen in the residuals (see Imbrie, 1987). This suggests some non-linear mechanism which was not accounted for in the model of insolation forcing. A number of additional casual mechanisms could be involved in producing deglaciation (Broecker, 1984, 1988b).

The Younger Dryas episode

The Younger Dryas (approximately 11 to 10 k.y. BP) is the most recent and abrupt return of cold conditions to western Europe, the North Atlantic and Greenland (Mott *et al.*, 1986; Peteet, 1987; Broecker *et al.*, 1988). Evidence suggests that it may have been initiated and terminated by changes in the mode of operation of the northern Atlantic Ocean (Broecker *et al.*, 1985). It is suggested that the predominant drainage of glacial meltwater into the Gulf of Mexico at first allowed the maintenance of efficient oceanic heat transport into the eastern North Atlantic via the Gulf Stream, and hence an abnormally rapid post-glacial warming in north-western Europe. When the drainage pattern switched to a high-latitude (Hudson Bay) run-off to the Atlantic, the anomalous warming was terminated and the Younger Dryas cold phase began. Rooth (1982) suggested that the impact of this diversion was to turn off deep-water production in the North Atlantic, and Rind *et al.* (1986) and Broecker *et al.* (1988) suggest mechanisms of regional climatic change caused by the shutdown of bottom water production. The response of the North Atlantic during the Younger Dryas was not limited to a cooling of surface waters but as shown by Boyle and Keigwin (1987) also involved deep-water production. This is consistent with the idea that the diversion of meltwater discharge into the region of deep-water formation caused salinity (and density) stratification of northern Atlantic waters and therefore slowed circulation.

Broecker *et al.* (1988) put forward the hypothesis that the Atlantic Ocean operates in a manner similar to a conveyor belt (Fig. 10.21). Warm and salty upper waters flow northward to the vicinity of Iceland and, during winter, cold air from North America sweeping over these warm waters is heated. The heat release from ocean to atmosphere cools surface waters which sink and form deep water. The warm air is transported towards the continents and causes amelioration of winter temperatures over land. Broecker *et al.* (1985) calculated that heat transferred from sea to air by this process accounts for 30% of the solar heating at the surface of the Atlantic north of 30° N latitude. Evidence from ocean cores (Broecker *et al.*, 1988) suggests that during the last glacial maximum and during the Younger Dryas, this conveyor belt was greatly weakened (Boyle and Keigwin, 1987; Bard *et*

Fig. 10.20. Patterns of orbital eccentricity, obliquity and index of precession over the past 800 k.y. The three upper time series are from Berger (1977, 1978). When normalized and combined they form an irregular curve that portrays variations in the amount of radiation reaching the Earth as a function of time. The two bottom curves represent $\delta^{18}O$ variations over the past 800 k.y. (from Imbrie et al., 1984). The upper of the two represents an average of normalized plankton data for five deep-sea cores. The bottom curve shows variance spectra calculated for the $\delta^{18}O$ time series and shows dominant variance at about 100 k.y., 41 k.y., 23 k.y. and 19 k.y. From Imbrie et al. (1984). With permission of Kluwer Academic Publishers.

Fig. 10.21. Schematic illustration of Broecker's idea of an oceanic conveyor belt. Freshwater and heat added at the Pacific end of the conveyor belt is carried into the Atlantic Ocean by surface currents. In the northern Atlantic heat and fresh water is removed by cold air moving off Canada, causing water to sink. Warm surface currents in the eastern Atlantic warm Europe when the belt is carrying a large load (present climate) and cool that area during times of decreased load (Younger Dryas episode). From Kerr (1988). Copyright © 1988 by AAAS.

al.,1987) due to extensive sea ice which limited northward heat transport in surface waters (for discussion see Berger, 1990).

Rapid cooling in the North Atlantic during the Younger Dryas can be correlated to the discharge of the Laurentide meltwater to the North Atlantic, but there are no known events which correlate to the abrupt termination of the Younger Dryas cool period and the abrupt northward migration of the Polar Front in the North Atlantic at 10.5 k.y. BP. Showers and Bevis (1988) suggested that changes in low-latitude circulation and subsequent rapid warming of the North Atlantic affected the Polar Front, reinitiated NADW production and ended the Younger Dryas cold period. Alternative explanations include a redirection of meltwater discharge back to the Mississippi drainage system due to a readvance of the Laurentide ice sheet into Lake Superior at ~ 10 k.y. BP (Broecker *et al.*, 1988).

Palynological and marine micropalaeontological data from the northern Atlantic and from Eurasia and North America at 10.5 k.y. BP led Grosvald,

Muratova and Shisharine (1985) to postulate a causal connection between the abrupt cooling of the late Dryas and glacier surges and massive discharge of icebergs into the ocean. The data reveal that maximum negative tempera-ture anomalies occupied the eastern subtropical North Atlantic between 35° and 45° N latitude (values of 8–9°C, double those recorded over the adjacent land masses). The area of maximum lowering of SSTs was located only some 5° south of the zone of highest accumulation rates of ice-rafted debris. It is suggested that ice sheets disintegrated by ice calving in the form of a series of massive discharges or glacier surges. The effect of glacier surges on climate involves absorption of heat due to the melting of ice in the ocean, expansion of the area of relatively fresh surface water, an increase in winter ice cover, and a sharp increase in the albedo of the ocean surface. On the basis of approximate calculations, Mercer (1969), Wilson (1969) and Flohn (1973) have postulated that the combined effect of these mechanisms could have led to a cooling of the atmosphere in the northern hemisphere by 5–10°C per century. The increase of ice cover on the ocean would have led to a reduction in evaporation and precipitation which, according to Ruddiman and McIntyre (1981a, 1984b), would have encouraged the starvation of the glaciers.

Grosvald *et al.* (1985) see a similar cooling episode at 16–13 k.y. BP in the Pacific, which was associated with the cooling influence of icebergs enter-ing the ocean from the destruction of the glacier complexes of British Columbia, south-eastern Alaska and the Bering and Okhotsk seas. Evidence of abrupt cooling is also found at ∼ 11 k.y. BP in the Gulf of California, in the far north-west Pacific, and off Japan (Keigwin and Jones, 1988).

The Holocene warming

Climate modelling of conditions at the beginning of the Holocene at 9 k.y. BP indicates that, relative to today, northern land winters were cooler and drier but with more snowfall; summers were warmer and wetter and there was a stronger monsoonal circulation (Kutzbach, 1981, 1984; Sellers, 1984). It is apparent that the Holocene warming was one of the more dramatic short-term events in Earth history.

Warming in the Holocene caused changes in frost conditions in the USSR, with the average annual temperature rising by 12°C. The southern boundary of permafrost shifted northward by 20° of latitude in the European part of the USSR, by 13° in western Siberia and by 6° in central Siberia (Baulin, Belopukhova and Danilova, 1984). Johnson (1984) calculated temperature changes in the Canadian Arctic (ice sheet nucleation centre) due to effects of orbital variations for the intervals 125–115 k.y. BP and 6 k.y. BP to 8 k.y. AP. The model temperatures for July are 4°C below present at 117 k.y. BP and 1.5°C above present at 8 k.y. AP. This temperature rise suggests a continua-tion of the present interglacial to at least 8 k.y. AP.

The record of aeolian deposition in the north-west Pacific presents an integrated picture of continental aridity and zonal westerly winds during the last 30 k.y. (Rea and Leinen, 1988). The data show the Holocene climatic optimum as the time of maximum aridity in China and the time of the northernmost shift of the zonal westerlies. This supports the conclusions from AGCM modelling of the Holocene climatic optimum (Kutzbach and Guetter, 1986; COHMAP, 1988), indicating enhanced aridity in central and western China and enhanced monsoonal circulation in south-east Asia.

11

Causes and chronology of climate change

Seeking the cause of climate change is both an exciting activity and a difficult task, as borne out by the long history of enquiry and the scarcity of firm conclusions. While numerous hypotheses have been advanced, either to explain local or temporally short changes or to provide a global framework for change throughout geologic time, it is safe to say that, although elements of truth exist in many of these hypotheses, no single one takes account of all the variables as they are known at present. To help us to understand the climate system, recent studies have utilized an integrated approach that considers the atmosphere, the hydrosphere, the biosphere and the solid Earth. This approach has tended, justifiably, to be historical in scope in order to make use of geological data bearing on the temporal and spatial variability of processes central to the system. The flood of new information means that there is a constant need to re-evaluate concepts and, given the growing awareness of the complexities of the climate system, it is likely that many more attempts will need to be made before we have a thorough understanding of how the system has worked in geological history.

The search for cyclical climate change

The palaeoclimate information collated here illustrates that the history of climate over the past 600 m.y. was not a simple trend of cooling or warming but was characterized by alternating periods of warm and cool climates. These periods we have defined as Cool or Warm Modes, based on the presence of cool or warm climate indicators as presented in the previous chapters.

The duration of Phanerozoic warm intervals ranges from 50 to 102 m.y. while that of cold intervals, excluding the late Cenozoic, lies in the range of 37–80 m.y. There is thus a fairly wide range of time for the duration of individual periods of warm or cool climate, which can be related in part to uncertainties in dating. Cyclicity is only apparent when one considers the length of a full cycle based on midpoints, that is, time between the midpoint

Table 11.1. *Approximate 150 m.y. climate cycles in the Phanerozoic and late Precambrian*

	Age range (Ma)	Midpoint (Ma)	Duration of cycles (m.y.) Warm	Cool
Early Eocene to present (cool)	~55–0	[~28]		
Early Cretaceous to early Eocene (warm)	~105–55	[~80]		
Late Jurassic to early Cretaceous (cool)	~183–105	[~144]		116+
Latest Permian to middle Jurassic (warm)	~253–183	[~218]	138	
Early Carboniferous to late Permian (cool)	~333–253	[~293]		149
Early Silurian to early Carboniferous (warm)	~421–333	[~377]	159	
Late Ordovician to early Silurian (cool)	~458–421	[~440]		147
Earliest Cambrian to late Ordovician (warm)	~560–458	[~509]	172	
Late Precambrian to earliest Cambrian (cool)	~615–560	[~588]		148
Late Precambrian (warm)		[~690]	181	
Late Precambrian (cool)		[~765]		177
Late Precambrian (warm)		[~852]	162	
Late Precambrian (cool)		[~940]		175

Note: Time scale from Palmer (1983).

of one Cool Mode and the midpoint of the next. A full cycle therefore consists of a Warm Mode and a Cool Mode, as shown in Table 11.1. The closer agreement gained by this method may reflect the difficulty in estimating the age of the beginning or end of a Cool or Warm Mode. This method is a means of rounding off errors of uncertainty in dating.

The duration of warm intervals is 138–172 m.y. for the Phanerozoic (or 138–181 m.y. if the late Precambrian is considered), which is not a good indication of very tightly controlled cyclicity. The range for (completed) cool cycles is 147–149 m.y. for the Phanerozoic, and 147–177 m.y. including the late Precambrian. The average length of a Warm–Cool cycle is 162 m.y. based on Warm Mode midpoints and 159 m.y. based on Cool Mode midpoints. The agreement with a 150 m.y. half Galactic Year cycle is quite good, given the uncertainties of dating.

It should be noted that the interval of glaciation represented in the late Devonian of Brazil and possibly North Africa (Caputo, 1985) is not included in this discussion. If it were, the apparent cyclicity at about 150 m.y. would vanish. However, the late Devonian glacial event (middle Famennian in age, i.e. less than 3 m.y. in duration) may be taken to reflect strictly regional conditions related to a high-latitude position for a short time. In this sense

the Brazilian late Devonian glacial deposits are consistent with the concept that the zones beyond 60° latitude have experienced cold (though not necessarily full glacial) polar climates throughout the Phanerozoic, irrespective of the global climate state (Frakes and Francis, 1988). Noteworthy also is the initiation of massive global volcanism in the late Devonian, which, through release of large quantities of CO_2 to the atmosphere, may have inhibited an earlier development of the Cool Mode.

It is also apparent that the climate state is governed by changing continental distributions on the globe. During cold episodes the distribution of continents within high-latitude zones played a critical role in providing sites for the growth of glaciers. The Palaeozoic and Cenozoic glaciations were polar glaciations, although their genesis may have differed somewhat. The Palaeozoic episodes originated in the southern hemisphere only as Gondwana drifted over the south pole of rotation. Extensive glaciation did not take place in the northern hemisphere apparently because the northern land masses occupied low-latitude positions, except for the Siberia and Kolyma plates. In contrast, the Cenozoic glaciation developed in high-latitude zones of both hemispheres. Also, further detailed work is required to document whether areas glaciated during the late Precambrian lay in tropical to subtropical latitudes, as is suggested by some palaeomagnetic evidence (for discussion see Frakes, 1979). If so, this interval of glaciation is unique in the history of the Earth. Continental distribution was not such an influential control during warm phases – the early Palaeozoic was characterized by dominantly low-latitude continents but during the Permian to late Eocene continents were distributed through all latitudinal zones.

An interesting aspect of the Warm and Cool Modes described above is that they appear to be independent of boundaries between major intervals of geological time, as defined on faunal, floral or lithological changes. For example, the late Palaeozoic Cool Mode both begins and ends *within* defined geologic Periods (Carboniferous and Permian respectively). Despite this, the effects of global climate change on both the evolution and the dispersal of organisms have been substantial, and clear examples are seen in the shorter subdivisions of geologic time.

The development of Cool Modes

There are certain similarities between modes of cool climate, including the better documented late Cenozoic cooling and glaciation. These include the following: (1) a considerable extent of high-latitude land, (2) long intervals of cooling leading to high-latitude glaciation, and (3) marked increase in $\delta^{13}C$ rising to a peak during extreme cooling. There also appears to have been a marked decrease in the abundance of evaporites. These four features can therefore be considered of vital importance in the development of Phanerozoic Cool Modes.

High-latitude land during Cool Modes

Two lines of reasoning support the idea that, during the Phanerozoic, land at high latitudes was a necessary precondition before extreme global cooling could occur. The first derives from the simple observation that glaciations at these times are recorded only in high-latitude sites; in other words, the land masses provided the seat on which glacial ice could form (but see Henderson-Sellers and Meadows, 1975; Hunt, 1984). Secondly, increased albedo and other positive feedbacks resulting from elevated masses of cold continental ice near the poles intensify global cooling trends much more effectively than feedbacks from sea ice.

Rapid relative movement between the poles and continents may, at times, have led to regional cooling but not full-blown refrigeration of the globe. Palaeomagnetic data from Australia (Embleton, 1984) indicate rapid relative motion between Gondwana and the pole in the Jurassic–Cretaceous, and although this interval displays evidence of high-latitude cooling, there is as yet no evidence of glaciation. Short-term development of glaciers in South America in the late Devonian similarly corresponds to a period of rapid relative motion between continents and the pole. Intervals of cooling with land at high latitudes for substantial periods appear to be necessary before the effects are felt globally.

Pre-glacial cooling

The Cenozoic Cool Mode had its initiation at about 50–55 Ma with marked cooling phases in the early Eocene. These took the form of sharp falls in temperature, as indicated by increases in $\delta^{18}O$ in the oxygen isotope signal, and by changes in faunal and floral palaeobiogeography (Chapter 9, Fig. 9.2). Falls in temperature were interspersed with gradual warming phases, which in some cases attained the nature of plateaux of several million years' duration. However, each succeeding temperature drop brought progressively cooler conditions. The Cenozoic saw a continued deterioration of Earth climates from the Eocene to the Pleistocene.

Although significant high-latitude glaciations began as early as the mid-Oligocene (\sim30 Ma) in Antarctica, the major ice sheets of both hemispheres did not originate until near the end of the Neogene (5–3 Ma). Considerable inertia in the climate system therefore is likely, with about 50 m.y. required for the transition from a Warm Mode to a Cool Mode. A similar pattern is seen in the late Palaeozoic Cool Mode, where the first clear signs of cooling are detectable no later than the Visean, and full development of ice sheets is recognizable in the Westphalian (i.e. an interval of about 40 m.y.). If cooling is taken to have begun in the late Devonian the interval would have been about 50 m.y.

The Ordovician–Silurian Cool Mode may have begun with intermittent

cooling in the early Llanvirnian: Spjeldnaes (1961) suggested (on fossil evidence) that at that time mean global temperatures were lower than present. Thus, pre-glacial cooling may have lasted for some 35 m.y. before an ice sheet grew in North Africa in the Ashgillian. The restriction of true glacial deposits (as opposed to ice-rafted deposits) to a narrow polar zone suggests the Ordovician–Silurian Cool Mode did not attain such globally extreme conditions as did the Cenozoic and late Palaeozoic episodes; neither did the late Jurassic to early Cretaceous cooling that persisted for some 70 m.y. without leading to a glaciation, at least as far as is presently known.

Increasing $\delta^{13}C$ and decreasing atmospheric CO_2

These relationships among Cool Mode histories lead to the conclusion that both the duration and the intensity of cooling varied during the intervals leading up to glaciation. Although the Cenozoic and late Palaeozoic apparently experienced similar conditions over similar time intervals, it seems that the early Palaeozoic and the mid-Mesozoic cool phases were less intense and lasted for shorter and longer time spans respectively. We suggest that Cool Modes developed when sufficient land existed at high latitudes to provide a seat for significant ice growth and over a time when relatively large amounts of carbon were locked up in sedimentary and/or biological reservoirs.

The development of ice sheets required a global cooling (at least cool summers) and for this we suggest there was a contribution from a marked decrease in atmospheric CO_2. The mechanism for this was provided through the sequestration of large quantities of organic carbon on land, on continental shelves and possibly in the deep ocean, as a result of the Warm Mode high stand of sea level; this resulted in the observed gradual increase in $\delta^{13}C$ during cooling (Fig. 11.1). The time lag between the major increase in $\delta^{13}C$ and development of glaciers in Cool Modes was between 10 and 30 m.y. and sequestration continued until well after glaciation began. Since the main episodes of high sea level appear to have occurred during previous Warm Modes, there is an element of autocyclicity in our model of climate change.

Early intensification of Cool Modes took place in concert with relatively high sea level (e.g. early Carboniferous, early Tertiary) and while ocean waters were still enriched in ^{13}C. A later stage saw the development of significant ice masses concurrent with large falls of sea level while the $\delta^{13}C$ signal continued to increase.

Why and how did the climate switch from a Warm to a Cool Mode? The carbon curve in Fig. 11.1 indicates that $\delta^{13}C$ increased throughout Warm Modes and into Cool Modes. This suggests that the transition into a cool climate phase was gradual, more gradual than from a Cool to a Warm Mode. The increase in $\delta^{13}C$ suggests either a gradual increase in biological produc-

Fig. 11.1. Estimated mean global temperature curve for the Phanerozoic, compared with 'eustatic' sea level curves (Hallam, 1984b; Vail et al., 1977), global abundances of volcanic rocks (Ronov, 1980; line in 'Geosynclinal Volcanism' column at 4m/m.y. represents mean rate of accumulation) and δ13C carbonate record. The carbon curves are as follows: thin solid line (Cenozoic) from Shackleton (1985); thick solid line (Mesozoic) from Weissert (1989); medium solid line (Phanerozoic) from Holser (1984); broken line (Phanerozoic) from Kump (1989). Warm and Cool Mode divisions are described in the text.

tivity or increased burial of ^{12}C in anoxic conditions. From Fig. 11.1 it appears that a critical threshold in $\delta^{13}C$ was reached several times, during which the Earth moved into a Cool Mode and glaciation. Perhaps the increased biological drawdown of atmospheric CO_2, plus decreased volcanic output during a quiet tectonic phase, allowed the sensitive polar regions to cool sufficiently for the development of ice.

The cooling may then have been enhanced by orbital influences of Milankovitch-type and the feedback effects of increased albedo etc. Changes in productivity in glacial–interglacial oceans appear to have been affected by orbital parameters (Sarnthein *et al.*, 1988). Changes in the orbital parameters warmed or cooled the polar oceans, changing the equator-to-pole temperature gradient. This would have affected the intensity of the meridional winds, strengthening them in cooler periods and weakening them during warm phases. Strong winds enhanced upwelling, nutrient supply and surface ocean productivity, whereas weak winds resulted in less well-mixed seas and lower productivity. The time scale of such change is obviously much smaller than that we are examining in our changes from Warm to Cool Modes in the geological record, but may have been an important contribution to the development of the longer time scales of change in the carbon cycle and climate.

The onset of Warm Modes

Four Phanerozoic Warm Modes have been recognized in this work: Cambrian to middle Ordovician, middle Silurian to early Carboniferous, latest Permian to early Jurassic, and late Cretaceous to early Tertiary. In contrast to Cool Modes, it appears that the advent of warm intervals was remarkably sudden, whereas the terminations were more gradual and geographically variable. Also consistent were the abrupt decreases in $\delta^{13}C$ shortly before the beginning of Warm Modes and the large but more gradual increases towards the end (Holser, 1984; Fig. 11.1). There is no apparent consistency in the distribution and volume of evaporites and marine carbonate rocks through Warm Modes, although both appear to be more abundant than in Cool Modes.

The initiation of Warm Modes

One of the most puzzling of palaeoclimatological problems is why Cool Modes terminated. The late Palaeozoic Cool Mode, for example, finished while much of the southern supercontinent of Gondwana was still in high latitudes and positioned near the pole. In these circumstances it is surprising that the glaciation terminated; the same can be said for the late Proterozoic and Ordovician–Silurian glaciations. It is not likely that abrupt shifts of the pole or the continents, as yet undetected in the palaeomagnetic

record, could have resulted in much warmer climates suddenly evolving. The polar and subpolar zones warmed substantially, to the extent that the ice masses melted. The driving force for termination may have been a one-off or random occurrence of either terrestrial or extraterrestrial origin, but it would seem most likely that it was of a related origin to processes that led to the initiation of Cool Modes.

Fig. 11.2 indicates that the onset of a Warm Mode was shortly followed by increased output of CO_2 resulting from tectonic activity. Such an increase was perhaps due to an upsurge of sea-floor spreading, as sea levels also rose in Warm Modes. This enhanced the warming effect. From Fig. 11.1 there is a very general trend of sea-level fall during a Cool Mode and a rise during a Warm Mode (this is clearer on the sea-level curve of Hallam). It is also interesting to note that during most Warm Modes the rise in sea level is interrupted by at least one small fall; these roughly coincide with a carbon peak. For example, the small peak in $\delta^{13}C$ at the Cambrian–Ordovician boundary is marked by a small drop in sea level, as is the carbon peak in the late Devonian (and, incidentally, also a smaller carbon peak in the early Devonian associated with a sea-level fall).

The best documented of the Warm Modes is the late Cretaceous to early Tertiary, the beginning of which was preceded by a sharp decrease in the $\delta^{13}C$ record in the late Aptian. Oxygen isotope and palynological data suggest that the late Jurassic and early Cretaceous Cool Mode was terminated by a warming phase in the middle Albian, coinciding with the termination of ice-rafting in high-latitude sites and of widespread Albian volcanic activity. The sea-level record, as deduced from coastal onlap relationships, suggests initial high levels, followed by a fall in the latest Cretaceous, an early Tertiary peak and a later fall (Figs 8.2, 11.1). Finally, there was an increase in volcanic activity in the late Cretaceous (Fig. 8.3). The sequence of events just prior to and in the early stages of this Warm Mode thus can be estimated as follows:

1 A sharp decrease in $\delta^{13}C$ of about 2‰) (late Aptian).
2 Warming and cessation of ice-rafting (mid-Albian).
3 Beginning of steady decline in accumulation rate of organic carbon.
4 Warming, rising sea level and volcanic activity.

The latest Permian to middle Jurassic Warm Mode appears to have differed in that the initial warming (as indicated by the termination of ice-rafting in the Kazanian) preceded a major decrease in $\delta^{13}C$ (~ 3‰) just before the Permian–Triassic boundary (Holser *et al.*, 1989); however, a significant fall of about 1.0‰ in the Sakmarian correlates with the youngest known glacial tillites. There was moderate accumulation of black shales and coals in intracratonic settings in the Artinskian–Kungurian interval, not long before strong warming terminated ice-rafting in high latitudes. A slight fall in

Fig. 11.2. The sequence of events characterizing the end of Cool Modes and the onset of Warm Modes for the four Warm Modes of the Phanerozoic. The end of a Cool Mode is characterized by a decrease in δ¹³C, representing decreasing levels of productivity and/or oxidation of organic matter. This caused an increase in CO_2 levels and climate warming, which terminated glaciation and ice-rafting. During the Warm Mode, input of CO_2 from volcanic outgassing reinforced the warming trend. For consistency the data of Ronov (1980) for volume of volcanics has been plotted for all Warm Modes, including the late Cretaceous/early Tertiary. However, other data (Arthur et al., 1985; Vogt, 1979) show that there was a relative increase in volcanic activity in the Aptian, late Campanian–Maastrichtian and the Palaeocene–Eocene boundary. Note also the carbon isotope decrease and sea-level drop at the end of the Devonian, which corresponds to the end of the short interval of glaciation in the southern hemisphere. Carbon isotope data from sources mentioned in Fig. 11.1.

197

sea level (latest Permian) was followed by a slight rise in the early Triassic, according to Vail *et al.* (1977), while Hallam (1984b) argues for a major fall at the end of the Permian. Continental volcanism was globally extensive during the middle and late Triassic but decreased markedly thereafter during the early Jurassic. The sequence of events during the initiation of this Warm Mode therefore was as follows:

1 Decrease in $\delta^{13}C$ by about 1‰ and end of tillite deposition (Sakmarian).
2 Black shale and coal accumulation; slight sea-level fall (Artinskian–Kungurian).
3 Warming and termination of ice-rafting (Kazanian).
4 Increase in $\delta^{13}C$ (decrease at end Permian).
5 Rise in sea level (early Triassic).
6 Extensive volcanism (middle and late Triassic).

In considering the next oldest Warm Mode, the late Silurian to early Carboniferous, a difficulty arises because of a shortage of carbon isotope data for the late Ordovician and the Silurian. However, Holser (1984) suggests a fall in $\delta^{13}C$ in the early Silurian. The youngest ice-rafting appears to be Wenlockian (late early Silurian) in age. The first rise in sea level occurred at the beginning of the Devonian. Volcanicity on the continents increased markedly through the Devonian to its Phanerozoic peak in the late Devonian. We suggest the following sequence of events:

1 Rapid decrease in $\delta^{13}C$ (age assumed; early Silurian).
2 Black shale accumulation (early Silurian).
3 Warming and termination of ice-rafting (Wenlockian).
4 Rise in sea level (early Devonian).
5 Rise in $\delta^{13}C$ but interrupted by a rapid decrease in $\delta^{13}C$ and black shale accumulation, coinciding with extensive volcanism (late Devonian).

The coincidence in time of events under (5) above may explain the short duration of the anomalous late Devonian glaciation in polar latitudes of South America. The late Famennian glaciation, which occurred as the region drifted near the pole, may have been inhibited by the initiation of intense volcanism. Continuation of extensive volcanism in the early Carboniferous may have similarly limited the development of glaciers in that interval. In this case, the early stages of a potential late Devonian Cool Mode were not sustained and the Warm Mode continued until the Namurian after only a brief interruption.

The final Warm Mode to be considered is the Cambrian to middle Ordovician interval. The age of the beginning of this Warm Mode is not known because it is uncertain whether, and for how long, the late Protero-

zoic glaciation extended into the Cambrian, and because identified climate indicators are entirely from low-latitude zones. Ice-rafting may have continued into the early Cambrian. There was also a rapid decrease in $\delta^{13}C$ at the Precambrian–Cambrian boundary (Magaritz, 1989). Holser (1984) also indicated a slight increase in $\delta^{13}C$ up to the middle Cambrian, and a sharp peak in the latest Cambrian followed rapidly by a significant decrease in the early Ordovician. Black shale events are known from the latest Cambrian and the late Ordovician and the second corresponds well with a sea-level high stand. Sea level apparently rose through most of the Warm Mode, though punctuated by a drop at the Cambrian–Ordovician boundary. The Ordovician was a time of major continental volcanism. On the basis of comparison with other Warm Modes, the following sequence of events is suggested:

1 Rapid decrease in $\delta^{13}C$ at the Precambrian–Cambrian boundary. (possible termination of ice-rafting?).
2 Rise in sea level through the Cambrian.
3 Decrease in $\delta^{13}C$ and fall in sea level (late Cambrian).
4 Black shale accumulation (late Cambrian and late Ordovician).
5 Warming (age uncertain).
6 Extensive volcanism (Ordovician).

A series of events thus appears to have characterized the beginning of Warm Modes (= the end of Cool Modes). These were as follows: (1) a rapid decrease in $\delta^{13}C$, (2) increasing deposition of abundant organic matter, and (3) warming and volcanism. Sea levels were initially high, fell just after the start of a Warm Mode, rose in the middle of a Mode, fell again and then rose at the end of the Warm Mode (Fig. 11.1). The preliminary stages suggest release of organic carbon from land, continental shelves, and/or the deep ocean floors, and perhaps initiation of significant carbonate sedimentation. Strong and abrupt decreases in the $\delta^{13}C$ record at approximately the time of termination of ice-rafting suggest widespread release of organic carbon to the ocean; in each case this was followed by an interval in which the accumulation of organic matter gradually increased. Each Warm Mode saw a gradual and steady increase in $\delta^{13}C$ following these events. The peak in this trend occurred in the succeeding Cool Mode.

The initial decrease in $\delta^{13}C$ can be interpreted as suggesting an initial rapid increase in atmospheric CO_2 coincident with the final phase of ice-rafting, and then a gradual decrease in atmospheric CO_2 through the Warm Mode as organic matter was again lost to the active system through burial. At this time the first warming effects and sea-level rises can be detected.

For the late Palaeozoic it is worth noting that the carbon signal would have been affected by the enormous amounts of organic matter now preserved as coal, particularly because marine Palaeozoic (and Mesozoic) organic matter is relatively depleted in ^{13}C (Lewan, 1986). The $\delta^{13}C$ values of

the oceans after the late Palaeozoic Cool Mode terminated are about the same as before it began ($\sim +0.5‰$), despite the increased sedimentary reservoir. This is perhaps explainable by the loss of the light isotope being balanced by a comparable loss of ^{13}C in enhanced limestone deposition on post-glacial Gondwana. Also, the decrease in $\delta^{13}C$ at the end of the Permian was the largest in the Phanerozoic, amounting to at least 3.5‰. The rebound of the carbon record during the Mesozoic and Cenozoic scarcely ever reached the values which prevailed from the early Carboniferous to the Permian, indicating either that the organic matter/limestone fractionation has continued to favour deposition of carbonate rocks (Popp *et al.*, 1989) or that the uptake of the light carbon by organisms has greatly decreased since the Permian.

Holland (1984) suggested that most of the organic carbon required to compensate for that lost through burial comes from release through weathering and metamorphic degassing, with about 4% possibly being from juvenile sources. However, these are figures which apply to the comparatively low modern burial rates for organic carbon; compensation rates required for the late Palaeozoic would have to have been up to about five times higher (Fig. 11.3). In the absence of a major orogenic episode at this time (post early Carboniferous), strong drawdown in both atmospheric and oceanic CO_2 levels is implied, which is consistent with the development of the late Palaeozoic Cool Mode.

Burial rates for carbon are notoriously difficult to estimate, since they involve both organic and carbonate carbon on land as well as in ancient seas. On the basis suggested by Holser (1984), it appears that Warm Mode burial rates also quite often exceeded the present rate (by factors of 2 to 3). It is possible that despite this tendency toward cooling owing to depleted atmospheric CO_2, increased weathering and outgassing maintained Warm Mode temperatures significantly above present levels for intervals averaging about 150 m.y. in duration, the length of the proposed Warm Modes.

An interesting observation is the occurrence of all major episodes of 'geosynclinal' volcanism either within Warm Modes or late in the Cool Modes (Figs 11.1, 11.2). These relationships suggest that the volcanic activity related to lithospheric plate mobility contributed to the termination of Cool Modes (Permian, early Cretaceous), as well as reinforcing warming trends (early Ordovician, Devonian, late Cretaceous). In both the late Ordovician and the mid-Carboniferous, cooling appears to have overwhelmed any CO_2 warming due to volcanism, perhaps because volcanic activity was waning. In all cases, improved control on the age and volume of volcanic sequences could alter these apparent relationships between volcanism and climate change.

During the development of a Warm Mode, there was an increase in continental volcanic activity, usually most enhanced towards the middle of the Warm Mode. Major peaks of 'geosynclinal' volcanism indicated by

Fig. 11.3. Variations over Phanerozoic time of (a) NaCl and CaSO₄ volumes in marine evaporites, (b) deposition rates of carbon in carbonate sediments and (c) deposition rates of carbon in organic matter. Shaded areas represent times of increased accumulation of organic-rich sediments. From Holser (1984).

Budyko and Ronov (1979) occur in the Ordovician, the Devonian to early Carboniferous, the early Permian, the late Triassic and the mid- to late Cretaceous (Aptian to Campanian) (Fig.11.1). Early Permian and Cretaceous volcanism thus began during the latest stages of the previous Cool Mode. Through increased release of CO_2 and aerosols, volcanism may have contributed to warming which led to termination of glacial activity in the late Permian and mid-Cretaceous. At the other times, the same process would appear to have sustained warmth. On Fig. 11.1, it can be seen that some 70% of above-average volcanism in tectonic regimes occurred during Warm Modes. Although the extent of greenhouse gases contributed by volcanic activity over geological time is uncertain, the timing of major episodes either late in Cool Modes or during Warm Modes (but not in the early stages of Cool Modes) is consistent with the hypothesis that global climates are influenced, and perhaps forced, by volcanic outgassing.

The carbon record in Warm and Cool Modes

A striking feature of the $\delta^{13}C$ record is the occurrence of an increase in $\delta^{13}C$ before each interval of Phanerozoic glaciation (Fig. 11.1). Several curves of $\delta^{13}C$ for the Phanerozoic are plotted on Fig. 11.1. Although the curves vary

somewhat in their details, the general trend is the same and, more importantly, the significant changes in the isotope record are in agreement.

It is apparent from this figure that many of the peaks in $\delta^{13}C$ are associated with (close to the end of or just after) the phases that we have delineated as Cool Modes; that is, periods when permanent or seasonal ice was abundant in high latitudes. These times include the early Silurian, the late Permian and the late early Cretaceous (Aptian). Other major peaks occur with our Warm Modes (late Devonian, mid-Jurassic, late Cretaceous and Palaeocene) but, as we will discuss later, we believe these represent brief cool intervals within a generally warmer time.

Of the increases in $\delta^{13}C$, the Ordovician–Silurian one represents the first major excursion of $\delta^{13}C$ into positive values (to 1.5–2‰), the late Palaeozoic exhibits a massive increase (to almost 4‰) which persists over an exceptionally long period of about 80 m.y., the middle Mesozoic is seen as a series of peaks followed by a low at the Jurassic–Cretaceous boundary, and the Cenozoic exhibits a short sharp spike in the earliest Tertiary (Holser, 1984). Another marked increase occurs in the middle Devonian; this is just older than glacial deposits in Brazil.

In Chapter 10 we have seen that in the late stages of a Cool Mode during the late Cenozoic, carbon isotope peaks are in some way related to nutrient supply in the world ocean, perhaps through the mechanism of oceanic upwelling. It is likely that sequestration of carbon on land in peat bogs and on the shelves has also played an important part. The strong late Miocene spike and deposition of organic carbon (~ 6.7 Ma) preceded the supposedly glacially induced drawdown of sea level which led to the first desiccation of the Mediterranean beginning at about 6.2 Ma. The abrupt fall in $\delta^{13}C$ that followed indicates a rapid shift of the light isotope ^{12}C from continental and/or oceanic sediments in amounts sufficient to account for at least $-1.0‰$ in the carbon signal.

As indicated, there appears to be a trend in the carbon isotope curve of a slow rise in values of $\delta^{13}C$ through a Warm Mode to a peak late in a Cool Mode. This peak is then followed by an abrupt drop in $\delta^{13}C$ values near the end of the Cool Mode. The longest and most extreme of this kind of cycle is seen in the mid Silurian to late Permian Warm–Cool Mode cycle. However, this pattern (slow rise in $\delta^{13}C$ followed by a sharp drop) is also present on shorter time scales; the climate change associated with shorter cycles still seems to be of the same pattern as that associated with the longer cycle (e.g. late Palaeocene to early Eocene peak). These will be discussed below.

At the Permian–Triassic boundary the beginning of the recovery stage (the shift back to more positive values of $\delta^{13}C$) was not instantaneous but occurred after an interval of about 1 m.y. (Magaritz *et al.*, 1988). Holser and Magaritz (1987) suggested that the peak in $\delta^{13}C$ was due to the expansion and high productivity of the land floras, especially the southern hemisphere

glossopterid flora, as a result of climate amelioration. In many sites extinction events are associated with the sharp decrease in $\delta^{13}C$ and new faunas appeared as old ones disappeared (Magaritz, 1989). The high positive values of $\delta^{13}C$ indicate that ocean waters were enriched in ^{13}C in the Permian; thus a large quantity of the light carbon fraction, ^{12}C, must have been buried out of the ocean system and would have formed large deposits of either black shales in the oceans or coals on land. Large sequestration of carbon lowers CO_2 concentrations in oceans and the atmosphere, and hence could contribute strongly to global heat loss and cooling.

Episodes of formation of oceanic carbon-rich shales in the Ordovician, the Jurassic and the Cretaceous correlate well with obvious peaks in $\delta^{13}C$. Further, both carbon peaks and shale intervals correlate reasonably well with sea-level highs. The late Palaeozoic $\delta^{13}C$ enrichment spanning the interval from the early Carboniferous to late in the Permian, on the other hand, corresponds precisely with the terrestrial coaly cyclothemic sequences of North America and Europe and to low sea levels related to the Gondwana glaciations. The supposed relationship (positive correlation of sea level with sedimentation of organic carbon) is thus not perfect, owing to trapping of important amounts of organic matter on land and in large marginal marine basins. The relationship of carbon isotope peaks to global cooling therefore may have different meanings at different times. For example, a cooling phase could diminish high-latitude vegetation, thus creating a negative feedback on climate.

The two late Palaeozoic rapid drops in $\delta^{13}C$ (Sakmarian, end Permian) were due, in part, to the erosion and oxidation of organic carbon in the coals, liberating large quantities of light-isotope carbon to the oceans. The first drop apparently represents an erosional and oxidation event (plus possible increased accumulation of carbonate sediments) which released sufficiently large amounts of CO_2 to bring about climatic warming. Holser and Magaritz (1987) attributed the erosion of the coals and extinction of marine faunas to a major drop in sea level caused by tectonic changes up to the end of the Permian (note Hallam curve in Fig. 11.1). Sea level rose again rapidly (as much as 240 ± 30 m/m.y.) in the earliest Triassic.

Within the late Palaeozoic glacial phase there was a major retreat of the ice masses in the Stephanian (Chapter 5). There was a slight rise in sea level at this time (Vail *et al.*, 1977), but this was followed by a major readvance of the ice and marine regression. The carbon isotopes reflect these changes, becoming suddenly lighter (by at least 1.5‰) in the Stephanian and heavier again as the climate cooled and ice fronts readvanced in the early Permian. The correlation between climate and the carbon isotope record suggested in our model is evident in this reversal within the late Palaeozoic Cool Mode.

Another carbon peak followed by a sharp fall is tentatively proposed by Holser near the Ordovician–Silurian boundary (Fig. 11.1), within the earliest

of our Cool Modes. This is also the time of black shale deposition (Llandover-ian to Wenlockian) and of many faunal extinctions. During the late Ashgil-lian (Hirnantian), sea levels dropped as glaciers developed, causing one episode of extinctions as epicontinental seas diminished in extent. The subsequent sea level rise as glaciers melted in the early Silurian may have caused another wave of extinctions (Brenchley, 1984).

There is also a significant peak and a sharp fall in the $\delta^{13}C$ record during the late Devonian, within our late Silurian to early Carboniferous Warm Mode. This is the approximate interval during which glaciation occurred in South America (Famennian) and extended to latitudes as low as 60° S (Chapter 4). The peak of $\delta^{13}C$ occurred at the Frasnian–Famennian bound-ary when large quantities of organic-rich shales were deposited. Similar to other times of high carbon peaks discussed above, this interval has been associated with marine extinctions (Wilde and Berry, 1984).

The peak of carbon fractionation appears, in cases discussed above, to have occurred during intervals when the climate was cold and glaciers were present or at least near-glacial conditions existed on Earth. What caused the ocean to be enriched in ^{13}C? Two mechanisms have been proposed to account for the formation of black shales (Chapter 7; discussion concerning Cretaceous black shales; also reviewed in Arthur *et al.*, 1990). One hypothe-sis is that conditions for the preservation of organic carbon were particu-larly favourable; that is, the deep oceans were depleted in oxygen and any organic carbon produced was therefore not oxidized but buried, thus leaving the ocean enriched in ^{13}C.

An alternative theory is that organic productivity was unusually high. Carbon isotope composition is related to the photosynthesis-respiration processes (Broecker and Peng, 1982), which result in the removal of the light isotope (^{12}C) from surface waters by organisms and its deposition in the deep ocean. If the deep waters contain sufficient oxygen, the organic carbon is oxidized and CO_2 released. However, if productivity in surface waters is very high, large amounts of ^{12}C will be buried; at some point the water will become too depleted in oxygen to remove all the organic matter present, and much of it will then be preserved.

The control on productivity in surface waters is the nutrient supply, particularly the availability of phosphorus and nitrogen (Broecker, 1982) and iron (Martin, 1990). These nutrients are abundant when ocean waters are well mixed and upwelling occurs to bring nutrients up into surface waters. Such vigorously mixed oceans are present on Earth during cold climate phases when the oceans are also cold. High productivity would have resulted in the extraction of large quantities of CO_2 from the atmosphere for photosynthesis by phytoplankton, thus causing a further positive feedback effect on climate cooling.

This kind of explanation was given by Vincent and Berger (1985; 'The

Monterey Hypothesis') to explain the excursion of $\delta^{13}C$ for a period of 4 m.y. during the latest early Miocene, when large quantities of black shales were deposited. They suggested that initial polar cooling caused an increase in productivity as a result of an increase in upwelling and nutrient supply. As organic carbon was buried atmospheric CO_2 levels became depleted, intensifying the cooling. The equator-to-pole temperature gradient increased as the poles became cooler, atmospheric circulation became more vigorous and the oceans became more strongly mixed, providing suitable conditions for further enhanced productivity.

A strong positive correlation between the oxygen and carbon isotope records was illustrated by Vincent and Berger for the latest early Miocene, suggesting that climate variations are closely linked to the carbon budget (Fig. 9.2). The build-up to the carbon peak started at about 17.5 Ma and lasted for about 4 m.y. After a lag of about 2 m.y. there was an increase in $\delta^{18}O$ at 15.5 Ma, interpreted as the result of polar cooling and ice build-up. This suggests that the cooling (which caused the initial increase in upwelling and productivity) needed enhancement from the positive feedback of CO_2 drawdown from increased productivity to push it into a major cooling and glaciation.

In the cases discussed above, the high $\delta^{13}C$ excursions correspond to cold climate phases. Do other carbon peaks signify similar climates, even within Warm Modes? The largest carbon peak in the Cretaceous (to about 3.5‰) occurred during the Aptian (Arthur *et al.*, 1990). This was related to Schlanger and Jenkins' (1976) Ocean Anoxic Event (OAE) 1 which spanned the latest Barremian to the earliest Albian, during which large quantities of black shales were deposited. In fact, because of this long time span there may have been more than one black shale event (early Aptian, early Albian, late Albian) or possibly rather patchy deposition. Debate still continues about whether the deposition of these organic-rich black shales was the consequence of sluggish circulation and deep water anoxia in warm oceans, or the result of increased biological productivity in ocean surface waters or on land (see review in Arthur *et al.*, 1990). The early Cretaceous shales appear to have been dominated more by terrestrial organic matter than by aquatic material (Tissot *et al.*, 1979), suggesting that productivity was enhanced on land rather than in the oceans at this time.

The Aptian–Albian was a time of rising sea levels and has been considered as a classic time of globally warm climates. However, the presence of cold climate indicators in sediments of these ages in both northern and southern high-latitude regions indicates that these areas were cold enough for the formation of river or shore ice during winter (Frakes and Francis, 1988) and that the early Cretaceous climates may have been cooler than originally thought (see Chapter 7). It is possible that, despite the lack of Cretaceous glacial deposits recorded at present, glaciers may have been present near

the poles (Frakes and Francis, 1990). The carbon peak in the Aptian occurs near the end of our mid-Jurassic to early Cretaceous Cool Mode, and we consider the decrease in $\delta^{13}C$ after this peak as signifying a decrease in productivity and thus increased levels of CO_2. In our view this was an important contribution to the development of the warm climate phase which characterized the late Cretaceous.

The late Cretaceous to early Tertiary does appear to have been a very warm interval (see Chapter 8) during which time there were several significant peaks of $\delta^{13}C$ (Fig. 11.1). One short-lived peak of $\delta^{13}C$, with an excursion of about 2‰, occurred within OAE 2 near the Cenomanian–Turonian boundary (Schlanger *et al.*, 1987; Arthur *et al.*, 1990). Organic-rich shales deposited during this short interval are widespread in the North Atlantic, the North Sea, in the Tethys, the Western Interior Seaway in America, and the western shelf area of Africa (Arthur *et al.*, 1988). The organic matter in the sediments is mainly of aquatic type and deposited independent of geographic setting or water depth.

Arthur *et al.* (1988) record that the positive excursion in the isotopic composition of the organic carbon ($\delta^{13}C_{org}$) at this time was even greater than that for carbonate ($\delta^{13}C_{carb}$). If the magnitude of the shift in isotopic values was the same in both $\delta^{13}C_{org}$ and $\delta^{13}C_{carb}$, then a change in the isotopic composition of the whole ocean reservoir of total dissolved carbon could be assumed. However, the different magnitudes of the isotopic shifts suggest that another factor was involved. Since the fractionation of carbon is affected by CO_2 levels (high CO_2 concentration leads to increased fractionation, resulting in more negative values of $\delta^{13}C_{org}$; Dean, Arthur and Claypool, 1986), the more positive shift for $\delta^{13}C_{org}$ suggests that CO_2 levels were low.

It was suggested that the burial rate of organic carbon was enhanced at this time and/or productivity was higher, causing a reduction in atmospheric CO_2 levels and a significant global cooling event (Arthur *et al.*, 1988). Volcanic activity was relatively low during this part of the Cretaceous so CO_2 output from volcanic outgassing may not have been sufficient to counteract the drawdown due to higher productivity (see Fig. 8.3). Cooler polar oceans would have encouraged upwelling and higher productivity, providing a positive feedback for the cooling. Evidence for cooler conditions cited by Arthur *et al.* include increases in $\delta^{18}O$ in inoceramids from chalk sequences in Europe, and the migration of boreal faunas into lower latitudes during the Turonian. These changes in ocean chemistry and a cooler climate may have been the cause of extinction events recorded for this time.

After this carbon peak there was a rapid decrease in values of $\delta^{13}C$ (see also Weissert, 1989). This suggests that productivity decreased markedly and/or there was a rapid input of ^{12}C into the ocean reservoir. Atmospheric CO_2 levels would have increased and the climate warmed. Another positive

(though smaller) carbon excursion has been recorded for the very early Santonian, within the Santonian to early Campanian OAE 3 (Arthur *et al.* 1990; and others therein), though it is not apparent whether there were any specific climate changes associated with this peak.

It is interesting to note that during the Toarcian in the early Jurassic, when there was a pronounced positive excursion of $\delta^{13}C$ to about 4‰ (Weissert, 1989), the shift in $\delta^{13}C_{org}$ was to more negative values (Jenkyns and Clayton, 1986). This could have been related to increased fractionation due to higher levels of CO_2 and therefore climate warming. The Toarcian carbon peak also occurs within a Warm Mode but there is no evidence for a climate cooling at this time. Carbon peaks within Warm Modes may not always be associated with higher productivity caused by cooler climates and upwelling, but some are perhaps related to enhanced burial in strongly anoxic conditions. Comparing the isotope records of both $\delta^{13}C_{carb}$ and $\delta^{13}C_{org}$ may provide an indication of environmental conditions at these times.

The late Cretaceous carbon peaks occur within a Warm Mode and therefore differ from other examples which correlate with the late stages of Cool Modes. It seems that the late Cretaceous was therefore rather an exceptional interval on Earth; Veizer (1985) also noted this in his studies of the carbon budget. A great deal of organic matter was buried and preserved, whether due to very anoxic deep oceans that preserved most of the organic matter produced, or due to high surface productivity (Arthur *et al.*, 1990). How could productivity have been enhanced in such warm conditions? Evidence discussed above (including Kemper, 1987) suggests that the late Cretaceous was not as uniformly warm as generally believed but punctuated by short intervals of cooler climate. During these times ocean circulation may have become somewhat more vigorous and/or more favourable for high productivity.

Despite general warmth over this interval, there is evidence of rapid climate change at the Cretaceous–Tertiary boundary. At the boundary there was a peak in $\delta^{13}C$ followed by a rapid drop of up to 3‰, recorded at many sites (Hsu and McKenzie, 1985; Zachos and Arthur, 1986). The carbon gradient within the ocean was also affected. From a normal gradient in the Cretaceous, where planktonics had about 1‰ more positive values of $\delta^{13}C$ than benthic foraminifera, the gradient then broke down. In the earliest Tertiary the benthic foraminifera values became even more positive than the planktonics. This was interpreted by Hsu and McKenzie as a complete breakdown of surface productivity, leaving a sterile ocean with no (or an inverse) carbon gradient ('The Strangelove Ocean').

Hsu and McKenzie considered that mass mortality at the ocean surface accounted for the lack of productivity, possibly as a consequence of a meteorite impact. The presence of blooms of nanoplankton in the earliest Tertiary indicates that the dead ocean lasted possibly for only a few

thousand years, and that ocean conditions were rather unstable for a while afterwards. Zachos and Arthur (1986) recorded that the carbon shift occurred before the Cretaceous–Tertiary boundary, suggesting that an impact was not the cause of the carbon change. Instead, they considered that changes in sea level and ocean structure were responsible.

The oxygen isotope curve of Hsu and McKenzie, which shows an increase in $\delta^{18}O$ at the boundary, indicates the Cretaceous–Tertiary transition was a time of significant climate cooling (though only a minor one was recorded by Zachos, Arthur and Dean, 1989). This may have been the trigger for many extinctions, both in the oceans and on land. A decrease in $\delta^{18}O$ just after the boundary records that the climate warmed again in the early Palaeocene. The restoration of higher levels of atmospheric CO_2, possibly due to a decrease in ocean productivity, would have caused climate warming.

A few million years later during the late Palaeocene, there was a very large peak in $\delta^{13}C$ (Shackleton, 1986a). Carbon isotope values were low (1‰) in the earliest Palaeocene and rose to a peak of about 3.5‰ in the late Palaeocene but then dropped rapidly (within about 5 m.y.) to about 1‰ in the early Eocene. The oxygen isotope curve follows this trend, showing a strong increase in $\delta^{18}O$ in the late Palaeocene and a decrease in the early Eocene (Shackleton, 1986a; see review in McGowran, 1990). Assuming a lack of diagenetic effects in the isotope signature, it is likely that the climate was cooling during a time of increased carbon productivity and warmed up again as productivity decreased. This again suggests that the enhanced productivity in the late Palaeocene caused a drawdown of atmospheric CO_2, resulting in cooling.

As productivity decreased to low levels in the early Eocene (possibly due to decreased nutrient supply), the CO_2 level in the atmosphere increased and the climate warmed. There is good evidence for these climate changes (late Palaeocene cooling followed by early Eocene warming) in the geological record (e.g. ice-rafting, microfossil changes etc.; see Chapter 8). In fact, the carbon curve suggests that productivity then increased slightly again during the rest of the Eocene–this occurred as the $\delta^{18}O$ signature of the ocean was increasing as a result of climate cooling and ice build-up.

Climate forcing

Orbital cycles

Do the major modes of climate as discussed in this book result from processes internal or external to the Earth? There is little doubt that external forces, in the form of the orbital parameters such as Milankovitch cycles, interact with the climate system to give rise to short-term (ice age) cyclicity. These external forcing functions, through mechanisms still not thoroughly understood, exert strong controls on the behaviour of climate.

However, orbital influence appears to affect climate strongly only during extremes of the Cool Modes, and although similar rhythmicities have been detected in sediments from non-glacial times (e.g. Cretaceous black shales; Herbert and Fischer, 1986), it is not certain that the observed effects result directly from significant climate change. Nor have the orbital parameters been called upon to explain the *initiation* of the late Cenozoic glacial episode, or indeed, any other major glaciation.

Climatic forcing factors can be divided into terrestrial and extraterrestrial. The first category includes such diverse mechanisms for controlling climates as atmospheric greenhouse gases, ocean geochemistry and biochemistry, land–sea distributions and their effects on global albedo, and geographic controls through changes in topography, oceanic circulation and water mass formation (see Pollack, 1979; Frakes, 1979; Barron, 1983; Saltzman, 1985; Worsley, Nance and Moody, 1986; Rea, 1986; Sheridan, 1987; Kasting, 1989). Extraterrestrial forcing can operate through changes in solar luminosity, passage of the solar system through clouds of matter in space, impacts of bolides, and changes in the Earth's rotation rate and its orbital parameters.

External processes, including orbital effects, cannot be eliminated as potential forcing functions for long-term change on the scale we are concerned with here. But in the present state of our knowledge, most long-term changes can only be convincingly ascribed to external forces if a rhythmicity can be discerned which corresponds in period to a known astronomical bandwidth. Patterns of change can be recognized within each of the 'cycles' delineated in Table 11.1. These include sedimentary events which appear to be cyclic themselves and some which are randomly distributed in time. Although cyclicity may be inferred from repetitive sedimentary sequences, in most cases it is not well established because the deposits cannot be dated precisely (see Klein, 1990). Normally, an average duration is calculated for each unit on the assumption that there was a constant sedimentation rate between dated horizons and that erosional or non-depositional events did not occur. At the end of the exercise a cyclic wavelength in years is derived and the figure is compared with known episodicities. However, this method does not establish cyclicity because a definite rhythm has not been documented. Whether cyclicity is expressed is important because its existence is usually necessary to demonstrate extraterrestrial forcing of Earth climates.

There are three important astronomical bandwidths. First is the possibility that bolide impacts on the Earth conform to a 300 m.y. cycle and force an 'Icehouse–Greenhouse' scenario, as described below. The second and third relate respectively to the Galactic Year (300 m.y.) and the half Galactic Year (150 m.y.), the latter being related to the position of the solar system with respect to the two spiral arms of the Galaxy (discussed below).

Steiner and Grillmair (1973) and Fischer (1981; 1982; 1986) proposed that climates have changed in cycles of 300 m.y. duration. Three phases of glaciation, or 'Icehouse' states, were identified, including the late Precambrian, late Carboniferous to early Permian and the Pleistocene to Recent. These alternated with 'Greenhouse' states, times of high concentrations of atmospheric CO_2 and warm, ice-free polar regions. These climate phases were linked to changes in sea level and volcanicity (high during 'Greenhouse' states), perhaps ultimately linked to cyclical changes in mantle convection. Biotic crises occurred at the changeovers from one state to another.

The 300 m.y. cyclicity may extend further back in time but present evidence does not allow a test beyond about 3 cycles (the beginning of the late Precambrian glaciation at about 900 Ma). Certainly there was a warmer phase before 900 Ma but on geologic evidence this warmth may have continued uninterrupted back as far as about 2300 Ma before changing significantly (resulting in the Huronian glaciation). It thus appears likely that cyclicity of 300 m.y. period may not have operated over the full lifetime of the Earth; confirmation of earlier cycles requires documentation of cool conditions in the intervals around 1200, 1500, 1800 and 2100 Ma. Unfortunately the 300 m.y. cycle does not fit perfectly for climate change during the Phanerozoic. The Ordovician–Silurian glaciation is an anomalous 'Icehouse' period within a 'Greenhouse' state and could not be accounted for by Fischer.

Cold and warm intervals in the 300 m.y. cycle are variable in their apparent characteristics. This results, in part, because different techniques are used in determining climate depending on the age of the rocks; that is, more aspects of climate can be delineated in young rocks than in relatively old ones because methods are more numerous (oxygen isotopes, etc.) and more reliable (palaeoecology, diversity of fossil forms, etc.).

Although a 300 m.y. cyclicity may be somewhat irregular, a periodicity at about 150 m.y. (the half Galactic Year), first suggested by Williams (1975), seems to fit most data on major climate changes in the Phanerozoic and latest Precambrian interval. This is now more clear (Table 11.1) because glaciation in the Ordovician–Silurian (Hambrey, 1985) falls halfway between the late Precambrian and the late Palaeozoic cold phases, and because a major cooling trend is now recognized in the late Mesozoic (Frakes and Francis, 1988). However, this proposed cycle is undetectable in the Proterozoic (no known warm intervals between ～570 and 900 Ma) and earlier in the Precambrian.

The five so-called Milankovitch periodicities relating to variations in the orbital parameters of the Earth are the ones most abundantly discerned in data bearing on climate. Wavelengths from these fall within the range between about 19 000 and 400 000 years. Their distribution appears to be widespread, without great regard to geological time or the gross climatic

state of the Earth, although their formation and preservation require rather special local conditions persisting over considerable intervals of time.

In relatively young deposits with good age control, Milankovitch cycles can be matched quite precisely to the chronological maxima and minima of the combined orbital effects (e.g. Hays *et al.*, 1976) and their absolute ages thus can be estimated. This is not possible for older deposits, however, and in many of these the calculated wavelengths correspond only approximately to Milankovitch periodicities (e.g. North American late Palaeozoic cyclothems average somewhere between 235 000 and 393 000 years in duration (Heckel, 1986) versus the Milankovitch cyclicity of 413 000 years; Imbrie and Imbrie, 1980). A further problem relates to variations in cyclicity; in sediments of the Brunhes palaeomagnetic epoch, for example, cyclicity corresponds well with the Milankovitch eccentricity parameter but slightly older strata more closely reflect the obliquity cycles.

Recognition of cycles with wavelengths greater than 400 000 years and less than 150 m.y. has so far been based mainly on palaeontological data (Fischer and Arthur, 1977; Raup and Sepkowski, 1984). Fischer and Arthur (1977) identified cyclical changes of 30 m.y. duration, caused by changes in ocean structure in response to climate change (possibly atmospheric CO_2). The cycles alternated between 'polytaxic' and 'oligotaxic' phases. During polytaxic phases oceans were warm and there was an expansion of organic diversity in the pelagic marine realm. They identified the late Cretaceous, Eocene and Miocene as polytaxic phases. Each of these states rapidly declined into 'oligotaxic' states, times of cool oceans and the spread of 'opportunist generalists'. Their oligotaxic states include the Maastrichtian to Danian, the Oligocene and the more recent temperature decline.

The 26–30 m.y. cycle of organic diversity in the Phanerozoic is perhaps influenced strongly by the geological time scale, as time units are defined on biological extinctions (Shaw, 1987). However, such variations, if climatically linked, could provide a valuable tool for understanding climatic evolution in general and assist in answering why and how cycles were generated in variable sediments of widely different ages. Further cycles have been recognized in Cretaceous and Tertiary sediments from the correlation of climatically related data. In the Tertiary the cycles were approximately 10 m.y. in duration (cool–cool) (Wolfe and Poore, 1982) and 20 m.y. duration (McGowran, 1990). In the mid Jurassic to early Cretaceous Cool Mode, cycles of about 20 m.y. appear to have been operating (Chapter 7).

Cycles with yet shorter wavelengths are also known. Sedimentological data locally show evidence of an 11-year cycle, possibly related to the sunspot cycle. Numerous studies on chemical and deep-sea sediments have identified peaks at 5000 years to 200 years under spectral analysis. It is not certain whether these sequences are related to climate variations or to some other control on sedimentation.

Random events can be important forcing factors. Many significant ones

are detectable in the geologic record and most are, except possibly in the case of bolide impacts, of a terrestrial nature. Such events are often of tectonic origin and confuse the issue in producing sedimentological changes which appear to be climatically induced fluctuations in the rocks (Molnar and England, 1990). When closely spaced in time, such as in deposits from turbidity currents, they can give rise to seemingly regular cycles in sediments, and because their cause might obviously not be associated directly with climate change, their occurrence should be taken as a warning against overinterpretation.

The problem is no less simple when dealing with events spaced at longer intervals, as with changes in palaeogeography. Changes in regional or global oceanic circulation patterns have been selected as causative agents in explaining many climatic changes (see Berggren, 1982). Alterations to land–sea distributions generally have tectonic origins, and it is possible that many of these in fact are cyclically controlled on time scales of 5–10 m.y. or more and may therefore be difficult to recognize. Although the truly random event may eventually prove to be rare, especially where tectonic activity is involved, there are large numbers of orogenic and related events presently interpreted as randomly spaced in time. Such events can interrupt, disrupt and change cyclic accumulation, and some undoubtedly constitute a major factor in climate change.

Did tectonic forces drive climatic change?

Analysis of climate forcing factors suggests that tectonic processes, particularly those related to volcanism, strongly govern the onset of cool and warm phases of the Earth's climate, while fluctuations *within* an ice age result from orbital forcing. The implication is that, even in the absence of orbital forcing, there would still be long-term glacial and non-glacial fluctuations. The available information suggests that the position of the Earth in the solar system leads to the situation where the energy imbalance between the equator and the poles favours the development of snow and ice in high latitudes. Intervals characterized by warm and equable climates and weak thermal gradients, on the other hand, probably represent times when increased tectonic activity has contributed to greater concentrations of atmospheric and oceanic CO_2 and quite different oceanic structure and circulation.

Several hypotheses have been proposed which link possible cyclical patterns of tectonic activity to climate change (including Hays and Pittman, 1973; Keith, 1982; Veizer, 1985; Worsley *et al.*, 1986; Sheridan, 1987; Loper, McCartney and Buzyna, 1988; Veevers, 1990). The duration of these proposed cycles is from 0.5 b.y. to 60–100 m.y. and they have been related to cyclical changes in the rate of sea-floor spreading, so they would then impact upon many facets of atmosphere and ocean properties.

Models of continental fragmentation and assembly, states which produced different climate regimes, were proposed by Worsley *et al.* (1986) and Veevers (1990). The tectonic cycle is driven by heat within the mantle and results in either large continental blocks or fragmented masses. While continents were fragmented and scattered, equable warm climates developed. Continental relief would have been low and levels of atmospheric CO_2 high. Sea levels rose during the latter part of the fragmentation stage, creating warm, salinity-stratified oceans. Oceanic anoxic events led to the deposition of black shales, even though rates of productivity were low due to nutrient- and oxygen-deficient waters. Conditions moderated as continents were at maximum dispersal with lower sea levels, cooler climates and better mixed oceans. While continents were being assembled into large land masses, sea levels fell and land relief was high. Oceans were well mixed and productive, so CO_2 levels were low, the climate cooled and glaciation occurred. The large scale of these cycles, in the order of 500 m.y., means that unfortunately only one cycle can be applied to the Phanerozoic. The continental mass of Pangaea represented the stage of continental assembly (Veevers, 1990), correlated with an 'Icehouse' climate, similar to that proposed by Fischer (1982).

Cycles of 60–100 m.y. in length generated by cyclical eruption of mantle plumes were proposed by Sheridan (1987). In essence, he proposed that during periods of slow sea-floor spreading, volcanic activity and sea levels would be low and output of CO_2 would decrease. Ocean waters would be less acidic, more calcite would be preserved on the sea floor and the CCD would deepen. Lower levels of CO_2 in the atmosphere would lead to cooler climates. Oxygen levels in the cool oceans would be higher, thus allowing oxidation of organic matter and preventing the formation of black shales.

During times of faster plate spreading, sea levels would be higher, atmospheric CO_2 levels would increase due to increased output of volcanic gases, and the climate would warm. Ocean acidity would increase and the CCD would be shallower. This warmer Earth would be more humid due to increased evaporation rates and this, along with higher levels of CO_2, would result in expansion of vegetation cover. Organic matter would be dumped in the oceans and black shale deposits would form (resulting in a peak of $\delta^{13}C$), the preservation enhanced by warm, saline, oxygen-poor waters in the deep ocean.

This proposal (high sea levels, high CO_2, warm climates and black shale deposition) is fitting for the warm late Cretaceous interval but does not explain the carbon peaks that occur during cooler times. In contrast, Veizer (1985) emphasized that during times of high sea level $\delta^{13}C$ values were low, and conversely, the peaks in $\delta^{13}C$ occurred at times of low sea levels (in other words, most organic carbon ^{12}C was buried during times of low sea levels). This is opposite to the situation proposed by Sheridan (1987) and, as

Veizer (1985) comments, conflicts with the high sea level/black shale scenario generally considered for the Cretaceous.

However, continental collision is likely to occur during rapid plate motion; this would produce high continental relief, suitable for alpine glaciation. Small quantities of land plants would source only small amounts of coal or non-marine shales. Climates would be slightly cooler and the oceans well oxygenated, preventing the preservation of organic-rich shales.

On an even shorter time scale Rea *et al.* (1990) consider that tectonic events were the driving mechanism for climate change (essentially changes in atmospheric CO_2) at the Palaeocene–Eocene boundary. At this time, about 57 Ma, there was an increase in hydrothermal activity (see also Vogt, 1979) and an associated increase in CO_2, causing climate warming. Ocean circulation may then have become salinity-driven by oxygen-poor warm saline bottom waters which carried warm waters to the poles, decreased the equator-to-pole temperature gradient and so reduced the intensity of atmospheric circulation. The weaker winds (indicated also by a decrease in grain size of aeolian dust) caused ocean circulation to be less vigorous, leading to a decrease in biological productivity and so a decrease in $\delta^{13}C$. Weaker atmospheric circulation resulted in enhanced seasonality on land because the oceans were less capable of moderating land climates.

Carbon cycling

It is well known that the radiation received at the top of the atmosphere, and thus the thermal state of the Earth, is determined to a considerable degree by the presence of optically active components such as carbon dioxide and related gases, water vapour, ozone and aerosols in the atmosphere. To a large extent these greenhouse gases prevent loss of heat from the surface through reradiation trapping. Kondratiev and Moskalenko (1984) calculated that the mean global temperature is maintained at about 32K above that which would prevail with a lack of atmospheric greenhouse gases. Also, because the spectral transmission of both CO_2 and water vapour exhibit a strong temperature dependence, the effectiveness of the greenhouse increases with increasing temperature. The single most important factor in keeping the Earth's surface temperature above zero over its history probably is the greenhouse effect and in particular that related to CO_2.

Computer models of the carbon cycle (Lasaga *et al.*, 1985; Berner, 1990) can calculate variations in atmospheric CO_2 over time, on the basis of weathering, burial and degassing rates for carbon. In the first model cited, weathering rates were assumed to be proportional to the exposed land area and the degassing rate taken as directly proportional to the rate of generation of new sea floor (for discussion, see Berner and Lasaga, 1989). Berner later assumed rates on the best available data. Although uncertainties about these rates somewhat limit the value of calculated CO_2 concentrations

through time, it is significant that realistic rates suggest that there were changes (up to at least a factor of about six) in the levels of atmospheric CO_2 during the Phanerozoic (Berner, 1990). Effective greenhousing is also required to explain the global warmth evident in the Precambrian, when the solar luminosity apparently was 10–35% lower than at present.

It appears that warm global climates were dominated by maritime climates, when high sea level and large shallow seas led to enhanced subtropical evaporation and hence alterations to oceanic water-mass structure by the creation of descending plumes of warm saline waters (Brass *et al.*, 1982). Polar land masses may have been warmed by high-latitude upwelling of these waters (alternatively, surface waters may have been systematically stripped of [18]O (Rea *et al.*, 1990), thus confusing the interpretation of the isotope records). Any such alterations in the water mass structure of the global ocean thus may have constituted a positive feedback during Warm Modes.

From the evidence we have compiled here, it seems that major peaks in $\delta^{13}C$ occurred during Cool Modes and that many large falls in $\delta^{13}C$ correspond to the initial stages of Warm Modes. At peak times, volcanic activity was comparatively low and the cold oceans were vigorously mixed and rich in nutrients, allowing high productivity of surface biotas. Carbon was fractionated and organic carbon ^{12}C transported to the deep ocean where later it was oxidized by oxygen-rich waters, thereby releasing nutrients back into the ocean. The process of photosynthesis at the ocean surface caused the drawdown of CO_2, resulting in climate cooling, thus enhancing cooling. On the occasions when productivity was very high and/or all the oxygen in the deep ocean was used up, the remaining organic carbon was preserved and organic-rich sediments formed.

The change in this cycle may have been due to depletion of the nutrient supply, causing a drastic drop in productivity, or it may have resulted from widespread oxidation of sea floor or other organic material. A consequent increase in atmospheric CO_2 led to a drop in $\delta^{13}C$. At times, conditions approached those of a 'Strangelove Ocean'. Climate began to warm and evaporites became more common. Oceans were less well mixed and may have been driven instead by warm saline bottom waters. Warm waters transported heat to high latitudes, reducing the equator-to-pole temperature gradient. This caused atmospheric circulation to become weaker. As bottom waters in the oceans were poor in oxygen, any organic carbon produced would have been preserved in bottom sediments.

During the Quaternary the removal of carbon from surface to deep-ocean waters is required to explain decreases in the concentration of CO_2 in surface waters and atmosphere, as detected in isotope records from sediment and ice cores. The carbon pump accomplishes this by settling carbonate and organic tissue to the ocean floor and by bringing carbonate

materials into the deep zone of dissolution. The mechanism also can thus be used to explain short-term depletion of atmospheric CO_2, as can storage of organic matter on land and continental shelves. Obviously such factors as thermal and dynamic structure of the water mass, oceanic fertility and chemistry of the water column influence the effectiveness of the carbon pump.

A type of reverse carbon pump operates to liberate carbon to the ocean and atmosphere, as presently perceived, from sediment reservoirs. Significant amounts of additional CO_2 might also be released through oxidation of large amounts of newly discovered dissolved CO_2 in ocean surface waters (Sugimura and Suzuki, 1988, cited in Toggweiler, 1990). Processes that govern the activity of the sedimentary reverse pump include oxidation and physical erosion of organic matter (which require either exposure to oxic environments or to the activity of currents, tides or waves), and chemical weathering of carbonate, other carbon-bearing rocks, and possibly even methane-rich tundra. It is commonly assumed that this type of reverse pump is engaged suddenly to deliver large amounts of dissolved carbon to the hydrosphere and the biosphere but the mechanism is almost entirely unknown and unspecified. Weathering of carbon from carbonate rocks is likely to work against this because of the concentration of the heavy isotope.

Also, the largest and most sudden changes observed in the carbon isotope record (up to at least 3‰) represent extremely rapid reverse pumping of carbon from the reservoirs, probably much too rapid to be explained by oxidation and erosion. It is unlikely, for example, that an amount of ^{12}C equal to that accumulated in coal and shallow marine sediments over some 115 m.y. in the Carboniferous–Permian, could be returned instantaneously (in less than 10 m.y.) to the ocean and atmosphere by processes such as these, even allowing for rapid structural oxidation of the deep ocean. Yet this is what is suggested by the carbon isotope record. The need for an effective reverse pump is also apparent in Cenozoic records, where short-term decreases of $\delta^{13}C$ can attain 1.5‰ in less than a million years.

Carbon can also be pumped into the hydrosphere–atmosphere system by outgassing from volcanism which, on volume, would be effective even though the $\delta^{13}C$ values are only moderate (~ -2 to $-10‰$), and from release of methane during diagenesis (~ -20 to $-50‰$). Fig. 11.4 shows how the various factors, operating alone, might give rise to the observed drops in $\delta^{13}C$.

A consequence of organic carbon deposition is the addition of oxygen to ocean waters. Moreover, oxidation of sea-floor organic material may lead to a decrease in the $\delta^{18}O$ signal, as suggested by Hoffman *et al.* (1989). Arthur *et al.* (1988) estimated that the oxidizing capacity of deep-ocean waters may have been increased by up to 50% towards the end of a lengthy period of organic carbon accumulation. After the Cenomanian–Turonian, this,

Beginning of
WARM MODE

End of
COOL MODE

Time →

-ve +ve increasing→ increasing→ increasing→

$\delta^{13}C$ ^{12}C oxidation Volcanic release ^{13}C burial
 (particulate and of CO_2 in carbonates
 dissolved C_{org})

Fig. 11.4. The transition from the end of a Cool Mode to the beginning of a Warm Mode, showing how fluxes in the carbon cycle may have contributed to the $\delta^{13}C$ signal. Dashed lines indicate possible trends before the transition. The $\delta^{13}C$ record reaches a peak near the end of a Cool Mode, then drops rapidly to lower values at the start of the next Warm Mode. Corresponding with the drop in $\delta^{13}C$, the rates of organic carbon (^{12}C) oxidation and ^{13}C burial in carbonates increase. An input of CO_2 from enhanced volcanic activity may reinforce the warming trend.

together with the effect of exposure of widespread carbonates due to a fall in sea level, may have resulted in greatly increased release of carbon. Whether this could have been sufficient to explain the subsequent fall in $\delta^{13}C$ and increased global warming is as yet uncertain, especially considering that carbonate sediments are strongly enriched in the heavy isotope.

Theoretically, consumption of the heavy isotope of oxygen during the deep-sea oxidation of organic matter will lead to an isotopically light oxygen signal. Railsback (1990) proposed that the heavy isotope ^{18}O was sequestered in the warm, deep saline waters and so would generate a light oxygen signal in surface waters. Together these models suggest that warm, deep saline waters in a stratified ocean are an agent of reverse carbon pumping (oxidation of organic carbon), and are at least a partial cause for light oxygen values observed, (and interpreted as climate warming), following large increases in $\delta^{13}C$. Parallel decreases in carbon and oxygen signals are the best indications that this process may have been operating (e.g. Palaeocene–Eocene; Rea *et al.*, 1990). In contrast, it is generally considered that warm saline bottom waters had a low oxygen content (Brass *et al.*, 1982) and were therefore unable to oxidize large quantities of organic carbon.

Effective reverse pumping might also function through the release of CO_2

by hydrothermal activity associated with large-scale metamorphism; short-term periodicity in this mechanism would be unlikely. Major events of a more prolonged nature could result from explosive volcanism near collisional plate boundaries, a type that yields substantial amounts of CO_2 and aerosols. The closing of the early Palaeozoic Cool Mode may have resulted in part from a contribution of CO_2 from such sources during the early stages of the assembly of Pangaea. This model suggests that volcanism and attendant metamorphism, being intermittent processes on long time scales, have altered the carbon budget over time by irregularly liberating greenhouse gases from both crustal and juvenile sources. For this to happen, a governing variable would be the quantity of carbon-rich rocks in the crustal reservoir. When the amount was small, cooling resulted because relatively small amounts of the greenhouse gases could be liberated to the ocean–atmosphere system; when it was large the global climate system was better able to retain heat through the increased greenhouse effect.

Our synthesis therefore indicates that cold climate phases during the Phanerozoic were correlated with low rates of volcanism and increased organic carbon accumulation resulting from high productivity, either on land or in the oceans. These were either full glacial phases (late Ordovician to early Silurian, mid-Carboniferous to late Permian, late Tertiary) or times of high-latitude seasonal ice (mid-Jurassic to mid-Cretaceous). Maxima of carbon accumulation also occurred during Warm Modes but new climate information suggests that these may also have been related to brief intervals of cooler climate during which productivity was enhanced, coupled with increased preservation of organic carbon in highly productive or anoxic oceans.

The carbon cycle therefore provided important forcing mechanisms and responses to the change from Warm to Cool Modes and vice versa, through a complicated cycle involving tectonics, atmospheric CO_2, sea-level changes, nutrient supply, and ocean circulation. The case we have presented here is obviously greatly simplified and does not consider in detail the positive and negative feedbacks of these mechanisms, which are known from more detailed studies of younger geological environments.

Conclusions

In this study Phanerozoic climate history has been divided into four Warm Modes and four Cool Modes (Fig.11.1). Among Warm Modes possibly the warmest was the late Cretaceous to early Tertiary, with the Silurian–Devonian and early Mesozoic ranked equal second. The position of the Cambrian–Ordovician is uncertain owing to the lack of information from high latitudes. The most extreme cold climates prevailed during the late Palaeozoic and the late Cenozoic, followed by the Ordovician–Silurian,

which was more intensely cold than both the middle Mesozoic and the anomalous late Devonian local cooling. An important conclusion to be drawn from this ranking is that the great variability of Phanerozoic climates has not seen a clear trend towards overall warming or cooling in the last 570 m.y, but rather can be characterized as alternating cool and warm intervals of long period. The base line in Fig. 11.1 is taken to be the mean annual global temperature of the present. As a consequence of the present interglacial state of the Earth, this results in most ancient climates being shown to be relatively warm.

In considering the mean global humidity over the Phanerozoic, the wettest times appear to have been the Carboniferous to early Permian and the early Tertiary; the driest likely occurred during the Triassic–Jurassic and the Devonian. Again there does not appear to be any Phanerozoic trend towards increasing or decreasing global aridity. In both the temperature and humidity estimations, however, it is necessary to remember that conclusions originate from diverse kinds of analyses of rock and fossil types and distributions, and that there are therefore many uncertainties.

With the recent large increases in information bearing on the palaeoclimatic history of the Earth, it has come to be recognized that climates have varied more often than previously accepted. This is particularly evident for the early Cretaceous and the late Cenozoic; the details are shown as completely as possible given the scale of Fig. 11.1. Similar variations very likely characterized the other Cool Modes, and several variations of comparatively long wavelength have been detected in Palaeozoic strata. Further work may also reveal greater variability on both short and long wavelengths in sequences deposited during Warm Modes.

The abrupt shifts from one mode to another signify enormous changes in the climate system, while the relative homogeneity of the internal record of each mode, persisting over tens of millions of years, indicates a large degree of inertia in the system. The search for the triggering mechanism of the large changes among the materials in the incomplete stratigraphic sequences deposited during each climate mode has long engaged palaeoclimatologists. Only future work will reveal whether mechanisms already indicated can stand the test of time.

Bibliography

Abelmann, A. and Gersonde, R. (1988) Plio-Pleistocene paleoceanography in the Weddell Sea: siliceous microfossil evidence. In *Geologic History of the Polar Oceans: Arctic versus Antarctic (Abstracts)*, Bremen, F. R. Germany: NATO Advanced Workshop, October 10–14 1988.

Ager, D. V. (1981) Major marine cycles in the Mesozoic. *J. Geol. Soc. Lond.* 138: 159–166.

Aharon, P. (1983) 140,000-yr climatic record from raised coral reefs in New Guinea. *Nature* 304: 720–723.

Aharon, P. (1984) Implications of the coral-reef record from New Guinea concerning the astronomical theory of ice ages. In *Milankovitch and Climate*, ed. A. Berger, J. Imbrie, J. Hays, G. Kukla and B. Saltzman, pp. 379–404. Dordrecht: Reidel.

Aharon, P. and Chappell, J. (1986) Oxygen isotopes, sea level changes and the temperature history of a coral reef environment in New Guinea over the last 10^5 years. *Palaeogeogr., Palaeoclimatol., Palaeoecol.* 56: 337–379.

Aitken, J. D. (1966) Middle Cambrian to Middle Ordovician cyclic sedimentation, southern Rocky Mountains of Alberta. *Can. Pet. Geol. Bull.* 14: 405–441.

Aksu, A. E. (1985) Planktonic foraminiferal and oxygen isotope stratigraphy of CESAR cores 102 and 103, preliminary results. Initial geological report on CESAR – the Canadian expedition to study the Alpha Ridge, Arctic Ocean. *Geol. Surv. Canada.* Paper 84–22: 115–124.

Aksu, A. E. and Mudie, P. J. (1985) Late Quaternary stratigraphy and paleoecology of northwest Labrador Sea. *Marine Micropalaeo.* 9: 537–557.

Allen, P. (1981) Pursuit of Wealden models. *J. Geol. Soc. Lond.* 138: 375–405.

Altermann, W. (1986) The Upper Palaeozoic pebbly mudstone facies of peninsular Thailand and western Malaysia; continental margin deposits of Palaeoeurasia. *Geol. Rundsch.* 75: 371–381.

Alvarez, W., Kauffman, E. G., Surlyk, F., Alvarez, L., Asaro, F. and Michel, H. V. (1984) Impact theory of mass extinctions and the vertebrate fossil record. *Science* 223: 1135–1140.

Anderson, R. Y. (1982) A long geoclimatic record from the Permian. *J. Geophys. Res.* 87: 7285–7294.

Andrews, J. T. and Fulton, R. J. (1987) Inception, growth and decay of the Laurentide ice sheet. *Episodes* 10: 13–15.

Andrews, J. T., Miller, G. M., Vincent, J. S. and Shilts, W. W. (1984) Quaternary correlations in Arctic Canada. *Geol. Surv. Canada.* Paper 84–10: 127–134.

Andrews, J. T., Aksu, A., Kelly, M., Klessen, R., Miller, G. H., Mode, W., and Mudie, P. (1985) Land-ocean correlations during the last interglacial-glacial transition, Baffin Bay, northwestern North Atlantic: a review. *Quat. Sci. Rev.* 4: 333–335.

Arhenius, G. (1952) Sediment cores from the east Pacific: properties of the sediments. *Reports of the Swedish Deep-Sea Expedition 1947–48.* 1: 1–288.

Arkhipov, S. A., Bespaly, V. G., Faustore, M. A., Glushkova, O. Y., Isaeve, L. L. and Velickko, A. A. (1986) Ice sheet reconstructions. *Quat. Sci. Rev.* 5: 475–484.

Armentrout, J. M. (1983) Distribution of organic carbon in the Gulf of Alaska; Neogene and Holocene sedimentary record. In *Coastal Upwelling, its Sedimentary Record*, ed. E. Suess and J. Theide, pp. 421–439. New York: Plenum Press.

Armentrout, J. M. (1987) *Recalibration of Miocene molluscan stages and consequent recorrelation of Alaskan formations (abstract)*. 4th International Congress on Pacific Neogene Stratigraphy. ed. J. A. Barron and J. R. Blueford. Berkeley, California. 5 pp.

Arthur, M. A. (1982) The carbon cycle: controls on atmospheric CO_2 and climate in the geologic past. In *Climate in Earth History*, ed. W. H. Berger and J.C. Crowell, pp. 55–67. Washington, D. C.: National Academy Press.

Arthur, M. A., Dean, W. E. and Pratt, L. M. (1988) Geochemical and climatic effects of increased marine organic carbon burial at the Cenomanian/Turonian boundary. *Nature* 335: 714–717.

Arthur, M. A., Dean, W. E. and Schlanger, S.O. (1985) Variations in the global carbon cycle during the Cretaceous related to climate, volcanism, and changes in atmospheric CO_2. In *The Carbon Cycle and Atmospheric CO_2: Natural Variations Archean to Present*, ed. E. T. Sundquist and W. S. Broecker, pp. 504–529. Washington D. C.: American Geophysical Union, Geophysical Monograph 32.

Arthur, M. A., Jenkyns, H. C., Brumsack, H.-J. and Schlanger, S. O. (1990) Stratigraphy, geochemistry, and paleoceanography of organic carbon-rich Cretaceous sequences. In *Cretaceous Resources, Events and Rhythms*, ed. R.N. Ginsburg and B. Beaudoin, pp. 75–119. Dordrecht: Kluwer Academic Publishers.

Arthur, M. A. and Schlanger, S. O. (1979) Cretaceous 'ocean anoxic events' as causal factor in development of reef-reservoired giant oil fields. *Am. Assoc. Pet. Geol. Bull.* 63: 870–885.

Arthur, M. A., Zachos, J. C. and Jones, D. S. (1987) Primary productivity and the Cretaceous/Tertiary boundary event in the oceans. *Cret. Res.* 8: 43–54.

Askin, R. A. (1983) Tithonian (uppermost Jurassic) – Barremian (Lower Cretaceous) spores, pollen and microplankton from the South Shetland Islands, Antarctica. In *Antarctic Earth Science*, ed. R. L. Oliver, P. R. James, and J. B. Jago, pp. 295–297. Australian Academy of Science, Canberra, and Cambridge University Press, Cambridge.

Bachtadse, V., Van der Voo, R. and Halbich, I. W. (1987) Paleomagnetism of the western Cape fold belt, South Africa, and its bearing on the Paleozoic apparent wander path for Gondwana. *Earth. Planet. Sci. Letters.* 84: 487–499.

Backman, J. (1979) Pliocene biostratigraphy of DSDP Sites 111 and 116 from the North Atlantic Ocean and the age of the Northern Hemisphere glaciation. *Stockholm Contributions to Geology.* 32: 115–137.

Baldauf, J. G. and Barron, J. A. (1990) Evolution of biosiliceous sedimentation patterns – Eocene through Quaternary: Paleoceanographic response to polar cooling. In *Geological History of the Polar Oceans: Arctic versus Antarctic,* ed. U. Bleil and J. Thiede, pp. 575–608. Dordrecht: Reidel.

Bambach, R., K., Scotese, C. R. and Ziegler, A. M. (1980) Before Pangea: The paleogeographies of the Paleozoic world. *Am. Sci.* 68: 26–38.

Bard, E. G., Arnold, M., Maurice, P., Duprat, J., Moyes, J. and Duplessey, J-C. (1987) Retreat velocity of the North Atlantic polar front during the last deglaciation determined by [14]C accelerator mass spectrometry. *Nature* 328: 791–794.

Bard, E. G, Hamelin, B., Fairbanks, R. G. and Zindler, A. (1990) U/Th ages obtained by mass spectrometry in corals from Barbados: calibration of [14]C for the last 30,000 years and sea level variations during the last 130,000 years. *Eos, Trans. Am. Geophys. Union.* 71: 543.

Bard, E. G. and Fairbanks, R. G. (1988) A [12]C deglacial SST rise west of Portugal 12.5 to 12 kyr B.P.: Local hydrographic changes or polar front migration? *Eos, Trans. Am. Geophys. Union.* 69: 378.

Bardossy, G. (1979) Growing significance of bauxites. *Episodes* 2: 22–25.

Barker, P. F. and Burrell, J. (1977) The opening of Drake Passage. *Marine Geol.* 25: 15–34.

Barnard, P. D. W. (1973) Mesozoic floras. *Special Papers in Palaeontology,* 12: 175–188.

Barnett, T. P. (1988) The effect of Eurasian snow cover on global climate. *Science* 239: 504–507.

Barnola, J. M., Reynaud, D., Korotkevitch, Y. S. and Lorius, C. (1987) Vostok ice core provides 160,000-year record of atmospheric CO_2. *Nature* 329: 408–414.

Barrera, E., Huber, B. T., Savin, S. M. and Webb, P. N. (1987) Antarctic marine temperatures: Late Campanian through Early Eocene. *Paleoceanography* 2: 21–47.

Barrett, P. J., Barrera, E., Harwood, D. M. and Webb, P. N. (1988) Middle Cenozoic glacial and sea level history from CIROS-1 drillhole, Victoria Land Basin, Antarctica (abs.). *Eos, Trans. Am. Geophys. Union.* 69: 383.

Barrett, P. J., Elston, D. P., Harwood, D. M., McKelvey, B. C. and Webb, P. N. (1987) Mid-Cenozoic record of glaciation and sea-level change on the margin of the Victoria Land Basin, Antarctica. *Geology* 15: 634–637.

Barron, E. J. (1983) A warm equable Cretaceous: the nature of the problem. *Earth Sci. Rev.* 19: 305–338.

Barron, E. J. (1985) Numerical climate modelling, a frontier in petroleum source rock prediction: results based on Cretaceous simulations. *Am. Ass. Petrol. Geol. Bull.* 69: 448–459.

Barron, E. J. (1987) Eocene equator-to-pole surface ocean temperatures: a significant climate problem? *Paleoceanography* 2: 729–739.

Barron, E. J., Arthur, M. A. and Kauffman, E. G. (1985) Cretaceous rhythmic bedding sequences: a plausible link between orbital variations and climate. *Earth Planet. Sci. Letters.* 72: 327–340.

Barron, E. J., Harrison, C. G. A., Sloan, J. L. and Hay, W. W. (1981) Paleogeography, 180 million years ago to the present. *Eclog. Geol. Helv.* 74: 443–470.

Barron, E. J. and Peterson, W. H. (1989) Model simulation of the Cretaceous ocean circulation. *Science* 244: 684–686.

Barron, E. J., Sloan, J. L. and Harrison, C. G. A. (1980) Potential significance of land-sea distribution and surface albedo variations as a climate forcing factor, 180 my to the present. *Palaeogeogr., Palaeoclimatol., Palaeoecol.* 30: 17–40.

Barron, E. J. and Washington, W. M. (1982) Cretaceous climate; a comparison of atmospheric simulations with the geologic record. *Palaeogeogr., Palaeoclimatol., Palaeoecol.* 40: 103–133.

Barron, E. J. and Washington, W. M. (1985) Warm Cretaceous climates: High atmospheric CO_2 as a plausible mechanism. In *The Carbon Cycle and Atmospheric CO$_2$: Natural Variations Archean to Present*, ed. E. T. Sundquist and W. S. Broecker, pp. 546–553. Washington, D.C.: American Geophysical Union.

Barron, J. A. (1987) *Identification of times of Miocene global sea-level falls through correlation of benthic foraminiferal carbon isotope records.* 4th International Congress on Pacific Neogene Stratigraphy. ed. J.A. Barron and J. R. Blueford. pp.6–7. Berkeley, California.

Barron, J. A. and Keller, G. (1982) Widespread Miocene deep-sea hiatuses: coincidence with periods of global cooling. *Geology* 10: 577–581.

Basinger, J. F. (1987) Our 'tropical' Arctic. *Can Geog.* 106: 28–37.

Batten, D. J. (1974) Wealden palaeoecology from the distribution of fossil plants. *Proc. geol. Ass.* 85: 433–458.

Batten, D.J. (1982) Palynofacies and salinity in the Purbeck and Wealden of southern England. In *Aspects of Micropalaeontology*, ed. F. T. Banner and A. R. Lord, pp. 278–308. London: Allen and Unwin.

Batten, D. J. (1984) Palynology, climate and the development of Late Cretaceous floral provinces in the Northern Hemisphere: a review. In *Fossils and Climate*, ed. P. Brenchley, pp. 127–164. New York: John Wiley & Sons.

Baulin, V. V., Belopukhova, Y. B. and Danilova, N. S. (1984) Holocene permafrost in the USSR. In *Late Quaternary Environments in the Soviet Union*, ed. A. A. Velichko, pp. 87–91. Minneapolis: University of Minnesota Press.

Bender, M. L. and Keigwin, L. D. (1979) Speculation about the upper Miocene change in abyssal Pacific dissolved bicarbonate delta ^{13}C. *Earth and Planet Sci. Letters.* 45: 383–393.

Benson, R. H., Chapman, R. E. and Deck, L. T. (1984) Palaeoceanographic events and deep-sea ostracods. *Science* 224: 1334–1336.

Berger, A. L. (1977) Support for astronomical theory of climatic change. *Nature* 269: 44–45.

Berger, A. L. (1978) Long-term variations of caloric insolation resulting from the Earth's orbital elements. *Quat. Res.* 9: 139–167.

Berger, A. L. (1984) Accuracy and frequency stability of the Earth's orbital elements during the Quaternary. In *Milankovitch and Climate*, ed. A. L. Berger, J. Imbrie, J. Hays, G. Kukla and B. Saltzman, pp. 3–39. Dordrecht: Reidel.

Berger, A. L., Galle, M., Fichefet, T., Marsiat, I. and Tricat, C. (1990) Testing the atronomical theory with a coupled climate–ice-sheet model. *Global and*

Planetary Changes 3: 125–141.

Berger, A. L., Guiat, J. and Kukla, G. (1981) Long-term variations of monthly insolation as related to climatic changes. *Geol. Rundsch.* 70: 748–758.

Berger, A. L. and Pestiaux, P. (1984) Accuracy and stability of the Quaternary terrestrial insolation. In *Milankovitch and Climate*, ed. A. L. Berger, J. Imbrie, J. Hays, G. Kukla and B. Saltzman, pp. 85–112. Dordrecht: Reidel.

Berger, W. H. (1982) Deep-sea stratigraphy: Cenozoic climate steps and the search for chemo-climatic feedback. In *Cyclic and Event Stratification*, ed. G. Einsele and A. Seilacher, pp. 121–157. Berlin: Springer-Verlag.

Berger, W. H. (1985) CO_2 increase and climate prediction: clues from deep-sea carbonates. *Episodes* 8: 163–168.

Berger, W. H. (1987) Ocean ventilation during the last 12,000 years: Hypothesis of counterpoint deep water production. *Marine Geol.* 78: 1–10.

Berger, W. H. (1990) The Younger Dryas cold spell – a quest for cause. *Global and Planetary Changes* 3: 214–238.

Berger, W. H., Killingley, J. S., Metzler, C. V. and Vincent, E. (1985) Two step deglaciation; ^{14}C dated high-resolution $\delta^{18}O$ records from the tropical Atlantic Ocean. *Quat. Res.* 23: 258–271.

Berger, W. H. and Mayer, L. A. (1987) Cenozoic paleoceanography 1986: An introduction. *Paleoceanography* 2: 613–624.

Berger, W. H. and Spitzy, A. (1988) History of atmospheric CO_2: constraints from the deep-sea record. *Paleoceanography* 3: 404–411.

Berggren, W. A. (1982) Role of ocean gateways in climatic change. In *Climate in Earth History*, ed. W. H. Berger and J. C. Crowell, pp. 118–125. Washington: National Academy Press.

Berggren, W. A. and Hollister, C. (1974) Paleogeography, paleobiogeography and history of circulation in the Atlantic Ocean. In *Studies in Palaeoceanography*, ed. W. W. Hay, pp. 126–186. Society of Economic Paleontologists Special Publication 20.

Berggren, W. A. and Olsson, R. K. (1986) North Atlantic Mesozoic and Cenozoic paleobiogeography. In *The Geology of the North America, The Western North Atlantic Region*, ed. P. R. Vogt and B. E. Tucholke, pp. 565–587. Boulder: Geological Society of America.

Berner, R. A. (1990) Atmospheric carbon dioxide levels over Phanerozoic time. *Science* 249: 1382–1386.

Berner, R. A. and Canfield, D. E. (1989) A new model for atmospheric oxygen over Phanerozoic time. *Am. J. Sci.* 289: 333–361.

Berner, R. A. and Lasaga, A. C. (1989) Modeling of the geochemical carbon cycle. *Sci. Am.* 260: 54–61.

Berner, R. A., Lasaga, A. C. and Garrels, R. M. (1983) The carbonate-silicate geochemical cycle and its effect on atmospheric carbon dioxide over the last 100 million years. *Am. J. Sci.* 283: 641–683.

Berner, W., Stauffer, B. and Oeschger, H. (1979) Past atmospheric composition and climate; gas parameters measured on ice cores. *Nature* 275: 53–55.

Berry, W. B. N. (1973) Silurian-early Devonian graptolites. In *Atlas of Paleobiogeography*, ed. A. Hallam, pp. 81–88. Amsterdam: Elsevier.

Berry, W. B. N. (1979) Graptolite biogeography: a biogeography of some Lower

Paleozoic plankton. In *Historical Biogeography, Plate Tectonics, and the Changing Environment*, ed. J. Gray and A. J. Boucot, pp. 105–115. Corvallis: Oregon State University Press.

Berry, W. B. N. and Wilde, P. (1978) Progressive ventilation of the oceans—an explanation for the distribution of the Lower Paleozoic Black Shales. *Am. J. Sci.* 278: 257–275.

Beuf, S., Biju-Dival, B., DeCharpal, O., Rognou, P., Gariel, O. and Bennoef, A. (1971) Les Grès du Paleozoique Inférieur au Sahara – Sédimentation et Discontinuités, Evolution Structurale d'un Craton. *Inst. Fr. Pétroles-Sci. Tech. Pétroles.* 18: 454 pp.

Biju-Duval, B., Delcourt, J. and LePichon, X. (1976) *From the Tethys Ocean to the Mediterranean Sea: a plate tectonic model of the evolution of the western alpine system.* ed. B. Biju-Duval and L. Montadert. International Symposium on the Structural History of the Mediterranean Basins. Split, Yugoslavia. pp.143–164. Paris: Editions Techniq.

Billard, A. and Orombelli, G. (1986) Quaternary Glaciations in the French and Italian Piedmont of the Alps. *Quat. Sci. Rev.* 5: 407–412.

Bird, K. J. (1985) The framework geology of the north slope of Alaska as related to oil-source rock correlations. In *Alaska North Slope Oil-Rock Correlation Study*, ed. L. B. Magoon and G. E. Claypool, pp. 3–18. Tulsa, Oklahoma: American Association of Petroleum Geologists Studies in Geology 20.

Birkenmajer, K. and Zastawaniak, E. (1990) Late Cretaceous-Early Tertiary floras of King George Island, West Antarctica: their stratigraphic distribution and palaeoclimatic significance. In *Origins and Evolution of the Antarctic Biota*, ed. J. A. Crame, pp. 227–240. London: The Geological Society.

Birkenmajer, K. J. (1987) Oligocene-Miocene glaciomarine sequences of King George Island (South Shetland Islands), Antarctica. *Palaeontol. Pol.* 49: 9–36.

Birkenmajer, K. J. (1991) Tertiary glaciations in the South Shetland Islands, West Antarctica: evaluation of data. In *The Geological Evolution of Antarctica*, ed. M. R. A. Thomson, J. A. Crame and J. W. Thomson, pp. 629–632. Cambridge University Press.

Birkenmajer, K. J. and Narebski, W. (1963) Dolerite drift blocks in marine Tertiary of Sorkapp Land and some remarks on the geology of the eastern part of this area. *Norsk Polarin. Arbok.* 1962 (1963), 68–79.

Boersma, A. and Premoli-Silva, I. (1983) Paleocene planktonic foraminiferal biostratigraphy and the paleoceanography of the Atlantic Ocean. *Micropalaeontology* 29: 355–381.

Boersma, A., Shackleton, N. J., Hall, M. and Given, Q. (1979) Carbon and oxygen records at DSDP Site 384 (North Atlantic) and some Paleocene paleotemperatures and carbon isotope variations in the South Atlantic. *Init. Repts. DSDP* 43: 695–718. Washington (Govt. Printing Office).

Bond, G. C., Nickeson, P. A. and Kominz, M. A. (1984) Breakup of a supercontinent between 625 Ma and 555 Ma: new evidence and implications for continental histories. *Earth Planet Sci. Lett.* 70: 325–345.

Bornhold, B. D. (1983) Ice-rafted debris in sediments from Leg 71, Southwest Atlantic Ocean. *Init. Repts. DSDP* 71: 307–316. Washington (Govt. Printing

Office).

Boucot, A. J. (1975) Malvinokaffric Devonian marine community distribution and implications for Gondwana. *An. Acad. Sci. Bras. Cienc.* 43: 23–49.

Boucot, A. J. (1985) Late Silurian-Early Devonian biogeography, provincialism, evolution and extinction. *Phil. Trans. Royal Soc. Lond.* B 309: 323–339.

Boucot, A. J. and Gray, J. (1983) A Palaeozoic Pangaea. *Science* 222: 571–581.

Boucot, A. J. and Johnson, J. G. (1973) Silurian brachiopods. In *Atlas of Palaeobiogeography*, ed. A. Hallam, pp. 59–65. Amsterdam: Elsevier.

Boulton, G. S. (1979) A model of Weschelian glacier variation in the North Atlantic temperate ice caps region. *Boreas* 8: 373–395.

Bowen, D. Q. (1980) Antarctic ice surges and theories of glaciation. *Nature* 283: 619.

Bowen, R. (1966) *Paleotemperature Analysis.* Amsterdam: Elsevier. 265 pp.

Boyle, E. A. (1984) Cadmium in benthic foraminifera and abyssal hydrography: evidence for a 41 kyr obliquity cycle. In *Climate Processes and Climate Sensitivity*, ed. J. E. Hansen and T. Takahashi, pp. 360–368. American Geophysical Union.

Boyle, E. A. (1986) Deep ocean circulation, performed nutrients and atmospheric carbon dioxide: Theories and evidence from oceanic sediments. In *Mesozoic and Cenozoic Oceans*, ed. K. J. Hsu, pp. 49–59. American Geophysical Union, Geodynamics Series 15.

Boyle, E. A. (1988a) Vertical oceanic nutrient fractionation and glacial/interglacial CO_2 cycles. *Nature* 275: 53–55.

Boyle, E. A. (1988b) Cadmium: chemical tracer of deep water paleoceanography. *Paleoceanography* 3: 471–489.

Boyle, E. A. and Keigwin, L. D. (1982) Deep circulation of the North Atlantic over the last 200,000 years: Geochemical evidence. *Science* 218: 784–787.

Boyle, E. A. and Keigwin, L. D. (1986) Comparison of Atlantic and Pacific paleochemical records for the last 215,000 years: changes in deep ocean circulation and chemical inventories. *Earth Planet. Sci. Letters.* 76: 135–150.

Boyle, E. A. and Keigwin, L. D. (1987) North Atlantic thermohaline circulation during the past 20,000 years linked to high-latitude surface temperature. *Nature* 330: 35–40.

Boyle, E. A. and Rosener, P. (1990) Further evidenc for a link between Late Pleistocene North Atlantic surface temperatures and North Atlantic Deep Water production. *Global and Planetary Changes* 3: 113–124.

Braackman, J. H., Levell, B. K., Martin, J. H., Potter, T. L. and van Vliet, A. (1982) Late Paleozoic Gondwana glaciation in Oman. *Nature* 299: 48–50.

Bradley, R. S. (1985) *Quaternary Paleoclimatology.* Boston: Allen and Unwin. 472 pp.

Bralower, T. J. and Thierstein, H. R. (1984) Low productivity and slow deep water circulation in mid-Cretaceous oceans. *Geology* 12: 614–618.

Brandt, K. (1986) Glacioeustatic cycles in the early Jurassic? *N. Jb. Geol. Palaont. Mh.* 1985(5): 257–274.

Brass, G. W., Southam, J. R. and Peterson, W. H. (1982) Warm saline bottom waters in the ancient ocean. *Nature* 296: 620–623.

Brenchley, P. J. (1984) Late Ordovician extinctions and their relationship to the Gondwana glaciation. In *Fossils and Climate*, ed. P. J. Brenchley, pp. 291–

315. New York: John Wiley and Sons Ltd.

Brenchley, P. J. and Newall, G. (1980) A facies analysis of Upper Ordovician regressive sequences in the Oslo region, Norway – A record of glacio-eustatic changes. *Palaeogeogr., Palaeoclimatol., Palaeoecol.*, 31: 1–38.

Briden, J. C. and Irving, E. (1964) Paleolatitude spectra of sedimentary paleoclimatic indicators. In *Problems in Paleoclimatology*, ed. A. E. M. Nairn, pp. 199–224. London: Interscience.

Broccoli, A. J. and Manabe, S. (1987) The influence of continental ice, atmospheric CO_2 and land albedo on the climate of the Last Glacial Maximum. *Climate Dynamics* 1: 87–99.

Broecker, W. S. (1982) Ocean chemistry during glacial time. *Geochim. Cosmochim. Acta.* 46: 1689–1750.

Broecker, W. S. (1984) Terminations. In *Milankovitch and Climate*, ed. A. L. Berger, J. Imbrie, J. Hays, G. Kukla and B. Saltzman, pp. 678–698. Dordrecht: Reidel.

Broecker, W. S. (1987) Greenhouse surprise. *Lamont-Doherty Geological Observatory Yearbook.* 1987: 11–15.

Broecker, W. S. (1988a) Can the Greenland climatic jumps be identified in records from the ocean and land? *Quat. Res.* 30: 1–6.

Broecker, W. S. (1988b) The salinity contrast between the Atlantic and Pacific Oceans during glacial times. *Paleoceanography* 4: 207–212.

Broecker, W. S., Andree, M., Klaus, M., Bonani, G., Wolfi, W. and Oeschger, H. (1988) New evidence from the South China Sea for an abrupt termination of the glacial period. *Nature* 333: 156–158.

Broecker, W. S., Andree, M., Wolfi, W., Oeschger, H., Bonani, G., Kennett, J. and Peteet, D. (1987) The chronology of the last deglaciation: Implications to the cause of the younger Dryas event. *Paleoceanography* 3: 1–20.

Broecker, W. S. and Denton, G. H. (1985) The role of ocean-atmosphere reorganisation in glacial cycles. *Geochim. Cosmochim. Acta* 53: 2465–2501.

Broecker, W. S. and Peng, T. H. (1982) *Tracers in the Sea.* Palisades, New York.: Lamont-Doherty Geological Observatory, Columbia University. 690 pp.

Broecker, W. S. and Peng, T. H. (1987) The role of $CaCO_3$ compensation in the glacial to interglacial atmospheric CO_2 change. *Global Biogeochem. Cycles* 1: 15–29.

Broecker, W. S. and Peng, T. H. (1989) The cause of the glacial to interglacial atmospheric carbon dioxide change: A polar alkalinity hypothesis. *Global Biogeochem. Cycles* 3: 215 – 239.

Broecker, W. S., Peteet, D. M. and Rind, D. (1985) Does the ocean-atmospheric system have more than one stable mode of operation? *Nature* 315: 21–26.

Brooks, J. and Fleet, A. J. (1987) Marine Petroleum Source Rocks. *Geol. Soc. London, Spec. Pub.* 26: 1–444.

Brouwers, E. M., Clemens, W. A., Spicer, R. A., Ager, T. A., Carter, L. D. and Sliter, W. V. (1987) Dinosaurs on the North Slope, Alaska: High latitude latest Cretaceous environments. *Science* 237: 1608–1610.

Budyko, M. I. and Ronov, A. B. (1979) The evolution of the atmosphere in the Phanerozoic. *Geochimiya.* 5: 643–653.

Bukry, D. (1985) Tropical Pacific silicoflagelate zonation and paleotemperature trends of the late Cenozoic. *Initial Reports DSDP.* 85: 477–497. Washington

(Govt. Printing Office).

Burckle, L. H. (1987) Diatom distribution in the Weddell Sea region during late winter. *Micropalaeontology.* 33: 177–184.

Burckle, L. H. and Abrams, N. (1986) Biotic response to two late Neogene intervals of global climatic change: evidence from the deep sea record. *S. Afr. J. Sci.* 82: 69–72.

Busch, R. M. and Rollins, H. B. (1984) Correlation of Carboniferous strata using a hierarchy of transgressive-regressive units. *Geology* 12: 471–474.

Caputo, M. V. (1985) Late Devonian glaciation in South America. *Palaeogeogr., Palaeoclimat., Palaeoecol.* 51: 291–317.

Caputo, M. V. and Crowell, J. C. (1985) Migration of glacial centers across Gondwana during Palaeozoic Era. *Geol. Soc. Am. Bull.* 96: 1020–1036.

Carlquist, S. (1988) Pliocene *Nothofagus* wood from the Transantarctic Mountains. *ALISO* 11: 571–583.

Carter, L. D., Brigham-Grette, J., Marinovich, L., Pease, V. L. and Millhouse, J.W. (1986) Late Cenozoic Arctic Ocean sea ice and terrestrial paleoclimate. *Geology* 14: 675–678.

Chafetz, H. S. (1980) Evidence for arid to semi-arid climate during deposition of the Cambrian System in central Texas. *Palaeogeogr., Palaeoclimatol., Palaeoecol.*, 30: 83–95.

Chaloner, W. G. and Creber, G. T. (1973) Growth rings in fossil wood as evidence of past climates. In *Implications of Continental Drift to the Earth Sciences*, ed. D. H. Tarling and S. K. Runcorn, pp. 425–437. New York: Academic Press.

Chaloner, W. G. and Lacey, W. S. (1973) The distribution of late Paleozoic floras. *Special Papers in Palaeontology, Paleontological Association*, London. 12: 271–289.

Chaloner, W. G. and Meyen, S. V. (1973) Carboniferous and Permian floras of the northern continents. In *Atlas of Paleobiogeography*, ed. A. Hallam, pp. 169–186. New York: Elsevier.

Chamley, H. (1979) North Atlantic clay sedimentation and paleoenvironment since the late Jurassic. In *Deep Sea Drilling Results in the Atlantic Ocean: Continental Margins and Paleoenvironment*, ed. M. Talwani, W. Hay and W. B. F. Ryan, pp. 342–361. Washington D.C.: American Geophysical Union.

Chamley, H., Debrabant, P., Candillier, A.-M. and Foulon, J. (1983) Clay mineralogy and inorganic geochemical stratigraphy of the Blake-Bahama Basin since the Callovian, Site 534, Deep Sea Drilling Project Leg 76. *Init. Repts. DSDP.* 76: 437–451. Washington (Govt. Printing Office).

Chappell, J. (1983) A revised sea level record for the last 300,000 yrs from Papua New Guinea. *Search* 4: 99–101.

Chappell, J. and Shackleton, N. (1986) Oxygen isotopes and sea level. *Nature* 324: 137–140.

Chinzei, K. (1987) *Faunal succession and geographic distribution of Neogene molluscan faunas in Japan (abs.)*. 4th International Congress on Pacific Neogene Stratigraphy. ed. J. A. Barron and J. R. Blueford, Abstract Volume p. 14. Berkeley, California.

Chow, N. and James, N. P. (1987) Cambrian grand cycles: a North Appalachian perspective. *Geol. Soc. Am. Bull.* 98: 418–429.

Christie, R. C. and McMillan, N. J. (1991) Tertiary fossil forests of the geodetic Hills, Axel Heiberg Island, Arctic Archipelago, *Geological Survey of Canada Bulletin* 403. 229 pp.

Chuey, J. M., Rea, D. K. and Pisias, N. G. (1987) Late Pleistocene paleoclimatology of the central equatorial Pacific: a quantitative record of eolian and carbonate deposition. *Quat. Rev.* 28: 323–339.

Chumakov, N. M. (1981) Upper Proterozoic glaciogenic rocks and their stratigraphic significance. *Precambrian Res.* 15: 373–395.

Ciesielski, P. and 18 members of the Leg 114 Shipboard Scientific Party (1987) Paleoceanography of the subantarctic South Atlantic. *Nature* 328: 671–672.

Ciesielski, P. F., Ledbetter, M. T. and Ellwood, B. B. (1982) The development of Antarctic glaciation and the Neogene paleonvironment of the Maurice Ewing Bank. *Mar. Geol.* 46: 1–51.

Ciesielski, P. F. and Weaver, F. M. (1974) Early Pliocene temperature changes in the Antarctic Seas. *Geology* 2: 511–551.

Ciesielski, P. F. and Weaver, F. M. (1983) Neogene and Quaternary paleoenvironmental history of DSDP Leg 71 sediments, southwest Atlantic Ocean. *Init. Repts. DSDP* 71:461–477. Washington (Govt. Printing Office).

Cita, M. B. and McKenzie, J. A. (1986) The terminal Miocene event. In *Mesozoic and Cenozoic Oceans*, ed. K. J. Hsu, pp. 123–140. Washington D.C.: American Geophysical Union, Geodynamics Series 15.

Clark, D. L. (1982) Origin, nature and world climate effect of Arctic Ocean ice-cover. *Nature* 300: 321–325.

Clark, D. L., Byers, C. W. and Pratt, L. M. (1986) Cretaceous black mud from the central Arctic Ocean. *Paleoceanography* 2: 265–271.

Clark, D. L., Whitman, R. R., Morgan, K. A. and Mackey, S. D. (1980) Stratigraphy and glacial-marine sedimentation of the Amerasian Basin, Central Arctic Ocean. *Geol. Soc. Am. Spec. Paper.* 181. 57 pp.

Clarke, M. J. and Farmer, N. (1976) Biostratigraphic nomenclature for late Palaeozoic rocks in Tasmania. *R. Soc. Tasm. Pap. Proc.* 110: 91–109.

Clarke, M. J. and Farmer, N. (1982) Late Paleozoic cold-water carbonate sedimentation in Tasmania; discussion. *J. Sed. Pet.* 52: 682–683.

Clauer, N. and Deynoux, M. (1987) New information on the probable isotopic age of the late Proterozoic glaciation in West Africa. *Precambrian Res.* 37: 89–94.

Claypool, G. E., Holser, W. T., Kaplan, I. R., Sakai, M. and Zak, I. (1980) The age curves of sulfur and oxygen isotopes in marine sulfate and their mutual interpretation. *Chem. Geol.* 28: 199–260.

CLIMAP (1976) The surface of the ice-age Earth. *Science* 191: 1131–1144.

CLIMAP (1981) Seasonal reconstructions of the Earth's surface at the last glacial maximum. *Geol. Soc. Amer. Map and Chart Series No. 36.*

CLIMAP (1984) The last glacial ocean. *Quat. Res.* 21: 123–224.

COHMAP (1988) Climatic changes of the last 18,000 years: observations and model simulations. *Science* 241: 1043–1052.

Collinson, M. E. (1986) Catastrophic vegetation changes. *Nature* 324: 112.

Collinson, M. E., Fowler, K. and Boulter, M. C. (1981) Floristic changes indicating a cooling climate in the Eocene of southern England. *Nature* 291: 315–317.

Cook, P. J. and McElhinny, M. W. (1979) A re-evaluation of the spatial and temporal

229

distribution of sedimentary phosphate deposits in the light of plate tectonics. *Econ. Geol.* 74: 315–340.

Cook, P. J. and Shergold, J. H. (eds) (1986) *Phosphate Deposits of the World, Vol 1.* Cambridge: Cambridge University Press. 386 pp.

Cooper, M. R. (1977) Eustacy during the Cretaceous: its implications and importance. *Palaeogeogr., Palaeoclimatol., Palaeoecol.* 22: 1060.

Cope, M. J. and Chaloner, W. G. (1980) Fossil charcoal as evidence of past atmospheric composition. *Nature* 284: 647–649.

Copper, P. (1977) Paleolatitudes in the Devonian of Brazil and the Frasnian-Famennian mass extinction. *Palaeogeogr., Palaeoclimatol., Paleoecol.* 21: 165–207.

Copper, P. (1986) Frasnian/Famennian mass extinction and cold water oceans. *Geology* 14: 835–839.

Copper, P. (1987) Frasnian/Famennian mass extinction and cold water oceans: Reply. *Geology* 15: 778.

Corliss, B. H. (1982) Linkage of North Atlantic and Southern Ocean deep water circulation during last glacial intervals. *Nature* 298: 458–460.

Corliss, B. H. and Keigwin, L. D. (1986) Eocene-Oligocene palaeoceanography. In *Mesozoic and Cenozoic Oceans*, ed. K. J. Hsu, pp. 101–118. Washington, D.C.: American Geophysical Union, Geodynamics Series 15.

Corliss, B. H., Martinson, D. G. and Keffer, T. (1986) Late Quaternary deep-ocean circulation. *Geol. Soc. Am. Bull.* 97: 1106–1121.

Cotillon, P. H. and Rio, M. (1984) Cyclic sedimentation in the Cretaceous of Deep Sea Drilling Project Sites 535 and 540 (Gulf of Mexico), 534 (Central Atlantic), and in the Vocontian Basin (France). *Init. Repts. DSDP*, 77:339–376. Washington (Govt. Printing Office).

Crame, J. S. and Howlett, P. J. (1988) Late Jurassic and Early Cretaceous biostratigraphy of the Fossil Bluff Formation, Alexander Island. *Brit. Antarct. Survey Bull.* 78: 1–35.

Crane, P. R. (1987) Vegetational consequences of angiosperm diversification. In *The Origin of Angiosperms and their Biological Consequences*, ed. E. M. Friis, W. G. Chaloner and P. R. Crane, pp. 107–144. Cambridge: Cambridge University Press.

Creber, G. T. and Chaloner, W. G. (1985) Tree growth in the Mesozoic and early Tertiary and the reconstruction of palaeoclimates. *Palaeogeogr., Palaeoclimatol., Palaeoecol.* 52: 35–60.

Cronin, T. M. (1988) Evolution of marine climate of the US Atlantic coast during the past four million years. *Phil. Trans. R. Soc. Lond.B.* 318: 661–678.

Crowell, J. C. (1978) Gondwana glaciation, cyclothems, continental positioning, and climate change. *Am. J. Sci.* 278: 1345–1372.

Crowell, J. C. (1981) Early Paleozoic glaciation and Gondwana drift. In *Paleoreconstruction of the Continents*, ed. M. W. McElhinny and D. A. Valencio. pp. 45–49. Washington, D.C.: American Geophysical Union, Geodynamics Series 2.

Crowell, J. C. (1983) Ice ages recorded on Gondwanan continents. *Geol. Soc. S. Africa Trans.* 86: 238–261.

Crowell, J. C. and Frakes, L. A. (1970) Phanerozoic glaciation and the causes of ice ages. *Amer. J. Sci.* 268: 193–224.

Crowley, T. J., Mengel, J. G. and Short, D. A. (1987) Gondwanaland's seasonal cycle. *Nature* 329: 803–807.

Crowley, T. J., Short, D. A., Mengel, J. G. and North, G. R. (1986) Role of seasonality in the evolution of climate during the last 100 million years. *Science* 231: 579–584.

Curry, W. B. and Crowley, T. J. (1987) The delta [13]C in equatorial Atlantic surface waters: Implications for Ice Age pCO_2 levels. *Paleoceanography* 2: 489–518.

Dalland, A. (1977) Erratic clasts in the Lower Tertiary deposits of Svalbard–evidence of transport by winter ice. *Norsk Polarin. Arbok.* 1976: 15–165.

Dansgaard, W. and Duplessy, J. C. (1981) The Eemian interglacial and its termination. *Boreas* 10: 219–228.

Dansgaard, W., Johnson, S. J., Clausen, H. B., Dahl-Jensen, D., Gundestrup, N., Hammer, C.V. and Oeschger, H (1984) North Atlantic climatic oscillations revealed by deep Greenland ice cores. In *Climatic Processes and Climate Sensitivity*, ed. J. E. Hansen and T. Takahashi, pp. 288–298. Washington D.C.: American Geophysical Union, Geophysical Monograph 32.

David, T. W. E. and Taylor, T. G. (1905) Occurrence of pseudomorph glendonite in New South Wales, Part 1. *NSW Geol. Surv. Rec.* VII: 162–172.

Davis, R. A. J. and Mallett, C. W. (1981) Sedimentation in a Permian subglacial channel. *Jour. Sed. Petrol.* 51: 185–190.

De Angelis, M., Barkov, N. I. and Petrov,V.N. (1987) Aerosol concentrations over the last climatic cycle (160kyr) from the Antarctic ice core. *Nature* 325: 318–321.

de Boer, P. L. (1982) Cyclicity and the storage of organic matter in Cretaceous pelagic sediments. In *Cyclic and Event Stratification*, ed. G. Einsele and A. Seilacher, pp. 456–475. New York: Springer Verlag.

de Boer, P. L. and Shackleton, N. (1984) Pre-Pleistocene evidence of orbital forcing. In *Milankovitch and Climate*, ed. A. Berger *et al.*, pp. 827–830. Dordrecht: Reidel.

de Vernal, A. and Hillaire-Marcel, C. (1987) Paleoenvironments along the eastern Laurentide ice sheet margin and timing of the last ice maximum and retreat. *Geographie. Phys. et Quat.* 41: 265–277.

Dean, W. and Gardner, J. (1985) Cyclic variations in calcium carbonate and organic carbon in Miocene to Holocene sediments, Walvis Ridge, South Atlantic Ocean. In *South Atlantic Paleoceanography*, ed. K. J. Hsu and H. J. Weissert, pp. 61–78. Cambridge: Cambridge University Press.

Dean, W. E. and Gardner, J. (1986) Milankovitch cycles in Neogene deep-sea sediments. *Paleoceanography* 1: 539–553.

Dean, W. E., Arthur, M. A. and Claypool, G. E. (1986) Depletion of [13]C in Cretaceous marine organic matter: Source, diagenetic, or environmental signal. *Mar. Geol.* 70: 119–157.

Deconinck, J. (1987) Identification de l'orogine détritique ou diagénétique des assemblages argilleux: le cas des alternances marne-calcaire du Crétacé inférieur subalpin. *Bull. Geol. Soc. France.* 3: 139–145.

Delaney, M. L. and Boyle, E. A. (1988) Tertiary paleoceanic chemical variability: unintended consequences of simple geochemical models. *Paleoceanography* 3: 137–156.

Denton, G. H. and Armstrong, R. L. (1969) Miocene-Pliocene glaciations in southern Alaska. *Am. J. Sci.* 267: 1121–1142.

Denton, G. H. and Hughes, T. J. (1981) *The Last Great Ice Sheets.* New York: Wiley Interscience. 484 pp.

Denton, G. H. and Hughes, T. J. (1986) Potential of floating ice shelves on the climate of an ice age. *S. Afric. J. Sci.* 82: 509–513.

Denton, G. H., Hughes, T. and Karlen, W. (1986) Global ice-sheet system interlocked by sea level. *Quat. Res.* 26: 3–26.

Denton, G. H., Prentice, M. L., Kellogg, D. E. and Kellogg, T.B. (1984) Late Tertiary history of the Antarctic ice sheet: evidence from the Dry Valleys. *Geology* 12: 263–267.

Dettman, M. E. (1986a) Early Cretaceous palynoflora of subsurface strata correlative with the Koonwarra Fossil Bed, Victoria. *Mem. Ass. Australas. Palaeontols.* 3: 79–110.

Dettman, M. E. (1986b) Significance of the Cretaceous-Tertiary spore genus *Cyatheacidites* in tracing the origin and migration of *Lophosoria* (Filicopsida). *Sp. pap. Palaeo.* 35: 63–94.

Dettman, M. E. (1990) Antarctica: Cretaceous cradle of austral temperate rainforests? In *Origins and Evolution of the Antarctic Biota*, ed. J. A. Crame, pp. 89–106. London: Geological Society of London.

Ditchfield, P. and Marshall, J. D. (1989) Isotopic variation in rhythmically bedded chalks: Palaeotemperature variation in the Upper Cretaceous. *Geology* 17: 842–845.

Donn, W. and Shaw, D. (1977) Model of climate evolution based on continental drift and polar wandering. *Geol. Soc. Amer. Bull.* 64: 1315–1326.

Dorf, E. (1964) The petrified forests of Yellowstone Park. *Sci. Amer.* 210: 106–114.

Dorofeyeva, L. A., Davydov, V. I. and Kashik, D. S. (1982) Mode of temperature variation in the Late Palaeozoic of the southwestern Darvaz. *Doklady Acad. Sci. USSR.* Earth Science section. 263: 81–89.

Douglas, J. G. and Williams, G. E. (1982) Southern polar forests: The Early Cretaceous floras of Victoria and their palaeoclimatic significance. *Palaeogeogr., Palaeoclimatol., Palaeoecol.* 39: 171–185.

Douglas, R. G. and Savin, S. M. (1973) Oxygen and carbon isotope analyses of Cretaceous and Tertiary foraminifers of the Central North Pacific. *Init. Repts. DSDP* 17: 591–605. Washington (Govt. Printing Office).

Douglas, R. G. and Savin, S. M. (1975) Oxygen and carbon isotope analyses of Cretaceous and Tertiary microfossils from Shatsky Rise and other sites in the North Pacific Ocean. *Init. Repts. DSDP* 32: 509–521. Washington (Govt. Printing Office).

Douglas, R. G. and Woodruff, F. (1981) Deep sea benthic foraminifera. In *The Sea*, vol. 7, ed. C. Emiliani, pp. 1233–1327. New York: Wiley Interscience.

Doyle, P. (1987) The Cretaceous Dimitobelidae (Belemnitida) of the Antarctic Peninsula region. *Palaeontology* 30: 147–177.

Doyle, P. (1988) The belemnite family Dimitobelidae in the Cretaceous of Gondwana. In *Cephalopods – Past and Present*, ed. J. Weidmann and J. Kullman, pp. 539–552. Stuttgart: Schweizerbart'she Verlagsbuchhandlung.

Drewry, D. J. (1975) Initiation and growth of the East Antarctic ice sheet. *J. Geol.*

Soc. Lond. 131: 255–273.

Drinnan, A. N. and Chambers, T. C. (1986) Flora of the Lower Cretaceous Koonwarra Fossil Bed (Korumburra Group), South Gippsland, Victoria, *Mem. Ass. Australas. Palaeontols.* 3: 1–77.

Duff, P. M. D. (1987) Mesozoic and Tertiary coals – a major world energy resource. *Modern Geol.* 11: 129–150.

Duplessy, J.-C. (1982) Glacial to interglacial contrasts in the northern Indian Ocean. *Nature* 295: 494–498.

Duplessy, J.-C., Arnold, M., Maurice, P., Berol, E., Duprat, J. and Moyes, J. (1986) Direct dating of oxygen isotope record of the last deglaciation by ^{14}C accelerator mass spectrometry. *Nature* 320: 350–352.

Duplessy, J.-C., Delibrias, G., Turon, J. L., Pujol, C. and Duprat, J. (1981) Deglacial warming of the northeastern Atlantic Ocean: correlation with the paleoclimatic evolution of the European continent. *Palaeogeogr., Palaeoclimatol., Palaeoecol.* 35: 121–144.

Duplessy, J.-C. and Pujol, C. (1983) Abrupt climatic events during the last glacial to interglacial transition. In *Paleoclimatic Research and Models*, ed. A. Ghazi, pp. 152–156. Dortrecht: D. Reidel.

Durante, M. V., Dmitriyev, V. Y. and Pavlova, Y. Y. (1985) The paleobiogeography of Mongolia and neighboring regions in the Late Paleozoic. *Int. Geol. Rev.* 27: 1396–1406.

Dyke, A. S. and Prest, V. K. (1987) Late Wisconsin and Holocene history of the Laurentide ice sheet. *Geographie phys. et Quater.* 41: 237–263.

Edwards, A. R. (1987) *Significant climatic events in the late Neogene marine strata of New Zealand (abstract)*. 4th International Congress on Pacific Neogene Stratigraphy. ed. J.A. Barron and J. R. Blueford. Berkeley, California. pp. 26–27.

Einarsson, T. and Albertsson, K. J. (1988) The glacial history of Iceland during the past three million years. *Phil. Trans. Roy. Soc. Lond.* B 318: 637–644.

Einsele, G. and Wiedmann, J. (1982) Turonian black shales in the Moroccan coastal basins: First upwelling in the Atlantic Ocean? In *Geology of the Northwest African Continental Margin*, ed. U. von Rad and others, pp. 396–414. Berlin: Springer Verlag.

Eldholm, O., Theide, J., Taylor, E. and others (1986) Leg 104 summary and preliminary conclusions. *Init. Repts. DSDP* 104: 759–771. Washington (Govt. Printing Office).

Elmstrom, K. M. and Kennett, J. P. (1986) Late Neogene palaeoceanographic evolution of DSDP Site 590: southwest Pacific. *Init. Repts. DSDP* 90: 1361–1381.

Embleton, B. J. (1984) Continental paleomagnetism. In *Phanerozoic Earth History of Australia*, ed. J. J. Veevers, pp. 11–16. Oxford: Clarendon Press.

Embleton, B. J. J. and Williams, G. E. (1986) Low palaeolatitude of deposition for late Precambrian periglacial varvites in South Australia: implications for palaeoclimatology. *Earth Planet. Sci. Lett.* 79: 419–430.

Embry, A. F. (1984) Upper Jurassic to lowermost Cretaceous stratigraphy, sedimentology and petroleum geology, Sverdrup Basin. *CSPG–CSEG Nat. Conv.* 49–50.

Ennever, F. K. and McElroy, M. B. (1985) Changes in atmospheric CO_2: factors

regulating the glacial to interglacial transition. In *The Carbon Cycle and Atmospheric CO₂: Natural Variations Archean to Present*, ed. E. T. Sundquist and W. S. Broecker, pp. 154–162. Washington, D. C.: American Geophysical Union, Geophysical Monograph 32.

Epshteyn, O. G. (1972) Upper Permian glacial marine deposits in the basin of the Kolyma River source. *Lithology and Mineral Resources* 7: 350–362.

Epshteyn, O. G. (1978) Mesozoic-Cenozoic climates of northern Asia and glacial-marine deposits. *Internat. Geol. Rev.* 20: 49–58.

Fairbridge, R. W. (1970) South Pole reaches the Sahara. *Science* 168: 880–881.

Faure, H. (1987) An amplifying mechanism for the climate cycle: the 'sea ice lid effect' controls the changes in atmospheric carbon dioxide. *C. R. Acad. Sci. Paris* 305: 523–527.

Finney, B. P., Lyle, M. W. and Heath, G. R. (1988) Sedimentation at MANOP site H (eastern equatorial Pacific) over the past 400,000 years: Climatically induced redox variations and their effects on transition metal cycling. *Paleoceanography* 3: 169–189.

Fischer, A. G. (1981) Climatic oscillations in the biosphere. In *Biotic Crises in Ecological and Evolutionary Time*, ed. M. Nitecki, pp. 103–131. New York: Academic Press.

Fischer, A. G. (1982) Long-term climatic oscillations recorded in stratigraphy. In *Climate in Earth History*, ed. W. H. Berger and J. C. Crowell, pp. 97–104. Washington, D.C.: National Academy Press.

Fischer, A. G. (1986) Climatic rhythms recorded in strata. *Ann. Rev. Earth Planet. Sci.* 14: 351–376.

Fischer, A. G. and Arthur, M. A. (1977) Secular variations in the pelagic realm. In *Deep Water Carbonate Environments*, ed. H. E. Cook and P. Enos, pp. 18–50. Soc. Econ. Paleontol. Mineral. Spec. Publ. 25.

Fischer, A. G., de Boer, P. L. and Premoli Silva, I. (1990) Cyclostratigraphy. In *Cretaceous Resources, Events and Rhythms*, ed. R. N. Ginsburg and B. Beaudoin, pp. 139–172. The Netherlands: Kluwer Academic Publishers.

Fitzgerald, P. G., Sandiford, M., Barrett, P. J. and Gleadow, A. J. W. (1986) Asymetric extension associated with uplift and subsidence in the Transantarctic Mountains and the Ross Embayment. *Earth Planet. Sci. Letters.* 81: 67–78.

Flexer, A., Rosenfeld, A., Lipson-Benitah, S. and Honigstein, A. (1986) Relative sea level changes during the Cretaceous in Israel. *Amer. Ass. Petrol. Geol. Bull.* 70: 1685–1699.

Flint, R. F. (1971) *Glacial and Quaternary Geology.* New York: John Wiley and Sons. 892 pp.

Flohn, H. (1973) Globale Energiebilanz und Klimaschwankungen. *Bonner Met. Abhandl.* 19: 75–117.

Fortey, R. A. (1984) Global earlier Ordovician transgressions and regressions and their biological implications. In *Aspect of the Ordovician System*, ed. D. L. Bruton, pp. 37–50. Paleontological Contributions from the University of Oslo.

Fotyanova, L. I. and Serova, M. Y. (1987) Late Miocene climatic optimum in northeast Pacific USSR. *Izvestiya Acad. Nauc. SSSR, seriya geol.* 5: 38–49.

Frakes, L. A. (1979) *Climates throughout Geologic Time.* Amsterdam: Elsevier. 310 pp.

Frakes, L. A. (1986) Mesozoic-Cenozoic climatic history and the causes of glaciation. In *Mesozoic and Cenozoic Oceans*, ed. K. J. Hsu, pp. 33–48. Washington, D.C.: American Geophysical Union, Geodynamics Series 15.

Frakes, L. A. and Francis, J. E. (1988) A guide to Phanerozoic cold polar climates from high-latitude ice-rafting in the Cretaceous. *Nature* 333: 547–549.

Frakes, L. A. and Francis, J. E. (1990) Cretaceous palaeoclimates. In *Cretaceous Resources, Events and Rhythms*, ed. R. N. Ginsburg and B. Beaudoin, pp. 273–287. Dordrecht: Kluwer Academic Publishers.

Frakes, L. A. and Kemp, E. M. (1972) Influence of continental positions on early Tertiary climates. *Nature* 240: 97–100.

Francois, L. M. and Gerard, J.-C. (1986a) Reducing power of ferrous iron in the Archean ocean: 1, Concentration of photosynthetic oxygen. *Paleoceanography* 1: 355–368.

Francois, L. M. and Gerard, J.-C. (1986b) A numerical model of the evolution of ocean sulfate and sedimentary sulfur during the last 800 million years. *Geochim. Cosmochim. Acta* 50: 2289–2302.

Francis, J. E. (1983) The dominant conifer of the Jurassic Purbeck Formation. *Palaeontology* 26: 277–294.

Francis, J. E. (1984) The seasonal environment of the Purbeck (Upper Jurassic) fossil forests. *Palaeogeogr., Palaeoclimatol., Palaeoecol.* 48: 285–307.

Francis, J. E. (1986a) The calcareous paleosols of the basal Purbeck Formation (Upper Jurassic), southern England. In *Paleosols*, ed. V. P. Wright, pp. 112–138. Oxford: Blackwell.

Francis, J. E. (1986b) Growth rings in Cretaceous and Tertiary wood from Antarctica and its palaeoclimatic implications. *Palaeontology* 48: 285–307.

Francis, J. E. (1987) The palaeoclimatic significance of growth rings in late Jurassic/ early Cretaceous fossil wood from southern England. In *Applications of Tree-Ring Studies*, ed. R. G. W. Ward, pp. 21–36. British Archaeological Reports.

Francis, J. E. (1988) A 50 million year old fossil forest from Strathcona Fiord, Ellesmere Island, Arctic Canada: evidence for a warm polar climate. *Arctic* 41: 314–318.

Francis, J. E. (1990) Polar fossil forests. *Geology Today* 6: 92–95.

Francis, J. E. and McMillan, N. J. (1987) Fossil forests of the far North. *Geos* 16: 6–9.

Fuji, N. (1988) Paleovegetation and paleoclimate changes around Lake Biwa, Japan, during the last ca. 3 million years. *Quat. Sci. Rev.* 7: 21–28.

Fullerton, D. S. and Prest, V. K. (1987) Introduction: The Laurentide ice sheet and its significance. *Geog. phys. et Quater.* 41: 181–186.

Fullerton, D. S. and Richmond, G. M. (1986) Comparison of the marine oxygen isotope record, the eustatic sea level record, and the chronology of glaciation in the United States. *Quat. Sci. Rev.* 5: 197–200.

Funder, S., Abrahamsen, N., Bennike, O. and Feyling-Hansen, R. W. (1985) Forested Arctic: evidence from North Greenland. *Geology* 13: 542–546.

Gates, W. L. (1976) Modelling the ice-age climate. *Science* 191: 1138–1144.

Geigtey, J. E. (1985) Correspondence between Ordovician conodont diversity and isotopically inferred marine paleotemperatures. *Geol. Soc. Amer., Abstracts with Program.* 21: 92.

Geldsetzer, H. H. J., Goodfellow, M. J., McLaren, D. J. and Orchard, M. J. (1987) Sulfur

isotope anomaly associated with the Frasnian-Famennian extinction, Medicine Lake, Alberta, Canada. *Geology* 15: 393–396.

Geldsetzer, H. H. J., Goodfellow, M. J., McLaren, D. J. and Orchard, M. J. (1988) Sulfur isotope anomaly associated with the Frasnian-Famennian extinction, Medicine Lake, Alberta, Canada: Reply. *Geology* 16: 87–88.

Genthon, C., Barnola, J. M., Raynaud, D., Lorius, C., Jouzel, J., Barkov, N. I., Korotevitch, Y. S. and and Kotlyakov, V. M. (1987) Vostok ice core: climatic response to CO_2 and orbital forcing changes over the last climatic cycle. *Nature* 329: 414–418.

Ghil, M. (1984) Climate sensitivity, energy balance models, and oscillatory climate models. *J. Geophys. Res.* 89: 1280–1284.

Ghil, M. (1988) Deceptively simple models of climate change. In *Climate and Geosciences*, ed. A. Berger, J. Duplessy and S. H. Schneider, pp. 211–240. Dordrecht: Kluwer Academic Publishers.

Gladenkov, Y. B. (1982) Neogene climate fluctuations in Northern Kamchatka. *Doklady Acad. Nauk. Sci. SSSR, Earth Science section* 265: 407–409.

Gladenkov, Y. B., Norton, P. and Spaink, G. S. (1980) Upper Cenozoic of Iceland, Stratigraphy of Pliocene-Pleistocene and palaeontological assemblages. *Trans. Acad. Sci. USSR.* 345: 1–116.

Glennie, K. W. (1987) Desert sediment environments, present and past; summary. *Sed. Geol.* 50: 135–165.

Goldhammer, R. K., Dunn, P. A. and Hardie, L. A. (1987) High frequency glacio-eustatic sea level oscillations with Milankovitch characteristics recorded in middle Triassic platform carbonates in northern Italy. *Am. J. Sci.* 287: 853–892.

Golovchenko, X. (1987) Milankovitch cycles detected from wireline logs: results from ODP Site 693, Antarctic continental margin (abs.). *Geol. Soc. Amer., Abstracts with Program* 19: 725.

Gradstein, F. M. and Sheridan, R. E. (1983) On the Jurassic Atlantic Ocean and a synthesis of results of DSDP Project Leg 76. *Init. Repts. DSDP* 76: 913–943. Washington (Govt. Printing Office).

Gregory, R. T., Douthitt, C. B., Duddy, I. R., Rich, P. V. and Rich, T. H. (1989) Oxygen isotope composition of carbonate concretions from the lower Cretaceous of Victoria, Australia: implications for the evolution of meteoritic waters on the Australian continent in a paleopolar environment. *Earth Planet. Sci. Letters.* 92: 27–42.

Grosvald, M. G. (1979) The last Eurasian ice sheet. *Polar Geogr. Geol.* 3: 169–188.

Grosvald, M. G. and Vozovik, Y. N. (1984) A marine ice cap in south Beringia (a working hypothesis). *Polar Geogr. Geol.* 8: 128–146.

Grosvald, M. G., Muratova, M. V. and Shisharine, Z. G. (1985) Climatic effects of late glacial surges based on the example of the 10.5 kyr BP cold phase. *Polar Geogr. Geol.* 9: 244–251.

Grube, F., Christensen, S., Vollmer, I., Duphorn, K., Klostermann, J. and Menke, B. (1986) Glaciations in North West Germany. *Quat. Sci. Rev.* 5: 347–358.

Gudjonsson, L. (1987) Local and global effects on the early Pliocene Mediterranean stable isotope records. *Mar. Micropalaeo.* 12: 241–254.

Hald, M. and Vorren, T. O. (1987) Stable isotope stratigraphy and paleoceanography

during the last deglaciation on the continental shelf off Troms, northern Norway. *Paleoceanography* 2: 583–600.

Hallam, A. (1975) *Jurassic Environments.* Cambridge: Cambridge University Press. 269 pp.

Hallam, A. (1984a) Continental humid and arid zones during the Jurassic and Cretaceous. *Palaeogeogr., Palaeoclimatol., Palaeoecol.* 47: 195–223.

Hallam, A. (1984b) Pre-Quaternary sea level changes. *Ann. Rev. Earth. Planet. Sci.* 12: 205–243.

Hallam, A. (1985) A review of Mesozoic climates. *J. Geol. Soc. Lond.* 142: 433–445.

Hallam, A. (1986a) The origin of minor limestone/shale cycles. *Geology* 14: 609–612.

Hallam, A. (1986b) Role of climate in affecting late Jurassic and early Cretaceous sedimentation in the North Atlantic. In *North Atlantic Palaeoceanography*, ed. C. P. Summerhayes and N. Shackleton, pp. 277–281. London: Geological Society of London, Special Publication.

Hallam, A. (1987a) End Cretaceous mass extinction event: argument for terrestrial causation. *Science* 238: 1237–1242.

Hallam, A. (1987b) Mesozoic marine organic shales. In *Marine Petroleum Source Rocks*, ed. J. Brooks and A. J. Fleet, pp. 251–262. Geological Society of London.

Hallberg, G. R. (1986) Pre-Wisconsin glacial stratigraphy of the Central Plains region in Iowa, Nebraska, Kansas. *Quat. Sci. Rev.* 5: 11–15.

Hambrey, M. J. (1985) The late Ordovician-early Silurian glacial period. *Palaeogeogr., Palaeoclimatol., Palaeoecol.* 51: 273–289.

Hambrey, M. J. and Harland, W. B. (1981) *Earth's Pre-Pleistocene Glacial Record.* Cambridge: Cambridge University Press. 1003 pp.

Hambrey, M. J. and Kluyver, H. M. (1981) Evidence of Devonian or Early Carboniferous glaciation in the Agades region of Nigeria. In *Earth's Pre-Pleistocene Glacial Record*, ed. M. J. Hambrey and W. B. Harland, pp. 188–190. Cambridge: Cambridge University Press.

Hambrey, M. J., Larsen, B., Ehrmann, W. U. *et al.* (1989) Forty million years of Antarctic glacial history yielded by Leg 119 of the ODP. *Polar Record* 25: 99–106.

Hamilton, T. D. (1986) Correlation of Quaternary glacial deposits in Alaska. *Quat. Sci. Rev.* 5: 171–180.

Hancock, J. M. and Kauffman, E. G. (1979) The great transgressions of the Late Cretaceous. *J. Geol. Soc. Lond.* 136: 175–186.

Hansen, J., Lacis, A., Rind, D., Russell, G., Stone, P., Fung, I., Rendy, R. and Lerner, J. (1984) Climate sensitivity: Analysis of feedback mechanisms. In *Climate Processes and Climate Sensitivity*, ed. J. E. Hansen and T. Takahashi, pp. 130–163. Washington, D.C.: American Geophysical Union.

Haq, B. U. (1982) Climatic acme events in the sea and on the land. In *Climates in Earth History*, ed. W. H. Berger and J. C. Crowell, pp. 126–132. Washington, D.C.: National Academy Press.

Haq, B. U. (1984) Paleoceanography: A synoptic overview of 200 million years of ocean history. In *Marine Geology and Palaeoceanography of the Arabian Sea and Coastal Pakistan*, ed. B. U. Haq and J. D. Milliman, pp. 201–231. New York: Van Nostrand Reinhold Company.

Haq, B. U., Hardenbol, J. and Vail, P. R. (1987) Chronology of fluctuating sea levels

since the Triassic (250 million years ago to present). *Science* 235: 1156–1167.

Harland, W. B. and Rudwick, M. J. S. (1964) The great Infra-Cambrian ice age. *Sci. Am.* 211: 28–36.

Harris, T. M. (1975) *The Yorkshire Jurassic Flora.* London: British Museum Natural History. 166 pp.

Hart, M. H. (1978) The evolution of the atmosphere of Earth. *Icarus* 33: 23 –39.

Hartmut, H. (1988) Origin and consequences of cyclic ice rafting in the northeast Atlantic during the past 130,000 years. *Quat. Res.* 29: 142–153.

Harvey, D. L. (1988) Climatic impact of ice-age aerosols. *Nature* 334: 333–335.

Harwood, D. M. (1986) Do diatoms beneath the Greenland ice sheet indicate interglacials warmer than present? *Arctic* 39: 304–308.

Harwood, D. M. (1987) Cenozoic Antarctic temperature, ice-volume and bottom water events, a comparison of Sirius Formation microfossil data to eustatic and oxygen isotope data (abs.). *Geol. Soc. Amer., Abstracts with Program.* 19: 695.

Hay, W. W. (1985) Potential errors in estimates of carbonate rock acumulating through geologic time. In *The Carbon Cycle and Atmospheric CO_2: Natural Variations Archean to Present,* ed. E. T. Sundquist and W. S. Broecker, pp. 573–583. American Geophysical Union, Geophysical Monograph 32.

Hay, W. W., Behensky, J. F., Barron, E. J. and Sloan, J. L. I. (1982) Late Triassic-Liassic paleoclimatology of the proto-central North Atlantic rift system. *Palaeogeogr., Palaeoclimatol., Palaeoecol.* 40: 13–30.

Hay, W. W., McClain, M. and Sloan, J. L. (1988) Fluxes to the ocean: Holocene, glacial and Pleistocene (abs.). *Eos, Trans. Amer. Geophys. Union* 69: 381.

Hayes, D. E. and Frakes, L. A. (1975) General synthesis. *Init. Repts. DSDP* 28: 919–942. Washington (Govt. Printing Office).

Hays, J., McIntyre, A., Bloom, A. and Duplessy, J.-C. (1984) Marine Pleistocene records of climatic response. In *Milankovitch and Climate,* ed. A. L. Berger, J. Imbrie, J. Hays, G. Kukla and B. Saltzman, pp. 831–835. Dordrecht: Reidel.

Hays, J. D., Imbrie, J. and Shackleton, N. J. (1976) Variations in the Earth's orbit: Pacemaker of the Ice Ages. *Science* 194: 1121–1132.

Hays, J. D. and Pittman, W. C. (1973) Lithospheric plate motion, sea level changes, and climatic and ecological consequences. *Nature* 246: 18–22.

Hecht, A. D. (1985) Paleoclimatology: A retrospective of the last 20 years. In *Paleoclimate Analysis and Modeling,* ed. A. D. Hecht, pp. 1–26. New York: John Wiley and Sons.

Heckel, P. H. (1986) Sea-level curve for Pennsylvanian eustatic marine transgressive-regressive depositional cycles along mid-continent outcrop belt, North America. *Geology* 14: 330–334.

Heckel, P. H. and Witzke, B. J. (1979) Devonian world palaeogeography determined from the distribution of carbonates and related lithic palaeoclimatic indicators. In *The Devonian System,* ed. M. R. House, C. T. Scrutton and M. G. Bassett, pp. 99–123. London: Palaeontological Association.

Hein, J. R. and Parrish, J. T. (1987) Distribution of siliceous deposits in space and time. In *Siliceous Sedimentary Rock-Hosted Ores and Petroleum,* ed. J. R. Hein, pp. 10–57. New York: Van Nostrand Reinhold.

Henderson-Sellers, A. and Meadows, A. J. (1975) Long-term variations in the albedo

and surface temperature of the Earth. *Nature* 254: 44.

Herbert, T. D. and Fischer, A. G. (1986) Milankovitch climatic origin of mid-Cretaceous black shales rhythms in central Italy. *Nature* 321: 739–743.

Herman, Y. and Hopkins, D. M. (1980) Arctic ocean climate in late Cenozoic time. *Science* 209: 557–562.

Hill, R. S. (1989) Fossil leaf. In *Antarctic Cenozoic History from the CSIROS-1 Drillhole, McMurdo Sound.*, ed. P. J. Barrett, *DSIR Bulletin*, 245: 143–144.

Hills, L. V., Klovan, J. E. and Sweet, A. R. (1974) *Juglans eocinerea* n.sp., Beaufort Formation (Tertiary), southwestern Banks Island, Arctic Canada. *Can. J. Bot.* 52: 65–90.

Hodell, D. A., Elmstrom, K. M. and Kennett, J. P. (1986) Late Miocene benthic $\delta^{18}O$ changes, global ice volume, sea level and the 'Messinian salinity crisis'. *Nature* 320: 411–414.

Hodell, D. A. and Kennett, J. P. (1986) Late Miocene–Early Pliocene stratigraphy of the south Atlantic and southwest Pacific Oceans. *Paleoceanography* 1: 285–311.

Hodell, D. A., Kennett, J. P. and Leonard, K. A. (1983) Climatically induced changes in vertical water mass structure of the Vema Channel during the Pliocene: Evidence from DSDP Holes 516A, 517, and 518. *Init. Repts. DSDP* 72: 907–919. Washington (Govt. Printing Office).

Hodell, D., Williams, D. and Kennett, J. P. (1985) Late Pliocene reorganization of deep vertical water mass structure in the western South Atlantic: Faunal and isotopic evidence. *Bull. Geol. Soc. Amer.* 96: 495–503.

Hoffman, A., Gruszczynski, M. and Malkowski, K. (1989) Stable isotope records of Permian–Triassic crisis and geosphere-biosphere interactions (abs.). 28th International Geological Congress. Washington DC. 2: 63.

Hollin, J. T. (1962) On the glacial history of Antarctica. *J. Glaciol.* 4: 247–280.

Hollin, J. T. (1980) Climate and sea level change in isotope stage 5: An East Antarctic ice surge at 95,000 BP? *Nature* 283: 629–633.

Holland, C. H. (1971) *Lower Paleozoic Rocks of the World. 1. Cambrian of the New World.* London: John Wiley and Sons. 456 pp.

Holland, H. D. (1984) *The Chemical Evolution of the Atmosphere and Oceans.* Princeton: Princeton University Press. pp. 582.

Holser, W. T. (1984) Gradual and abrupt shifts in ocean chemistry during Phanerozoic time. In *Patterns of Change in Earth Evolution*, ed. H. D. Holland and A. F. Trendall, pp. 123–143. New York: Springer–Verlag.

Holser, W. T. and Magaritz, M. (1987) Events near the Permian–Triassic boundary. *Modern Geology* 11: 155–180.

Holser, W. T., Magaritz, M. and Clark, D. L. (1986) Carbon isotope stratigraphic correlation in the late Permian. *Amer. J. Sci.* 286: 390–402.

Holser, W. T., Schonlaub, H. P., Attrep, M., Boeckelmann, K., Klein, P., Margaritz, M., Orth, C. J., Fenninger, A., Jenny, C., Kralik, M., Mauritsh, H., Pak, E., Schramm, J. M., Stattegger, K. and Schmoller, R. (1989) A unique geochemical record at the Permian–Triassic boundary. *Nature* 337: 39–43.

Hooghiemistra, H. (1984) A palynological registration of climatic change of the last 3.5 Ma. In *Milankovitch and Climate*, ed. A. L. Berger, J. Imbrie, J. Hays, G. Kukla and B. Saltzman, pp.371–377. Dordrecht: Reidel.

Hooghiemistra, H., Bechler, A. and Beug, H.-J. (1987) Isopollen maps for 18,000 yr BP of the Atlantic offshore of northwest Africa: evidence for palaeo-wind circulation. *Paleoceanography* 2: 561–582.

Horrell, M. A. (1987) Apparent oceanic O^{18} fluctuations and their effect on paleo-sea surface temperature estimates: Cretaceous to present (abs.). *Eos, Trans. Amer. Geophys. Union.* 68: 333.

Houghton, S. D. and Jenkins, D. G. (1988) Subtropical microfossil indicators from the late Pliocene Celtic Sea. *Marine Geol.* 79: 119–126.

Hovan, B. A., Rea, D. K. and Pisias, N. G. (1988) Late Pleistocene eolian and $CaCO_3$ deposits in the Northwest Pacific. (abs.). *Eos, Trans. Amer. Geophys. Union.* 69: 382.

Howard, W. R. and Prell, W. L. (1984) A comparison of radiolarian and foraminiferal paleoecology in the Southern Indian Ocean: New evidence for the inter-hemispheric timing of climatic change. *Quat. Res.* 21: 244–263.

Hsu, K. J. (1986) Cretaceous/Tertiary boundary event. In *Mesozoic and Cenozoic Oceans*, ed. K. J. Hsu, pp. 75–84. Washington, D.C.: American Geophysical Union Geodynamics Series.

Hsu, K. J. and McKenzie, J. A. (1985) A "Strangelove" ocean in the earliest Tertiary. In *The Carbon Cycle and Atmospheric CO_2: Natural Variations Archean to Present*, ed. E. T. Sundquist and W. S. Broecker, pp. 487–492. Washington, D.C.: American Geophysical Union, Geophysical Monograph 32.

Hsu, K. J., McKenzie, J. A. and Weissert, H. J. (1986) Cenozoic carbon-isotope record in South Atlantic sediments. In *South Atlantic Paleoceanography*, ed. K. J. Hsu and H. J. Weissert, pp. 188–198. Cambridge: Cambridge University Press.

Hsu, K. J., Oberhansli, H., Gao, J. Y., Shu, S., Haihang, Ch. and Krakenbühl, V. (1985) "Strangelove ocean" before the Cambrian explosion. *Nature* 316: 809–811.

Hubbard, R. N. L. B. and Boulter, M. C. (1983) Reconstruction of Paleogene climate from palynological evidence. *Nature* 301: 147–150.

Hudson, J. D. and Anderson, T. F. (1989) Ocean temperatures and isotopic compositions through time. *Trans. Roy. Soc. Edinburgh.* 80: 183–192

Hughes, T. J. (1982) Did the west Antarctic ice sheet create the east Antarctic ice sheet? *Annals of Glaciology* 3: 138–145.

Hughes, T. J. (1987) The marine ice transgression hypothesis. *Geogr. Ann.* 69A: 237–249.

Hughes, T. J., Denton, G. M., Anderson, B. G., Schiling, D. M., Fasthook, J. L. and Lingle, C. S. (1981) The last great ice sheets: A global view. In *The Last Great Ice Sheets*, ed. G. H. Denton and T. J. Hughes, pp. 263–317. New York: Wiley–Interscience.

Hunt, B. G. (1984) Polar glaciation and the genesis of ice ages. *Nature* 308: 48–50.

Hyde, W. T. and Peltier, W. R. (1987) Sensitivity experiments with a model of the ice age cycle: the response to Milankovitch forcing. *J. Atmos. Sci.* 44: 1351–1374.

Ibaraki, M. (1987) Planktonic foraminiferal event stratigraphy of the Neogene Kakegawa Basin, the Pacific coast of central Japan (abs.). 4th International Congress on Pacific Neogene Stratigraphy. ed. J. A. Barron and J. R. Blueford. Berkeley, California. Abstract Volume: 48–49.

Imbrie, J. (1985) A theoretical framework for the Pleistocene ice ages. *J. Geol. Soc.*

London. 142: 417–432.

Imbrie, J. (1987) Abrupt terminations of late Pleistocene ice ages: a simple Milanko-vitch explanation. In *Abrupt Climatic Change*, ed. W. H. Berger and L. D. Labeyrie, pp. 365–367. Dordrecht: D. Reidel.

Imbrie, J. (1988) Climatic response to radiation forcing over the Milankovitch frequency band (abs.). *Eos, Trans Amer. Geophys. Union*. 69: 299.

Imbrie, J., Hayes, J., Martinson, D., McIntyre, A., Mix, A., Horley, J., Pisias, N., Prell, W. and Shackleton, N. (1984) The orbital theory of Pleistocene climate: support from a revised chronology of the marine $\delta^{18}O$ record. In *Milankovitch and Climate*, ed. A. Berger, J. Imbrie, J. Hays, G. Kukla and B. Saltzman, pp. 269–306. Dordrecht: Reidel.

Imbrie, J. and Imbrie, J. Z. (1980) Modeling the climatic response to orbital variations. *Science* 207: 943–953.

Imbrie, J., McIntyre, A., Pisias, M., Shackleton, N., Ruddiman, W., Mix, A., Horley, J., Prell, W., Howard, W. and Boyle, E. (1986) *Global patterns in Pleistocene Climate Records*. 2nd International Conference on Paleoceanography, Abstracts with program. Woods Hole, Massachusetts: Woods Hole Oceanographic Institution.

Irving, E., North, F. K. and Couillard, R. (1974) Oil, climate and tectonics. *Can. J. Earth Sci.* 11: 1–15.

Ives, J. D., Andrews, J. T. and Barry, R. G. (1975) Growth and decay of the Laurentide Ice Sheet and comparison with Fenno-Scandinavia. *Natturwissenschaften*. 62: 118–25.

Janecek, T. R. and Rea, D. (1985) Quaternary fluctuations in the Northern Hemisphere tradewinds and westerlies. *Quat. Res.* 24: 150–163.

Janecek, T. R. and Rea, D. (1984) Pleistocene fluctuations in Northern Hemisphere tradewinds and westerlies. In *Milankovitch and Climate*, ed. A. L. Berger, J. Imbrie, J. Hays, G. Kukla and B. Saltzman, pp. 331–347. Dordrecht: D. Reidel.

Jansen, E. and Sejrup, H. P. (1987) Stable isotope stratigraphy and amino-acid epimerization for the last 2.4 Ma at Site 610, Holes 610 and 610A. *Init. Repts. DSDP*. 94: 879–887. Washington (Govt. Printing Office).

Jansen, J. H. F., Woensdegt, C. F., Kooistra, M. J. and van der Gaast, S. J. (1987) Ikaite pseudomorphs in the Zaire deep-sea fan: An intermediate between calcite and porous calcite. *Geology* 15: 245–248.

Jefferson, T. H. (1982) Fossil forests from the Lower Cretaceous of Alexander Island, Antarctica. *Palaeontology* 25: 681–708.

Jefferson, T. H. and Taylor, T. N. (1983) Permian and Triassic woods from the Transantarctic Mountains: Paleoenvironmental indicators. *Ant. J. US* 18: 55–57.

Jell, P. A. (1974) Faunal provinces and possible planetary reconstruction of the middle Cambrian. *J. Geol.* 82: 319–350.

Jell, P. A. and Roberts, J. (1986) Plants and invertebrates from the Lower Cretaceous Koonwarra Fossil Bed, South Gippsland, Victoria. *Ass. Australas. Palaeontols., Memoir* 3: 1–205.

Jenkins, D. G., Bowen, D. Q., Adams, C. G., Shackleton, N. J. and Brassell, S. C. (1985) The Neogene: Part 1. *Mem. Geol. Soc. London* 10: 199–210.

Jenkyns, H. C. (1980) Cretaceous anoxic events: From continents to oceans. *J. Geol.*

Soc. Lond. 137: 171–188.

Jenkyns, H. C. (1988) The Early Toarcian (Jurassic) Anoxic Event: stratigraphic, sedimentary and geochemical evidence. *Amer. J. Sci.* 288: 101–151.

Jenkyns, H. C. and Clayton, J. C. (1986) Black shales and carbon isotopes in pelagic sediments from the Tethyan lower Jurassic. *Sedimentology* 33: 87–106.

Johnson, D. A. (1983) Palaeocirculations of the South Atlantic. *Init. Repts. DSDP.* 72: 977–994. Washington (Govt. Printing Office).

Johnson, D. A. (1985) Abyssal teleconnections II. Initiation of the Antarctic bottom water flow in the southwestern Atlantic. In *South Atlantic Paleoceanography*, ed. K. J. Hsu and H. J. Weissert, pp. 243–281. Cambridge: Cambridge University Press.

Johnson, J. G., Klapper, G. and Sandberg, C. A. (1985) Devonian eustatic fluctuations in Euramerica. *Geol. Soc. Amer. Bull.* 96: 567–587.

Johnson, R. G. (1982) Bruhnes-Matuyama magnetic reversal dated at 790,000 yr BP by marine-astronomical correlations. *Quat. Res.* 17: 135–147.

Johnson, R. G. (1984) Summer temperature variations in the Arctic during interglacials. In *Milankovitch and Climate*, ed. A. L. Berger, J. Imbrie, J. Hays, G. Kukla and B. Saltzman, pp. 715–720. Dordrecht: D. Reidel.

Johnson, R. G. and Andrew, J. T. (1986) Glacial terminations in the oxygen isotope record of deep sea cores: Hypothesis of massive Antarctic ice-shelf destruction. *Palaeogeogr., Palaeoclimatol., Palaeoecol.* 53: 107–138.

Jones, G. N. (1987) The timing of the last deglaciation in the Northern Norwegian-Greenland Sea: Direct dating by C^{14} accelerator mass spectrometry (abs.). *Eos, Trans. Amer. Geophys. Union.* 68: 1331.

Jouzel, J., Lorius, C., Petit, J. R., Genthon, C., Barkov, M. I., Korathevitch, Y. S. and Kotlyakov, V. M. (1988) Climatic interpretation of a continuous deuterium profile obtained from the Vostok ice core, Antarctica (160,000 years). *Annals of Glaciology* 10: 206–207.

Jouzel, J., Lorius, C., Petit, J. R., Genthon, C., Barkov, M. I., Kotlyakov, V. M. and Petrov, V. M. (1987) Vostok ice core: A continuous isotope temperature record over the last climatic cycle (160,000 years). *Nature* 329: 403–408.

Juignet, P. (1980) Transgressions-regressions, variations eustatiques et influences tectoniques de l'Aptien au Maastrichtian dans le basin de Paris Occidental et sur la bordure au Massif Amoricain. *Cretaceous Research.* 1: 341–357.

Kallel, N., Labeyrie, L., Juillet–Leclerc, A. and Duplessy, J.-C. (1988) A deep hydrological front between intermediate and deep-water masses in the glacial Indian Ocean. *Nature* 333: 651–655.

Kanari, S., Fuji, N. and Horie, S. (1984) The paleoclimatological constituents of paleotemperature in Lake Biwa. In *Milankovitch and Climate*, ed. A. L. Berger, J. Imbrie, J. Hays, G. Kukla and B. Saltzman, pp. 405–414. Dordrecht: D. Reidel.

Karhu, J. and Epstein, S. (1986) The implication of the oxygen isotope records in coexisting cherts and phosphates. *Geochim. Cosmochim. Acta.* 50: 1745–1756.

Kasting, J. F. (1987) Theoretical constraints on oxygen and carbon dioxide concentrations in the Precambrian atmosphere. *Precambrian Res.* 34: 205–229.

Kasting, J. F. (1989) Long-term stability of the Earth's climate. *Glob. Planet. Change.*

1: 83 –96.

Kauffman, E. G. (1977) *Cretaceous facies, faunas and palaeoenvironments across the Western Interior Basin.* Field Guide. Boulder, Colorado: Rocky Mountain Association of Geologists. pp. 75–100.

Kauffman, E. G. (1984) The fabric of Cretaceous marine extinctions. In *Catastrophes in Earth History*, ed. W. A. Bergren and J. A. Van Couvering, pp. 151–245. Princeton: Princeton University Press.

Keany, J. (1978) Palaeoclimatic trends in early and middle Pliocene deep-sea sediments of the Antarctic. *Marine Micropalaeo.* 3: 35–49.

Keffer, T., Martinson, D. G. and Corliss, B. H. (1988) The position of the Gulf Stream during Quaternary glaciations. *Science* 241: 440–442.

Keigwin, L. D. (1979) Late Cenozoic stable isotope stratigraphy and paleoceanography of DSDP sites from the east equatorial and central North Pacific Ocean. *Earth Planet. Sci. Letters.* 45: 361–382.

Keigwin, L. D. (1986) North Atlantic record of late Neogene climatic change. *S. Afr. J. Sci.* 82: 499–500.

Keigwin, L. D. (1987a) North Pacific deep water formation during the latest deglaciation. *Nature* 330: 362–363.

Keigwin, L. D. (1987b) Pliocene stable-isotope record of DSDP Site 606: sequential events of ^{18}O enrichment beginning at 3.1 Ma. *Init.Repts.DSDP* 94: 911–917. Washington (Govt. Printing Office).

Keigwin, L. D. (1987c) Towards a high-resolution chronology for latest Miocene paleoceanographic events. *Paleoceanography* 2: 639–660.

Keigwin, L. D., Aubury, M.-P. and Kent, D. V. (1987) North Atlantic late Miocene stable isotope stratigraphy, biostratigraphy, and magnetostratigraphy. *Init. Repts. DSDP* 94: 935–956. Washington (U.S. Govt. Printing Office).

Keigwin, L. D. and Boyle, E. A. (1985) Carbon isotopes in deep-sea benthic foraminifera: Precession and changes in low-latitude biomass. In *The Carbon Cycle and Atmospheric CO_2: Natural Variations Archean to Present.*, ed. E. T. Sundquist and W. S. Broecker, pp. 319–328. Washington, D.C.: American Geophysical Union, Geophysical Monograph 32.

Keigwin, L. D. and Jones, G. A. (1988) Global evidence for abrupt climate events in the marine record of deglaciation (abs.). *Eos, Trans. Amer. Geophys. Union.* 69: 299.

Keigwin, L. D. and Thunell, R. C. (1979) Middle Pliocene climatic change in the western Mediterranean from faunal and oxygen isotopic trends. *Nature* 282: 294–296.

Keir, R. S. (1988) On the late Pleistocene ocean geochemistry and circulation. *Paleoceanography* 3: 413–445.

Keith, M. L. (1982) Violent volcanism, stagnant oceans and some inferences regarding petroleum, strata-bound ores and mass extinctions. *Geochim. Cosmochim. Acta.* 46: 2621–2637.

Keller, G. (1983) Palaeoclimatic analysis of Middle Eocene through Oligocene planktic foraminiferal faunas. *Palaeogeogr., Palaeoclimatol., Palaeoecol.* 43: 73–94.

Keller, G. (1986) Stepwise mass extinctions and impact events: Late Eocene to early Oligocene. *Mar. Micropalaeo.* 10: 267–294.

Keller, G., Herbert, T., D'Hondt, S., Johnsson, M. and Chi, W. R. (1987) Global distribution of late Paleogene hiatuses. *Geology* 15: 199–203.

Kemp, E. M. and Barrett, P. J. (1975) Antarctic glaciation and early Tertiary vegetation. *Nature* 258: 507–508.

Kemp, E. M., Balme, B. E., Helby, R. J., Kyle, R. A., Playford, G. and Price, P. L. (1977) Carboniferous and Permian palynostratigraphy in Australia and Antarctica: A review. *Australian Bur. Min. Resources, Jour. Aust. Geol. Geophys.* 2: 177–208.

Kemper, E. (1987) Das Klima der Kreide-Zeit. *Geol. Jb.* A96: 5–185.

Kennett, J. P. (1977) Cenozoic evolution of Antarctic glaciation, the circum-Antarctic ocean, and their impact on global palaeoceanography. *J. Geophys. Res.* 82: 3843–3860.

Kennett, J. P. (1982) *Marine Geology.* Englewood Cliffs, New Jersey: Prentice–Hall. 813 pp.

Kennett, J. P. (1983) Paleoceanography: Global ocean evolution. *Rev. Geophys. Space Phys.* 21: 1258–1274.

Kennett, J. P. (1985) Neogene paleoceanography and plankton evolution. *S. Afr. J. Sci.* 81: 1251–1253.

Kennett, J. P. (1986) Miocene to early Pliocene oxygen and carbon isotope stratigraphy in the southwest Pacific, DSDP Leg 90. *Init. Repts. DSDP* 90: 1388–1412. Washington (US Govt. Printing Office).

Kennett, J. P. and Hodell, D. A. (1986) Major events in Neogene oxygen isotopic records. *S. Afr. J. Sci.* 82: 497–498.

Kennett, J. P., Houtz, R. E., Andrew, P. B., Edwards, A. R., Gostin, V. A., Hajos, M., Hampton, M., Jenkins, D. G., Margolis, S. V., Ovenshine, A. F. and Perch-Nielsen, K. (1975) Cenozoic paleoceanography in the southwest Pacific Ocean, Antarctic glaciation and the development of the Circum-Antarctic current. *Init. Repts. DSDP* 29: 1155–1169. Washington (Govt. Printing Office).

Kennett, J. P., Shackleton, N. J., Margolis, S. V., Goodney, D. E., Dudley, W. C. and Kroopotnick, P. M. (1979) Late Cenozoic oxygen and carbon isotopic history and volcanic ash history: DSDP Site 284, South Pacific. *Amer. J. Sci.* 279: 52–69.

Kennett, J. P. and von der Borch, C. C. (1986) Southwest Pacific Cenozoic paleoceanography. *Init. Repts. DSDP* 90: 1493–1517.Washington (Govt. Printing Office).

Kerr, R. A. (1987) Ocean drilling details steps to an icy world. *Science* 236: 912–913.

Kerr, R. A. (1988) Linking Earth, ocean, and air at the AGU. *Science* 232: 259–260.

Killingley, J. S. (1983) Effects of diagenetic recrystallization on O^{18}/O^{16} values of deep-sea sediments. *Nature* 301: 594–597.

Kimura, T. and Sekido, S. (1975) *Nilssoniocladus* n. gen. (Nilssoniaceae n. fam.), newly found from the early Lower Cretaceous of Japan. *Palaeontographica, B.* 153: 111–118.

Klein, G. de V. (1990) Pennsylvanian time scales and cycle periodicity. *Geology* 18: 455–457.

Kohl, H. (1986) Pleistocene glaciations in Austria. *Quat. Sci. Rev.* 5: 421–428.

Kolodny, Y. and Raab, M. (1988) Oxygen isotopes in phosphatic fish remains from

Israel: Paleothermometry of tropical Cretaceous and Tertiary shelf waters. *Palaeogeogr., Palaeoclimatol., Palaeoecol.* 64: 59–67.

Kondratiev, K. Y. and Moskalenko, N. I. (1984) The role of carbon dioxide and other minor gaseous components and aerosols in the radiation budget. In *The Global Climate*, ed. J. T. Houghton, pp. 225–233. New York: Cambridge University Press.

Krassilov, V. A. (1973) Climate changes in eastern Asia as indicated by fossil floras. 1 Early Cretaceous. *Palaeogeogr., Palaeoclimatol., Palaeoecol.* 13: 261–275.

Krassilov, V. A. (1981) Changes in Mesozoic vegetation and the extinction of the dinosaurs. *Palaeogeogr., Palaeoclimatol., Palaeoecol.* 34: 207–224.

Kruck, W. and Thiele, J. (1983) Late Palaeozoic glacial deposits in the Yemen Arab Republic. *Geol. Jahr.* 46: 3–29.

Kuhn, W. R., Walker, J. C. G. and Marshall, H. G. (1989) The effect on Earth's surface temperature from variations in rotation rate, continent formation, solar luminosity, and carbon dioxide. *J. Geophys. Res.* 94: 11129–11136.

Kukla, G. (1977) Pleistocene land-sea correlations; 1 Europe. *Earth Sci. Rev.* 13: 307–374.

Kukla, G. (1987) Loess stratigraphy in Central China. *Quat. Sci. Rev.* 6: 191–219.

Kukla, G. and Aharon, P. (1984) Plio-Pleistocene land records. In *Milankovitch and Climate*, ed. A. L. Berger, J. Imbrie, J. Hays, G. Kukla and B. Saltzman, pp. 835–838. Dordrecht: D. Reidel.

Kukla, G., Heller, F., Lin, X. M., Xu, T. C., Liu, T. S. and An, Z. S. (1988) Pleistocene climate in China dated by magnetic susceptibility. *Geology* 16: 811–814.

Kump, L. R. (1989) Alternative modeling approaches to the geochemical cycles of carbon, sulfur, and strontium isotopes. *Amer. J. Sci.* 289: 390–410.

Kutzbach, J. E. (1981) Monsoon climate of the early Holocene; Climatic experiment using the Earth's orbital parameters for 9,000 years ago. *Science* 214: 59–61.

Kutzbach, J. E. (1983) Monsoonal rains of the late Pleistocene and early Holocene; Patterns, intensity, and possible causes of changes. In *Variations in the Global Water Budget*, ed. F. A. Street–Perrott, M. Beran and R. A. S. Ratcliffe, pp. 371–389. Dordrecht: D. Reidel.

Kutzbach, J. E. (1984) The sensitivity of monsoon climates to orbital parameter changes for 9,000 yr BP; Experiments with the NCAR GCM. In *Milankovitch and Climate*, ed. A. L. Berger, J. Imbrie, J. Hays, G. Kukla and B. Saltzman, pp. 418–429. Dordrecht: D. Reidel.

Kutzbach, J. E. (1985) Modeling of paleoclimates. *Adv. Geophys.* 28A: 159–196.

Kutzbach, J. E. (1987) Model simulations of the climatic patterns during the deglaciation of North America. In *North America and Adjacent Oceans During the Last Deglaciation*, ed. W. F. Ruddiman and H. E. Wright Jr., pp. 425–446. Boulder: Geol. Soc. Amer.

Kutzbach, J. E. and Gallimore, R. G. (1989) Pangaean climates: Megamonsoons of the megacontinent. *J. Geophys. Res.* 94: 3341–3357.

Kutzbach, J. E., Guetter, P. J. and Washington, W. M. (1990) Simulated circulation of an idealized ocean for Pangaean time. *Paleoceanography* 5: 299–317.

Kutzbach, J. E. and Guetter, P. J. (1986) The influence of changing orbital parameters and surface boundary conditions on climate simulation for the past 18,000 years. *J. Atmos. Sci.* 43: 1726–1759.

Kutzbach, J. E. and Otto-Bliesner, B. L. (1982) The sensitivity of the African-Asian monsoonal climate to orbital parameters changes for 9,000 years BP in a low-resolution general circulation model. *J. Atmos. Sci.* 39: 1177–1188.

Kutzbach, J. E. and Street-Perrot, F. A. (1985) Milankovitch forcing of fluctuations in the tropical lakes for 18 Kyrs BP. *Nature* 317: 130–134.

Kutzbach, J. E. and Wright, H. E. J. (1985) Simulation of the climate of 18,000 yr BP; Results for the North American, North Atlantic/European Sector. *Quat. Sci. Rev.* 4: 147–187.

Kvasov, D. P. and Verbitsky, M. Y. (1981) Causes of Antarctic glaciation in the Cenozoic. *Quat. Res.* 15: 1–17.

Labeyrie, L. D., Duplessy, J.-C. and Blanc, P. L. (1987) Variations in the mode of formation and temperature of oceanic deep waters over the past 125,000 years. *Nature* 327: 477–482.

Labeyrie, L. D., Pichon, J.-J., Labracherie, M., Ippolito, P., Duprat, J. and Duplessey, J.-C. (1986) Melting history of Antarctica during the past 60,000 years. *Nature* 322: 701–706.

Labracherie, M., Bard, M. E., Labeyrie, L., Pichon, J.-J., Arnold, M., Duprat, J. and Duplessey, J.-C. (1988) First direct dating of the last deglaciation in the South Ocean 14C accelerator mass spectrometry (abs.). *Eos, Trans. Amer. Geophys. Union.* 69: 382.

Lasaga, A. C., Berner, R. A. and Garrels, R. M. (1985) An improved geochemical model of atmospheric CO_2 fluctuations over the past 100 million years. In *The Carbon Cycle and Atmospheric CO_2: Natural Variations Archean to Present.*, ed. E. T. Sundquist and W. S. Broecker, pp. 397–411. Washington, D.C.: American Geophysical Union, Geophysical Monograph 32.

Laukhin, S. A. (1986) First climatic traces of Cenozoic glaciation in northeastern Asia. *Doklady Akademii Nauk SSSR.* 291: 1445–1449.

Lawver, L. A., Schlater, J. G. and Meinke, L. (1985) Mesozoic and Cenozoic reconstructions of the South Atlantic. *Tectonophysics* 114: 233–254.

Le Treut, H., Portes, J., Jouzel, J. and Ghil, M. (1988) Isotopic modeling of climatic oscillations: Implication for a comparative study of marine and ice-core records. *J. Geophys. Res.* 93: 9365–9383.

Leblanc, M. (1981) The Late Precambrian Tiddiline tilloid of the Anti-Atlas, Morroco. In *Earth's pre-Pleistocene Glacial Record*, ed. M. J. Hambrey and W. B. Harland, pp. 120–122. Cambridge: Cambridge University Press.

Ledbetter, M. T., Williams, D. F. and Ellwood, B. B. (1978) Late Pliocene climate and southwest Atlantic abyssal circulation. *Nature* 272: 237–239.

Leggett, J. K., McKerrow, W. S., Cocks, L. R. M. and Rickards, R. B. (1981) Periodicity in the early Palaeozoic marine realm. *J. Geol. Sci.* 138: 167–176.

Legrand, M. R., Delmas, R. J. and Charlson, R. J. (1988) Climate forcing implications from Vostock ice-core sulphate data. *Nature* 334: 418–420.

Lehman, T. M. (1987) Late Maastrichtian paleoenvironment and dinosaur biogeography in the Western Interior of North America. *Palaeogeogr. Palaeoecol. Palaeoclimat.* 60: 189–217.

Leinen, M. (1985) Quartz content of northwestern Pacific Hole 576A and implication for Cenozoic eolian transport. *Init. Repts. DSDP* 86: 581–588. Washington (Govt. Printing Office).

Leinen, M. (1988) Eolian sedimentation in the northwest Pacific and its implications for Pleistocene climate changes (abs.). *Eos, Trans. Amer. Geophys. Union.* 69: 299.

LeMasurier, W. E. and Rex, D. C. (1982) Volcanic record of Cenozoic glacial history in Marie Byrd Land and Western Ellsworth Land, II Revised chronology and evaluation of tectonic factors. In *Antarctic Geosciences,* ed. C. Craddock, pp. 725–734. Madison: University of Wisconsin Press.

Leonard, K. A., Williams, D. F. and Thunell, R. C. (1983) Pliocene paleoclimatic and paleoceanographic history of the South Atlantic Ocean: stable isotopic records from Leg 72 DSDP Holes 516A and 517. *Init. Repts. DSDP* 72: 895–906. Washington (Govt. Printing Office).

Lewan, M. D. (1986) Stable carbon isotopes of amorphous kerogens from Phanerozoic sedimentary rocks. *Geochim. Cosmochim. Acta.* 50: 1583–1591.

Lindstrom, M. (1984) The Ordovician climate based on the study of carbonate rocks. In Aspects of the Ordovician System. *Palaeont. Contrib. Univ. Oslo, Univeritetsforlaget.* pp.81–88.

Lindzen, R. S. (1988) CO_2 feedbacks and the 100-K year cycle in climate. *Met. Atmo. Phys.* 36: 42–49.

Lindström, D. R. and MacAyeal, D. R. (1986) Paleoclimatic constraints on the maintenance of possible ice-shelf cover in the Norwegian and Greenland seas. *Paleoceanography* 1: 313–337.

Lister, G. S. (1988) A 15,000 year isotopic record from Lake Zurich of deglaciation and climatic change in Switzerland. *Quat. Res.* 29: 129–141.

Livermore, R. A., Smith, A. G. and Briden, J. C. (1985) Palaeomagnetic constraints on the distribution of the continents in the late Silurian and early Devonian. *Phil. Trans. Roy. Soc. Lond.* B 309: 29–56.

Lloyd, C. R. (1982) The mid-Cretaceous Earth: Palaeogeography; ocean circulation and temperature; atmospheric circulation. *J. Geol.* 90: 393–413.

Locke, S. and Thunell, R. C. (1988) Paleoceanographic record of the last glacial-interglacial cycle in the Red Sea and Gulf of Aden. *Palaeogeogr., Palaeoclimatol., Palaeoecol.* 64: 163–187.

Lohman, K. C. and Walker, J. G. C. (1989) The delta [18]O record of Phanerozoic abiotic marine calcite cements. *Geophys. Res. Letters.* 16: 319–322.

Loper, D. E., McCartney, K. and Buzyna, G. (1988) A model of correlated episodicity in magnetic-field reversals, climate, and mass extinctions. *J. Geol.* 96: 1–15.

Lorius, C., Barkov, M. I., Jouzel, J., Koratkevitch, Y. S., Kotlyakov, V. M. and Raynaud, D. (1988) Antarctic ice-core: CO_2 and climatic change over the last climatic cycle. *Eos, Trans. Amer. Geophys. Union* 69: 681–684.

Lorius, C., Jouzel, J., Ritz, C., Merlivat, L., Barkov, M. I., Kotlyakov, V. M. and Petrov, V. M. (1985) A 150,000–year climatic record from Antarctic ice. *Nature* 316: 591–596.

Lorius, C., Raynaud, D., Petit, J., Jouzel, L. and Merlivat, L. (1984) Late-glacial maximum Holocene atmospheric and ice-thickness changes from Antarctic core studies. *Annals of Glaciology,* 5: 88–94.

Louberie, P. (1988) Gradual Late Pliocene onset of glaciation: a deep-sea record from the northeast Atlantic. *Palaeogeogr., Palaeoclimatol., Palaeoecol.* 63: 327–334.

Louberie, P. and Moss, K. (1986) Late Pliocene climatic change and the onset of Northern Hemisphere glaciation as recorded in the northeast Atlantic Ocean. *Geol. Soc. Am. Bull.* 97: 818–828.

Loughnan, F. C. (1975) Laterites and flint clays in the early Permian of the Sydney Basin, Australia, and their palaeoclimatic implications. *J. Sed. Petrol.* 45: 591–598.

Loughnan, F. C. and Evans, P. R. (1978) The Permian and Mesozoic of the Merriwa-Binnaway-Ballimore area, New South Wales. *Roy. Soc. New South Wales, J. and Proc.* 111: 107–119.

Loughnan, F. C., Goldberry, R. and Holland, W. N. (1974) Kaolinite clayrocks in the Triassic Banks Wall Sandstone of the western Blue Mountains, New South Wales. *Geol. Soc. Aust.* 21: 393–402.

Loutit, T. S., Kennett, J. P. and Savin, S. M. (1984) Miocene equatorial and southwest Pacific paleoceanography from stable isotope evidence. *Marine Micropalaeo.* 8: 215–233.

Lowenstam, H. A. (1964) Paleotemperatures of the Permian and Cretaceous periods. In *Problems in Palaeoclimatology*, ed. A. E. M. Nairn, pp. 227–248. London: Wiley Interscience.

Lowenstam, H. A. and Epstein, S. (1954) Paleotemperatures of the post-Aptian Cretaceous as determined by the oxygen isotope method. *J. Geol.* 62: 207–248.

Lu, S., Ma, G., Gao, Z. and Lin, W. (1985) Sinica ice ages and glacial sedimentary facies-areas in China. *Precambrian. Res.* 29: 53–63.

Ludwig, W. J., Krashennikow, V. A. and Wise, S. (1983) Introduction and explanatory notes. *Init. Reports. DSDP.* 71: 205–249. Washington (Govt. Printing Office).

Magaritz, M. (1989) ^{13}C minima follow extinction events: A clue to faunal radiation. *Geology* 17: 337–340.

Magaritz, M., Bar, R., Baud, A. and Holser, W. T. (1988) The carbon-isotope shift at the Permian-Triassic boundary in the southern Alps is gradual. *Nature* 331: 337–339.

Malmgren, B. A. and Berggren, W. A. (1987) Evolutionary changes in some late Neogene planktonic foraminiferal lineages and their relationships to paleoceanographic changes. *Paleoceanography* 2: 445–456.

Manabe, S. and Broccoli, A. J. (1984) Influence of CLIMAP ice sheet on the climate of a GCM: Implications for the Milankovitch theory. In *Milankovitch and Climate*, ed. A. L. Berger, J. Imbrie, J. Hays, G. Kukla and B. Saltzman, pp. 789–799. Dordrecht: D. Reidel.

Manabe, S. and Broccoli, A. J. (1985) The influence of continental ice sheets on the climate of the ice age. *J. Geophys. Res.* 90: 2167–2190.

Mankinen, E. A. and Dalrymple, G. B. (1979) Revised geomagnetic polarity time scale for the interval 0–5 m.y.B.P. *J. Geophys. Res.* 84: 615–626.

Manspeizeer, W. (1982) Triassic-Liassic basins and climate of the Atlantic passive margins. *Geol. Rundsch.* 71: 895–917.

Margolis, S. V. and Herman, Y. (1980) Northern Hemisphere sea-ice and glacial development in the late Cenozoic. *Nature* 286: 145–149.

Marsiat, I., Berger, A., Gallee, H., Fichefet, T. and Tricot, C. (in press) Modeling the transient response of a coupled model to the astronomical forcing over the

last glacial to interglacial cycle. In *The Beginning of an Inland Glaciation, Facts and Problems of Climate Dynamics*, ed. B. Frenzel, Mainz.

Martin, J. H. (1990) Glacial-Interglacial CO_2 change: the iron hypothesis. *Paleoceanography.* 5: 1–14.

Martinson, D., Pisias, N. G., Hays, J. D., Imbrie, J., Moore, T. C. and Shackleton, N. J. (1987) Age dating and the orbital theory of the ice ages: development of a high resolution 0 to 300,000–year chronology. *Quat. Res.* 27: 1–29.

Matsch, C. L. and Schneider, A. F. (1986) Stratigraphy and correlation of the glacial deposits of the Lobe Complex in Minnesota and northwestern Wisconsin. *Quat. Sci. Rev.* 5: 59–64.

Matsuoka, N. (1987) Malacofaunal succession in Pliocene to Pleistocene non-marine sediments in the Omi and Ueno Basins, Central Japan. *J. Earth Sci., Nagoya Univ.* 35: 91–101.

Matthews, J. V. (1987) Plant microfossils from the Neogene Beaufort Formation on Banks and Meighen Islands, District of Franklin. *Geol. Surv. Can. Pap.* 87–1A: 73–87.

Matthews, R. K. and Poore, R. Z. (1980) Tertiary δO^{18} record and glacio-eustatic sea-level fluctuations. *Geology* 8: 501–504.

Matthews, S. C. and Cowie, J. W. (1979) Early Cambrian transgression. *J. Geol. Soc. London.* 136: 133–135.

McClure, H. A. (1980) Permian-Carboniferous glaciation in the Arabian Peninsula. *Geol. Soc. Amer. Bull.* 91: 707–712.

McGowran, B. (1978) Stratigraphic record of the early Tertiary oceanic and continental events in the Indian Ocean region. *Mar. Geol.* 26: 1–39.

McGowran, B. (1979) Comments on Early Tertiary tectonism and laterization. *Geol. Mag.* 116: 227–230.

McGowran, B. (1986) Cainozoic oceanic and climatic events: the Indo-Pacific foraminiferal biostratigraphic record. *Palaeogeogr., Palaeoclimatol., Palaeoecol.* 55: 247–265.

McGowran, B. (1990) Fifty million years ago. *Amer. Sci.* 78: 30–39.

McGowran, B. and Beecroft, A. S. (1986) Foraminiferal biofacies in a silica-rich neritic sediment, Late Eocene, South Australia. *Palaeogeogr., Palaeoclimatol., Palaeoecol.* 52: 321–345.

McKenna, T. E., Levche, I., Williams, D. F. and Full, W. E. (1988) Quantitative techniques in isotope chronostratigraphy. *Palaeogeogr., Palaeoclimatol., Palaeoecol.* 64: 241–264.

McKenzie, J. A. and Oberhansli, H. (1985) Paleoceanographic expressions of the Messinian salinity crisis. In *South Atlantic Paleoceanography*, ed. K. J. Hsu and H. S. Weissert, pp.99–129. Cambridge: Cambridge University Press.

McKerrow, W. S. (1979) Ordovician and Silurian changes in sea level. *J. Geol. Soc.* 136: 137–145.

Mercer, J. H. (1969) The Allerod oscillation: A European climatic anomaly? *Arctic and Alpine Res.* 1: 37–43.

Mercer, J. H. (1983) Cenozoic glaciation in the southern hemisphere. *Ann. Rev. Earth Sci.* 11: 99–132.

Mercer, J. H. (1987) The Antarctic ice sheet during the late Neogene. *Paleoecology of Africa.* 18: 21–32.

Mercer, J. H. and Sutter, J. F. (1982) Late Miocene-earliest Pliocene glaciation in southern Argentina: Implications for global ice-sheet history. *Palaeogeogr., Palaeoclimatol., Palaeoecol.* 38: 185–206.

Meyerhoff, A. A. (1970) Continental drift II. High-latitude evaporite deposition and geologic history of Arctic and North Atlantic oceans. *J. Geol.* 78: 406–444.

Middleton, P. D., Marshall, J. D. and Brenchley, P. J. (1988) *Isotope evidence for oceanographic changes associated with the late Ordovician glaciation (abs.).* International Symposium on the Ordovician System, p. 59. Memorial University, St. Johns, Newfoundland.

Miller, G., Mountain, G. S. and Christie-Blick, N. (1988) Cenozoic eustatic fluctuations: Constraints from isotope and seismic stratigraphy (abs.). *Eos, Trans. Amer. Geophys. Union.* 69: 300.

Miller, J. R. and Russell, G. L. (1988) Ocean heat transport during the last glacial maximum. *Paleoceanography* 4: 141–156.

Miller, K. G. and Fairbanks, R. G. (1985) Oligocene to Miocene carbon isotopic cycles and abyssal circulation changes. In *The Carbon Cycle and Atmospheric CO_2: Natural Variations Archean to Present*, ed. E. T. Sundquist and W. S. Broecker, pp. 504–529. Washington, D.C.: American Geophysical Union.

Miller, K. G., Fairbanks, R. G. and Mountain, G. S. (1987) Tertiary oxygen isotope synthesis, sea level history and continental margin erosion. *Paleoceanography* 1: 1–20.

Mix, A. C. and Ruddimann, W. F. (1984) Oxygen isotope analyses and Pleistocene ice volumes. *Quat. Res.* 21: 1–20.

Mix, A. C. and Ruddimann, W. F. (1985) Structure and timing of the last deglaciation; Oxygen isotope evidence. *Quat. Sci. Rev.* 4: 59–108.

Mix, A. C., Ruddimann, W. F. and McIntyre, A. (1986a) Late Quaternary paleoceanography of the tropical Atlantic. 1, Spatial variability of annual mean sea surface temperatures 0–20,000 years BP. *Paleoceanography* 1: 43–66.

Mix, A. C., Ruddimann, W. F. and McIntyre, A. (1986b) Late Quaternary paleoceanography of the tropical Atlantic. 2, The seasonal cycle of sea surface temperatures. *Paleoceanography* 1: 339–353.

Molfino, B., Heusser, L. H. and Woillard, G. M. (1984) Frequency components of a Grande Pile pollen record: Evidence of precessional orbital forcing. In *Milankovitch and Climate*, ed. A. L. Berger, J. Imbrie, J. Hays, G. Kukla and B. Saltzman, pp. 391–405. Dordrecht: D. Reidel.

Molnar, P. and England, P. (1990) Late Cenozoic uplift of mountain ranges and global climate change: chicken or egg? *Nature* 346: 29–34.

Moore, T. C., Lohman, G. P., Srivastava, M. and Woodruff, F. (1981) Variability of the Miocene Ocean (abs.). *Geol. Soc. Amer. Abstracts with Program.* 13: 514.

Moore, T. C., Loutit, T. S. and Greenie, S. M. (1987) Estimating short-term changes in eustatic sea level. *Paleoceanography* 2: 625–638.

Moore, T. C., Pisias, N. G. and Dunn, D. A. (1982) Carbonate time series of the Quaternary and Late Miocene sediments in the Pacific Ocean: A spectral comparison. *Marine Geol.* 45: 217–233.

Morel, P. and Irving, E. C. (1978) Tentative paleocontinental maps for the early Phanerozoic and Proterozoic. *J. Geol.* 86: 535–561.

Morley, J. J. (1987) *Pliocene-Pleistocene climatic changes in the northwest Pacific*

(abs.). 4th International Congress on Pacific Neogene Stratigraphy. ed. J.A. Barron and J. R. Blueford, Abstract Volume, p. 79. Berkeley, California.

Morley, J. J., Pisias, N. G. and Leinen, M. (1987) Late Pleistocene time series of atmospheric and oceanic variables recorded in sediments from the subarctic Pacific. *Paleoceanography* 2: 49–62.

Morley, J. J., Prell, W. L. and Howard, W. R. (1988) Responses of the Southern Ocean over the Milankovitch frequency band (abs.). *Eos, Trans. Amer. Geophys. Union* 69: 299.

Morley, J. J. and Shackleton, N. J. (1984) The effect of accumulation rate on the spectrum of geologic time series: Evidence from two South Atlantic sediment cores. In *Milankovitch and Climate*, ed. A. L. Berger, J. Imbrie, J. Hays, G. Kukla and B. Saltzman, pp. 467–480. Dordrecht: Reidel.

Mott, R. J., Grant, D.R., Stea, R. and Occhietti, S. (1986) Late glacial climatic oscillation in Atlantic Canada equivalent to the Allerod/younger Dryas event. *Nature* 323: 247–250.

Mudie, P. J., de Vernal, A. and Head, M. J. (1990) Neogene to Recent palynostratigraphy of circum-Arctic Basins: Results of ODP Leg 104, Norwegian Sea, Leg 105, Baffin Bay, and DSDP Site 611, Irminger Sea. In *Geological History of the Polar Oceans: Arctic Versus Antarctic*, ed. U. Bleil and J. Thiede, pp. 609–646. Dordrecht: Kluwer Academic Publ.

Mudie, P. J. and Helgason, J. (1983) Palynological evidence for Miocene climate cooling in eastern Iceland about 9.8 Myr ago. *Nature* 303: 689–692.

Muller, D. W. and Hsu, K. J. (1987) Event stratigraphy and paleoceanography in the Fortuna Basin (Southeast Spain): A scenario for the Messinian salinity crisis. *Paleoceanography* 2: 679–698.

Murray, J. W. (1984) Paleogene and Neogene benthic foraminifers from Rockall Plateau. *Init. Repts. DSDP.* 81: 503–534. Washington (Govt. Printing Service).

Murray, J. W. (1988) Neogene bottom water masses and benthic foraminifera in the NE Atlantic Ocean. *J. Geol. Soc. Lond.* 145: 125–132.

Nance, R. D., Worsley, T. R. and Moody, J. B. (1988) The supercontinent cycle. *Sci. Amer.* 7: 44–51.

Nelson, C. S., Mildenhall, D. C., Todd, A. J. and Pochnall, D. T. (1988) Subsurface stratigraphy, paleoenvironments, palynology, and deposition history of the late Neogene Tauronge Group at Ohinewai, Lower Waikato Lowland, South Auckland, New Zealand. *N. Z. Jnl. Geol. Geophys.* 31: 21–40.

Nikolayev, S. D., Blyum, N. S. and Nikolayev, V. I. (1986) Reconstruction of the past thermal structure of surface water masses in the oceans from oxygen isotope data: The Late Miocene. *Doklady Akademii Nauk SSSR.* 291: 250–253.

Norris, G. (1982) Spore-pollen evidence for early Oligocene high-latitude cool climate episode in Northern Canada. *Nature* 297: 387–389.

Norton, I. O. and Sclater, J. G. (1979) A model for the evolution of the Indian Ocean and the break-up of Gondwanaland. *J. Geophys. Res.* 84: 6803–6830.

Oberhansli, H. (1986) Latest Cretaceous-early Neogene oxygen and carbon isotopic record at DSDP sites in the Indian Ocean. *Marine Micropalaeo.* Special Issue on Stable Isotopes: 10: 91–116.

Oberhansli, H. and Hsu, K. J. (1986) Paleocene-Eocene Paleoceanography. In *Mesozoic and Cenozoic Oceans*, ed. K. J. Hsu, pp. 85–100. Washington, D.C.:

American Geophysical Union Geodynamics Series 15.

Oerlemans, J. (1984) On the origin of ice ages. In *Milankovitch and Climate*, ed. A. L. Berger, J. Imbrie, J. Hays, G. Kukla and B. Saltzman, pp. 607–611. Dordrecht: D. Reidel.

Oeschger, H., Beer, J., Siegenthaler, U., Stauffer, B., Dansgaard, W. and Longway, C. (1984) Late glacial climatic history from ice cores. In *Climatic Processes and Climate Sensitivity*, ed. J. E. Hansen and T. Takahashi, pp. 299–326. Washington D. C.: American Geophysical Union, Geophysical Monograph 29.

Officer, C. B., Hallam, A., Drake, C. L. and Devine, J. D. (1987) Late Cretaceous and paroxysmal Cretaceous/Tertiary extinctions. *Nature* 326: 143–149.

Olausson, E. (1988) On the Cenozoic oceans: evidence of the calcium carbonate content, $\delta^{13}C$ and $\delta^{18}O$. *Palaeogeogr., Palaeoclimatol., Palaeoecol.* 67: 103–118.

Olsen, P. E. (1986) A 40–million-year lake record of early Mesozoic orbital climatic forcing. *Science* 234: 842–848.

Olson, E. C. (1979) Nonmarine Vertebrates and Late Paleozoic Climates. 9th Congress Internationale de Stratigraphie et Geologie du Carbonifer. *Compte Rendu* 5: 403–414.

Oppo, D. W. and Fairbanks, R. G. (1987) Variability in the deep and intermediate water circulation of the Atlantic Ocean during the past 25,000 years: Northern hemisphere modulation of the Southern Ocean. *Earth Planet. Sci. Letters.* 86: 1–15.

Palmer, A. R. (1972) *Problems of Cambrian Biogeography*. 24th Session International Geological Congress. Montreal. Section 7, Paleontology: 310–315.

Palmer, A. R. (1979) Cambrian. In *Treatise on Invertebrate Paleontology. Part A. Introduction*, ed. R. A. Robinson and C. Teichert, pp. A119–A135. Boulder and Lawrence: Geological Society of America and the University of Kansas.

Palmer, A. R. (1981) On the correlatability of grand cycle tops. *U.S. Geol. Survey Open–File Report.* 81–743: 156–159.

Palmer, A. R. (1983) The Decade of North American Geology 1983 Geological Time Scale. *Geology* 11: 503–504.

Parrish, J. T. (1982) Upwelling and petroleum source beds with reference to Paleozoic. *Amer. Assoc. Petrol. Geol. Bull.* 66: 750–774.

Parrish, J. T. and Curtis, D. L. (1982) Atmospheric circulation, upwelling and organic-rich rocks in the Mesozoic and Cenozoic eras. *Palaeogeogr., Palaeoclimatol., Palaeoecol.* 40: 31–66.

Parrish, J. T. and Spicer, R. A. (1988) Late Cretaceous vegetation: A near-polar temperature curve. *Geology* 16: 22–25.

Parrish, J. T., Ziegler, A. M. and Humphreville, R. G. (1983) Upwelling in the Paleozoic era. In *Coastal Upwelling; Its Sediment Record. Part B: Sedimentary Record of Ancient Coastal Upwelling*, ed. J. Thiede and E. Suess, pp. 553–557. New York: Plenum Press.

Parrish, J. T., Ziegler, A. M. and Scotese, C. R. (1982) Rainfall patterns and the distribution of coals and evaporites in the Mesozoic and Cenozoic. *Palaeogeogr., Palaeoclimatol., Palaeoecol.* 40: 67–101.

Pavlov, S. F. (1979) New data on Upper Paleozoic glacial deposits in the Siberian platform. *Lithology and Mineral Resources (English translation from*

Russian Litologia i Paleznye Iskopaemye). 15: 254–262.

Perroud, H., Van der Voo, R. and Bonhommet, N. (1984) Paleozoic evolution of the Armorica plate on the basis of paleomagnetic data. *Geology* 12: 579–582.

Peteet, D. M. (1987) Paleoenvironmental shift in maritime Canada parallels Europe. *Tree* 2: 86–87.

Petersen, G. M., Webb, T., Kutzbach, J. E., Von der Hammen, T., Wimjstra, T. A. and Street, F. R. (1979) The continental record of environmental conditions at 18,000 yr BP: An initial evaluation. *Quat. Rev.* 12: 47–82.

Peterson, J. A. (1980) Deposition history and petroleum geology of the Permian Phosphoria, Park City and Shedhorn formations, Wyoming and southeastern Idaho. *U.S. Geol. Surv. Open-File Rep.* 80–667: 51.

Pewe, T. L. (1976) Late Cenozoic history of Alaska. In *Quaternary Stratigraphy of North America*, ed. W. C. Mahoney, pp. 493–504. New York: Dowden Hutchinson.

Pfefferkorn, H. W. and Thomson, M. C. (1982) Changes in dominance patterns in Upper Carboniferous plant-fossil assemblages. *Geology* 10: 641–644.

Phillips, T. L. and Peppers, R. A. (1984) Changing patterns of Pennsylvanian coal-swamp vegetation and implications of climatic control on coal occurrence. *Internat. J. Coal. Geol.* 3: 205–255.

Pickard, J., Adamson, D. A., Harwood, D. M., Miller, G. H., Quilty, P. G. and Dell, R. K. (1988) Early Pliocene marine sediments, coastline and climate of East Antarctica. *Geology* 16: 158–161.

Piper, J. D. A. (1987) *Paleomagnetism and the Continental Crust*. New York: J. Wiley & Sons. 434 pp.

Pirrie, D. and Marshall, J. D. (1990) High-paleolatitude Late Cretaceous paleotemperatures: New data from James Ross Island, Antarctica. *Geology* 18: 31–34.

Pisias, N. G. and Leinen, M. (1984) Milankovitch forcing of the oceanic system. In *Milankovitch and Climate*, ed. A. L. Berger, J. Imbrie, J. Hays, G. Kukla and B. Saltzman, pp. 307–330. Dordrecht: Reidel.

Pisias, N. G., Mix, A., Shackleton, N.J. and Imbrie, J. (1988) Response of the carbon system over the Milankovitch frequency band (abs.). *Eos, Trans. Amer. Geophys. Union.* 69: 299.

Pisias, N. G. and Moore, T. C. (1981) The evolution of Pleistocene climate: a time series approach. *Earth Planet. Sci. Rev.* 4: 59–108.

Pisias, N. G. and Prell, W. L. (1985) High resolution carbonate records from the hydraulic piston cored section of Site 572. *Init. Repts. DSDP* 85: 711–722. Washington (Govt. Printing Office).

Pisias, N. G. and Rea, D. K. (1988) Late Pleistocene paleoclimatology of the central equatorial Pacific: Sea surface response to the southeast trade winds. *Paleoceanography* 3: 21–38.

Pisias, N. G. and Shackleton, N. J. (1984) Modeling the global climate response to orbital forcing and atmospheric carbon dioxide changes. *Nature* 310: 757–759.

Plafker, G. (1971) Possible future petroleum resources of Pacific margin Tertiary basins, Alaska. In *Future Petroleum Provinces of North America*, ed. I. H. Cram, pp. 120–135. Am. Assoc. Pet. Geol. Mem. 15.

Pokras, E. M. and Mix, A. C. (1985) Eolian evidence for spatial variability of Late

Quaternary climates in Tropical Africa. *Quat. Res.* 24: 137–149.

Pokras, E. M. and Mix, A. C. (1987) Earth's precession cycle and Quaternary climatic change in tropical Africa. *Nature* 326: 486–487.

Pollack, J. B. (1979) Climatic change on the terrestrial planets. *Icarus* 37: 479–553.

Pollard, D. (1983) A coupled climate-ice sheet model applied to Quaternary ice ages. *J. Geophys. Res.* 88: 7705–7718.

Pollard, D. (1984) Some ice-age aspects of a calving ice-sheet model. In *Milankovitch and Climate*, ed. A. L. Berger, J. Imbrie, J. Hays, G. Kukla and B. Saltzman, pp. 541–564. Dordrecht: Reidel.

Pomerol, C. and Premoli-Silva, I. (1986) The Eocene-Oligocene transition events and boundary. In *Terminal Eocene Events*, ed. C. Pomerol and others, pp. 1–24. Amsterdam: Elsevier.

Poore, R. Z. (1981) Temporal and spatial distribution of ice-rafted mineral grains in Pliocene sediments of the North Atlantic: Implication for Late Cenozoic climatic history. *Soc. Econ. Pet. Min., special publication* 32: 505–515.

Poore, R. Z. and Matthews, R. K. (1984) Late Eocene-Oligocene oxygen and carbon isotope record from the South Atlantic DSDP site 522. *Init. Repts. DSDP* 73: 725–735.

Popp, B. N., Anderson, T. F. and Sandberg, P. A. (1986) Brachiopods as indicators of original isotopic compositions in some Paleozoic limestones. *Geol. Soc. Amer. Bull.* 97: 1262–1269.

Popp, B. N., Takigiku, R., Hayes, J. M., Louda, J. W. and Baker, E. W. (1989) The post-Paleozoic chronology and mechanisms of ^{13}C depletion in primary organic matter. *Amer. J. Sci.* 289: 436–454.

Powell, C. M. and Veevers, J. J. (1987) Namurian uplift in Australia and South America triggered the main Gondwanan glaciation. *Nature* 326: 177–179.

Prell, W. L. (1982) A reevaluation of the initiation of northern hemisphere glaciation at 3.2 Ma: new isotopic evidence (abs.). *Geol. Soc. Amer, Abstracts with Program.* 592.

Prell, W. L. (1984a) Covariance patterns of foraminiferal $\delta^{18}O$: an evaluation of Pliocene ice volume near 3.2 Ma. *Science* 226: 692–693.

Prell, W. L. (1984b) Monsoonal climate of the Arabian Sea during the late Quaternary: a response to solar radiation. In *Milankovitch and Climate*, ed. A. L. Berger, J. Imbrie, J. Hays, G. Kukla and B. Saltzman, pp. 349–366. Dordrecht: Reidel.

Prell, W. L. and Kutzbach, J. E. (1987) Monsoon variability over the past 15,000 years. *J. Geophys. Res.* 92: 8411–8425.

Prell, W. L. and McIntyre, A. (1988) Response of the monsoon system over Milankovitch frequency band (abs.). *Eos, Trans. Amer. Geophys. Union.* 69: 299.

Prell, W. L., Hutson, W., Williams, D., Be, A., Geitzenaueur, G. and Molfino, B. (1980) Surface circulation of the Indian Ocean during the last glacial maximum, approximately 18,000 yr BP. *Quat. Res.* 14: 309–336.

Pujol, C. and Duplessy, J.-C. (1983) The ocean surface during the last interglacial transition: A review of the available data. In *Paleoclimatic Research and Models*, ed. A. Ghazi, pp. 145–151. Dordrecht: D. Reidel.

Railsback, L. B. (1990) Influence of changing deep ocean circulation on the Phanerozoic oxygen isotope record. *Geochim. Cosmochim. Acta.* 54: 1501–1509.

Railsback, L. B. and Anderson, T. F. (1987) Control of Triassic seawater chemistry and temperature on the evolution of post-Paleozoic aragonite-secreting faunas. *Geology* 15: 1002–1005.

Railsback, L. B., Ackerly, S. C., Cisne, J. L. and Anderson, T. F. (1990) Paleontologic and isotopic evidence for warm saline deep waters in Ordovician oceans. *Nature* 343: 156–159.

Rampino, M. R. and Stothers, R. B. (1988) Flood basalt volcanism during the past 250 million years. *Science* 241: 663–668.

Ramsbottom, W. H. C. (1979) Rates of transgression and regression in the Carboniferous of NW Europe. *J. Geol. Soc. Lond.* 136: 147–153.

Rao, C. P. and Green, D. C. (1982) Oxygen and carbon isotopes in early Permian cold-water carbonates. *J. Sedim. Petrol.* 52: 1111–1125.

Rao, C. P. and Green, D. C. (1983) Oxygen- and carbon-isotope composition of cold shallow-marine carbonates of Tasmania, Australia. *Marine Geol.* 53: 117–129.

Raup, D. M. and Sepkowski, J. J., Jr. (1984) Periodicity of extinctions in the geologic past. *Proc. Natl. Acad. Sci. U.S.A.* 81: 801–805.

Raymo, M. E., Ruddiman, W. F., Backman, J., Clement, B. M. and Martinson, D. G. (1989) Late Pliocene variation in northern hemisphere ice sheets and North Atlantic deep water circulation. *Paleoceanography* 4: 413–446.

Raymo, M. E., Ruddimann, W. F. and Bradford, B. M. (1987) Pliocene-Pleistocene paleoceanography of the North Atlantic at DSDP Site 609. *Init. Repts. DSDP* 94: 895–901. Washington D.C. (Govt. Printing Office).

Raymo, M. E., Ruddimann, W. F. and Froelich, P. N. (1988) Influence of late Cenozoic mountain building on ocean geochemical cycles. *Geology* 16: 649–653.

Raymond, A. (1985) Floral diversity, phytogeography, and climatic amelioration during the Early Carboniferous (Dinantian). *Paleobiology* 11: 293–309.

Raymond, A. (1987) Paleogeographic distribution of Early Devonian plant traits. *Palaios* 2: 113–132.

Raymond, A., Kelly, P. H. and Lutken, C. B. (1987) Frasnian-Famennian mass extinction and cold water oceans: Comment. *Geology* 15: 777–778.

Raymond, A., Kelly, P. H. and Lutken, C. B. (1989) Polar glaciers and life at the equator: the history of Dinantian and Namurian (Carboniferous) climate. *Geology* 17: 408–411.

Raynaud, D., Chappellar, J., Barnola, J.M., Korotkevich, Y.S. and Lorius, C. (1988) Climatic and CH_4 cycle implications of glacial interglacial CH_4 change in the Vostok ice core. *Nature* 333: 655–667.

Rea, D. and Vallier, T. (1983) Two Cretaceous volcanic episodes in the western Pacific Ocean. *Geol. Soc. Amer. Bull.* 94: 1430–1937.

Rea, D. K. (1986) 420,000 year record of cyclicity in oceanic and atmospheric processes from the eastern equatorial Pacific. *Paleoceanography* 1: 577–586.

Rea, D. K. and Janecek, T. R. (1982) Late Cenozoic changes in atmospheric circulation deduced from North Pacific eolian sediments. *Marine Geol.* 49: 149–167.

Rea, D. K. and Leinen, M. (1988) Asian aridity and the zonal westerlies: Late Pleistocene and Holocene record of eolian deposition in the northwest Pacific Ocean. *Palaeogeogr., Palaeoclimatol., Palaeoecol.* 66: 1–8.

Rea, D. K., Zachos, J. C., Owen, R. M. and Gingerich, P. D. (1990) Global change at the Paleocene-Eocene boundary: climatic and evolutionary consequence of tectonic events. *Palaeogeogr. Palaeoclimatol., Palaeoecol.* 79: 117–128.

Read, J. and Francis, J. E. (in press) Responses of some southern hemisphere tree species to a prolonged dark period and their phytogeographic and palaeoecological implications for Cretaceous and Tertiary floras. *Palaeogeogr., Palaeoclimatol., Palaeoecol.*

Renard, M. (1986) Pelagic carbonate chemostratigraphy (Sr, Mg, ^{18}O, ^{13}C). *Marine Micropalaeo.* 10: 117–164.

Retallack, G. J. (1977) Reconstructing Triassic vegetation of eastern Australasia: a new approach for the biostratigraphy of Gondwanaland. *Alcheringa* 1: 247–277.

Retallack, G. J. (1986) The fossil record of soils. In *Paleosols: Their Recognition and Interpretation*, ed. V. P. Wright, pp. 1–57. Princeton: Princeton University Press.

Rich, P. V., Rich, T. H., Wagstaff, B. E., McEwen-Mason, J., Douthitt, C. B., Gregory, R. T. and Felton, E. A. (1988) Evidence for low temperatures in Cretaceous high latitudes of Australia. *Science* 242: 1403–1406.

Rich, T. H., Rich, P. V., Wagstaff, B., McEwan-Mason, J., Douthitt, C. B. and Gregory, R. T. (1989) Early Cretaceous biota from the northern side of the Australo–Antarctic rift valley. ed. J. A. Crame. *Origins and Evolution of the Antarctic Biota.* pp. 121–130. London. Geological Society of London Special Publication 47.

Richmond, G. M. and Fullerton, D. S. (1986) Introduction to Quaternary glaciations in the United States of America. *Quat. Sci. Rev.* 5: 3–10.

Ridd, M. F. (1971) Faults in South-East Asia and the Andaman Rhombochasm. *Nature* 229: 51–52.

Rind, D. (1987) Components of the ice age circulation. *J. Geophys. Res.* 92: 4241–4281.

Rind, D. and Peteet, D. (1985) Terrestrial conditions at the last glacial maximum and CLIMAP sea-surface temperature estimates: Are they consistent? *Quat. Res.* 24: 1–22.

Rind, D., Peteet, D., Broecker, W., McIntyre, A. and Ruddiman, W. (1986) The impact of cold North Atlantic sea surface temperatures on climate: Implications for the Younger Dryas cooling (11–10K). *Climate Dynamics* 1: 3–33.

Ritchie, J. C. (1985) Quaternary pollen records from the Western Interior and the Arctic of Canada. In *Pollen Record of Late Quaternary North American Sediments*, ed. V. M. Bryan, pp. 327–52. Am. Assoc. Stratigr. Palynol. Found., United States.

Roberts, J. D. (1976) Late Precambrian dolomites, Vendian glaciation and synchroneity of Vendian glaciations. *J. Geol.* 84: 47–63.

Roberts, J., Hunt, J. W. and Thompson, D. M. (1976) Late Carboniferous marine invertebrate zones of eastern Australia. *Alcheringa* 1: 197–225.

Robin, G. de Q. (1987) The Antarctic ice sheet, its history and response to sea level and climatic changes over the past 100 million years. *Palaeogeogr., Palaeoclimatol., Palaeoecol.* 67: 31–50.

Robinson, P. L. (1973) Paleoclimatology and continental drift. In *Implications of*

Continental Drift to the Earth Sciences, vol. 2, ed. D. H. Tarling and S. K. Runcorn, pp. 451–476. New York: Academic Press.

Rocha-Campos, A. C. and dos Santos, P. R. (1981) Late Paleozoic tillites of the Lutoe Series, Angola. In *Earth's Pre-Pleistocene Glacial Record,* ed. M. J. Hambrey and W. B. Harland, pp. 48–51. Cambridge: Cambridge University Press.

Ronov, A. B. (1976) Global carbon geochemistry, volcanism, carbonate accumulation, and life. *Geoch. Int.* 13: 172–195.

Ronov, A. B. (1980) A comparative estimate of volcanic intensity on continents and in oceans. *Internat. Geol. Rev.* 22: 1383–1389.

Ronov, A. B., Khain, V. and Seslavinsky, K. (1984) *Atlas of Lithological-Palaeontological Maps of the World (Late Precambrian and Paleozoic of Continents)* . Leningrad: USSR Academy of Science. 70 pp.

Rooth, C. (1982) Hydrology and ocean circulation. *Progress in Oceanography* 11: 131–149.

Ross, C. A. (1979) Carboniferous. In *Treatise on Invertebrate Paleontology, Part A,* ed. R. A. Robinson and C. Teichert, pp. 254–290. Boulder: Geol. Soc. Am. and University of Kansas Press.

Ross, C. A. and Ross, J. R. P. (1985) Late Paleozoic depositional sequences are synchronous and worldwide. *Geology* 13: 194–197.

Rossignol-Strick, M. (1983) African monsoon, an immediate climate response to orbital insolation. *Nature* 295: 105–110.

Rossignol-Strick, M. (1984) Immediate climate response to orbital insolation: Mediterranean sapropels and the African monsoon. In *Milankovitch and Climate,* ed. A. L. Berger, J. Imbrie, J. Hays, G. Kukla and B. Saltzman, pp. 367–368. Dordrecht: D. Reidel.

Rossignol-Strick, M. (1987) Insolation variation, formation of Mediterranean Quaternary sapropels: Significance to Cretaceous black shales. *Mem. Soc. geol. France, N. S.* 151: 7–26.

Roth, P. H. (1986) Mesozoic palaeoceanography of the North Atlantic and Tethys Oceans. In *North Atlantic Palaeoceanography,* ed. C. P. Summerhayes and N. J. Shackleton, pp. 299–320. London: Geological Society of London Special Publication.

Roth, P. H. (1987) Mesozoic calcareous nannofossil evolution: relation to paleoceanographic events. *Paleoceanography* 2: 601–611.

Royer, J. F., Deque, M. and Pestiaux, P. (1983) A GCM simulation of the importance of insolation forcing for the initiation of the Laurentide ice sheet. In *Palaeoclimatic Research and Models,* ed. A. Ghazi, pp. 164 –167. Dordrecht: Reidel.

Royer, J. F., Deque, M. and Pestiaux, P. (1984) A sensitivity experiment to astronomical forcing with spectral GCM: Simulation of the annual cycle at 125,000 BP and 115,000 BP. In *Milankovitch and Climate,* ed. A. L. Berger, J. Imbrie, J. Hays, G. Kukla and B. Saltzman, pp. 733–764. Dordrecht: Reidel.

Ruddiman, W. F. and Duplessy, J.-C. (1985) Conference on the last deglaciation: Timing and mechanism. *Quat. Res.* 23: 1–17.

Ruddiman, W. F. and Kutzbach, J. E. (1989) Forcing of the late Cenozoic northern hemisphere climate by plateau uplift in southeast Asia and the American Southwest. *J. Geophys. Res.* 94: 18409–18427.

Ruddiman, W. F. and McIntyre, A. (1979) Warmth of the subpolar North Atlantic

Ocean during northern hemisphere ice-sheet growth. *Science* 204: 173–175.

Ruddiman, W. F. and McIntyre, A. (1981a) Oceanic mechanisms for the amplification of the 23,000–year ice-volume cycle. *Science* 212: 617–627.

Ruddiman, W. F. and McIntyre, A. (1981b) The mode and mechanism of the last deglaciation: Oceanic evidence. *Quat. Res.* 16: 125–134.

Ruddiman, W. F. and McIntyre, A. (1984a) Ice-age thermal response and climatic role of the surface North Atlantic Ocean, 40° to 63°N. *Geol. Soc. Amer. Bull.* 95: 381–396.

Ruddiman, W. F. and McIntyre, A. (1984b) An evaluation of ocean-climate theories on the North Atlantic. In *Milankovitch and Climate*, ed. A. L. Berger, J. Imbrie, J. Hays, G. Kukla and B. Saltzman, pp. 671–686. Dordrecht: D. Reidel.

Ruddiman, W. F., McIntyre, A. and Raymo, M. (1986) Paleoenvironmental results from North Atlantic sites 607 and 609. *Initial Reports DSDP* 94: 855–878. Washington (Govt. Printing Office).

Ruddiman, W. F., Molfino, B, Esmay, A. and Pokras, E. (1980) Evidence bearing on the mechanism of rapid deglaciation. *Climate Change.* 3: 65–87.

Ruddiman, W.F. and Raymo, M. (1988) Northern hemisphere climatic regimes during the past 3 Ma: Possible tectonic connections. *Phil. Trans. Roy. Soc. London,* Ser. B 318: 411–430.

Ruddiman, W. F., Raymo, M. and McIntyre, A. (1986) Matuyama 41,000-year cycles, North Atlantic Ocean and Northern Hemisphere ice sheets. *Earth Planet. Sci. Letters.* 80: 117–129.

Ruddiman, W. F., Raymo, M. E., Martinson, D. G., Clements, B. M. and Backman, J. (1989) Pleistocene evolution of northern hemisphere climate. *Paleoceanography* 4: 353–412.

Ruddiman, W. F., Shackleton, N. J. and McIntyre, A. (1986) North Atlantic sea-surface temperatures for the last 1.1 million years. In *North Atlantic Paleoceanography* ed. C. P. Summerhayes and N. J. Shackleton, pp. 155–173. Geol. Soc. London, Spec. Publ.

Runnegar, B. and Campbell, K. S. W. (1976) Late Paleozoic faunas of Australia. *Earth Sci. Rev.* 12: 235–257.

Rust, I. C. (1975) The evolution of the Paleozoic Cape Basin, southern margin of Africa. In *The Ocean Basins and Margins*, ed. A. E. M. Nairn and F. J. Stehli, pp. 247–276. New York: Plenum.

Saltzman, B. (1985) Paleoclimate modeling. In *Paleoclimate Analysis and Modeling*, ed. A. Hecht, pp. 341–396. New York: John Wiley and Sons Ltd.

Saltzman, B. (1987) Carbon dioxide and the $\delta^{18}O$ record of Quaternary climatic change: a global model. *Climate Dynamics* 1: 77–85.

Saltzman, B., Hansen, A. R. and Maasch, K. A. (1984) The Late Quaternary glaciations as the response of a three-component feedback system to Earth-orbital forcing. *J. Atmos. Sci.* 41: 3380–3389.

Saltzman, B. and Maasch, K. A. (1988) Carbon cycle instability as a cause of the late Pleistocene ice age oscillations: modeling the asymmetric response. *Global Biogeochem. Cycles* 2: 177–185.

Saltzman, B. and Sutera, A. (1987) The mid-Quaternary climatic transition as the free response of a three-variable dynamical model. *J. Atmos. Sci.* 44: 236–241.

Saltzman, E. S. and Barron, E. J. (1982) Deep circulation in the late Cretaceous:

oxygen isotope paleotemperatures from Inoceramus remains in DSDP cores. *Palaeogeogr., Palaeoclimatol., Palaeoecol.* 40: 167–181.

Sancetta, C. (1979) Paleogene Pacific microfossils and paleoceanography. *Marine Micropal.* 4: 363–398.

Sancetta, C. (1987) Cycles, steps and CO_2; paleoceanography in the United States from 1983–1986. *Reviews of Geophysics* 25: 1363–1375.

Sandberg, P. A. (1985) Nonskeletal aragonite and pCO_2 in the Phanerozoic and Proterozoic. In *The Carbon Cycle and Atmospheric CO_2:Natural Variations Archean to Present*, ed. E. T. Sundquist and W. S. Broecker, pp. 585–594. Washington, D.C.: American Geophysical Union.

Sarmiento, J. L., Herbert, T. D. and Toggweiler, J. R. (1988) Causes of anoxia in the world ocean. *Global Biogeochemical Cycles* 2: 115–128.

Sarnthein, M. (1978) Sand deserts during glacial maximum and climatic optimum. *Nature* 272: 43–46.

Sarnthein, M., Winn, K., Duplessy, J.-C. and Fontugne, M. R. (1988) Global variations of surface ocean productivity in low and mid latitudes: influence on CO_2 reservoirs of the deep ocean and atmosphere during the last 21,000 years. *Paleoceanography* 3: 361–399.

Savin, S. M. (1977) The history of the Earth's surface temperature during the past 100 Ma. *Ann. Rev. Earth Planet. Sci.* 5: 319–355.

Savin, S. M., Douglas, R. G. and Stehli, F. G. (1975) Tertiary marine paleotemperatures. *Geol. Soc. Am. Bull.* 86: 1499–1510.

Schlanger, S. O. (1986) High-frequency sea-level fluctuations in Cretaceous time: an emerging geophysical problem. In *Mesozoic and Cenozoic Oceans*, ed. K. J. Hsu, pp. 61–74. Washington, D.C.: American Geophysical Union, Geodynamics Series 15.

Schlanger, S. O., Arthur, M. A., Jenkyns, H. C. and Scholle, P. A. (1987) The Cenomanian-Turonian oceanic anoxic event, I. Stratigraphy and distribution of organic carbon-rich beds and the marine $\delta^{13}C$ excursion. In *Marine Petroleum Source Rocks*, ed. J. Brooks and A. Fleet, pp. 371–399. Geological Society of London Special Publication.

Schlanger, S. O. and Cita, M. B. (1982) *Nature and Origin of Cretaceous Carbon-Rich Facies*. London: Academic Press. 229 pp.

Schlanger, S. O. and Jenkyns, H. C. (1976) Cretaceous ocean anoxic events: causes and consequences. *Geol. Mijnb.* 55: 179–184.

Schluchter, C. (1986) The Quaternary glaciations of Switzerland, with special reference to the northern Alpine foreland. *Quat. Sci. Rev.* 5: 413–420.

Schmidt, P. W., Cudahy, T. J. and Powell, C. M. (1986) Prefolding and Premegakinking magnetizations from the Devonian Comerong Volcanics, NSW, Australia and their bearing on the Gondwana pole path. *Tectonics* 5: 135–150.

Schneider, S. H. (1984) Energy balance models (EBM). In *Milankovitch and Climate*, ed. A. L. Berger, J. Imbrie, J. Hays, G. Kukla and B. Saltzman, pp. 845–854. Dordrecht: D. Reidel.

Schneider, S. H., Thompson, S. L. and Barron, E. J. (1985) Mid-Cretaceous continental surface temperatures: are high CO_2 concentrations needed to simulate above freezing winter conditions? In *The Carbon Cycle and Atmospheric CO_2: Natural Variations Archean to Present*, ed. E. T. Sundquist and W. S.

Broecker, pp. 554–560. Washington, D.C.: American Geophysical Union.

Scholl, D. W. and Stevenson, A. J. (1987) *Ocean margin tectonics and North Pacific paleoceanography: Wonderments about the Pacific-Beringa connections.* 4th International Congress on Pacific Neogene Stratigraphy. ed. J.A. Barron and J. R. Blueford. Berkeley, California. Abstract Volume: 95–98.

Scholle, P. A. and Arthur, M. A. (1980) Carbon isotope fluctuations in Cretaceous pelagic limestones: Potential stratigraphic and petroleum exploration tool. *Am. Assoc. Pet. Geol. Bull.* 64: 67–87.

Schrader, H. J., Bjorklund, K., Manum, S., Martini, E. and van Hinte, J. (1976) Cenozoic biostratigraphy, physical stratigraphy and paleoceanography in the Norwegian-Greenland Sea, DSDP Leg 38. Paleoceanographical synthesis. *Init. Repts. DSDP* 38: 1197–1211. Washington (Govt. Printing Office).

Scotese, C. R. (1974) The evolutionary biogeography of lower Paleozoic crinoids in relation to the tectonic history of the proto-Atlantic (abs.). *Geol. Soc. Amer, Abstracts with Program* 6: 543–44.

Scotese, C. R., Bambach, R. K., Van der Voo, R. and Ziegler, A. M. (1979) Paleozoic base maps. *J. Geol.* 87: 217–227.

Scotese, C. R., Barrett, S. F. and Van der Voo, R. (1985) Silurian and Devonian base maps. *Phil. Trans. Roy. Soc. London.* B 309: 57–77.

Scotese, C. R., Gahagan, L. M. and Larson, R. L. (1988) Plate tectonic reconstructions of the Cretaceous and Cenozoic ocean basins. *Tectonophysics* 155: 27–48.

Scotese, C. R. and McKerrow, W. S. (1990) Revised world maps and introduction. *Mem. Geol Soc. London* 12: 1–21.

Scotese, C. R., Van der Voo, R. and Ross, W. C. (1981) Mesozoic and Cenozoic base maps (abs.). *Amer. Assoc. Petrol. Geol. Bull.* 65: 989.

Sellers, W. D. (1984) The response of a climate model to orbital variations. In *Milankovitch and Climate*, ed. A. L. Berger, J. Imbrie, J. Hays, G. Kukla and B. Saltzman, pp.765–788. Dordrecht: D. Reidel.

Sepkoski, J. J. (1979) A kinetic model of Phanerozoic taxonomic diversity. 2. Early Phanerozoic families and multiple equlibria. *Paleobiology* 5: 222–251.

Sepkoski, J. J., (1981) A factor analytic description of the Phanerozoic marine fossil record. *Paleobiology* 7: 36–53.

Serova, M. Y. (1985) Marine Oligocene in stratigraphic section of Paleogene from Western Sakhalin. *Izvestiya Acad. Nauk. SSSR. Seriya Geol.* 11: 86–89.

Seslavinskiy, K. B. (1978) Ordovician and Silurian climates and global climatic belts. *Internat. Geol. Rev.* 21: 140–151.

Shackleton, N. J. (1967) Oxygen isotope analyses and Pleistocene temperatures reassessed. *Nature* 215: 15–17.

Shackleton, N. J. (1977) The oxygen isotope stratigraphic record of the Late Pleistocene. *Phil. Trans. R. Soc. Lond., Ser. B.* 280: 169–182.

Shackleton, N. J. (1985) Oceanic carbon isotope constraints on oxygen and carbon dioxide in the Cenozoic atmosphere. In *The Carbon Cycle and Atmospheric CO_2: Natural Variations Archean to Present*, ed. E. T. Sundquist and W. S. Broecker, pp. 412–417. Washington, D.C.: American Geophysical Union, Geophysical Monograph 32.

Shackleton, N. J. (1986a) Paleogene stable isotope events. *Palaeogeogr., Palaeoclimatol., Palaeoecol* 57: 91–102.

Shackleton, N. J. (1986b) The Plio-Pleistocene Ocean: Stable isotope history. In *Mesozoic and Cenozoic Oceans*, ed. K. J. Hsu, pp. 141–153. Washington, D.C. : American Geophysical Union, Geodynamics Series, 15.

Shackleton, N. J. (1987a) The carbon isotope record of the Cenozoic: History of organic carbon burial of oxygen in the ocean and atmosphere. In *Marine Petroleum Source Rocks*, ed. J. Brooks and A. J. Fleet, pp. 319–322. Geological Society London Special Publication.

Shackleton, N. J. (1987b) Oxygen isotopes, ice volume and sea level. *Quat. Sci. Rev.* 6: 183–190.

Shackleton, N. J. and Boersma, A. (1981) The climate of the Eocene Ocean. *J. Geol. Soc. London.* 138: 153–157.

Shackleton, N. J. and Cita, M. B. (1979) Oxygen and carbon isotope stratigraphy of benthic foraminifers at Site 597: detailed history of climatic change during the late Neogene. *Init. Repts. DSDP* 29: 743–755. Washington (Govt. Printing Office).

Shackleton, N. J. and Duplessy, J.-C. (1986) Temperature changes in ocean deep waters during the late Pleistocene (abs.). 2nd International Conference on Paleoceanography. Woods Hole, Massachusetts: Woods Hole Oceanographic Institution.

Shackleton, N. J., Hall, M. A., Line, J. and Shuxi, C. (1983) Carbon isotope data in core V19–30 confirm reduced carbon dioxide concentration in the ice age atmosphere. *Nature* 306: 319–322.

Shackleton, N. J. and Kennett, J. P. (1975) Paleotemperature history of the Cenozoic and the initiation of Antarctic glaciation: Oxygen and carbon isotope analysis in DSDP sites 277, 279 and 281. *Init. Repts. DSDP* 29: 743–755. Washington (Govt. Printing Office).

Shackleton, N. J. and Opdyke, N. D. (1973) Oxygen isotope and paleomagnetic stratigraphy of equatorial Pacific core V28–239: oxygen isotope temperature and ice volume on a 10^5-year scale. *Quat. Res.* 3: 39–55.

Shackleton, N. J. and Opdyke, N. D. (1977) Oxygen isotope and paleomagnetic evidence for early northern hemisphere glaciation. *Nature* 270: 216–219.

Shackleton, N. J. and Pisias, N. G. (1985) Atmospheric carbon dioxide, orbital forcing and climate. In *The Carbon Cycle and Atmospheric CO₂: Natural Variations Archean to Present*, ed. E. T. Sundquist and W. S. Broecker, pp. 303–317. Washington, D.C.: American Geophysical Union.

Shackleton, N. J., van Andel, T. H., Boyle, E. A., Jansen, E. and Sundquist, E. (1990) Contributions from the oceanic record to the study of global change on three time scales – Report of Working Group 1, Interlaken Workshop for Past Global Changes. *Global and Planetary Changes* 2: 5–38.

Shackleton, N. J. and 16 others (1984) Oxygen isotope calibration of the onset of ice-rafting and history of glaciation in the North Atlantic region. *Nature* 307: 620–623.

Shaw, H. R. (1987) The periodic structure of the natural record, and nonlinear dynamics. *Eos, Trans. Amer. Geophys. Union.* 68: 1651–1662.

Sheard, M. J. (1991) Glendonites from the southern Eromanga Basin in South Australia: palaeoclimatic indicators for Cretaceous ice. *Geol. Surv. S. Aust. Geol. Notes* 114: 17–23.

Shearman, D. J. and Smith, A. J. (1985) Ikaite, the parent mineral of jarrowite-type pseudomorphs. *Proc. Geol. Ass.* 96: 305–314.

Sheridan, R. E. (1986) Pulsation tectonics as the control of North Atlantic paleoceanography. In *North Atlantic Paleoceanography*, ed. C. P. Summerhayes and N. J. Shackleton, pp. 255–275. Geological Society of London Special Publication.

Sheridan, R. E. (1987) Pulsation tectonics as the control of long-term stratigraphic cycles. *Paleoceanography* 2: 97–118.

Shibuya, M. (1977) A study of the Gen'no-ishi; its mode of occurrence, form, and texture. *J. Geol. Soc. Jap.* 83: 19–26.

Short, D. A. and Mengel, J. G. (1986) Tropical climatic phase lags and Earth's precession cycle. *Nature* 323: 48–50.

Showers, W. J. and Bevis, M. (1988) Amazon cone isotopic stratigraphy: Evidence for the source of the tropical freshwater spike. *Palaeogeogr., Palaeoclimatol., Palaeoecol.* 64: 189–199.

Showers, W. J. and Genna, B. (1988) The deglacial tropical freshwater spike on the the middle and lower Amazon Fan (abs.). *Eos, Trans. Amer. Geophys. Union* 69: 730.

Sigleo, W. (1983) Soil stratigraphy and Middle Pennsylvanian climatic change: Evidence from the Pennsylvania system stratotype, Appalachian Basin. *Geol. Soc. Amer.* 15: 687.

Sigleo, W. and Reinhardt, J. (1988) Paleosols from some Cretaceous environments in the southeastern United States. *Geol. Soc. Amer. Spec. Pap.* 216: 123–142.

Sladen, C. P. (1983) Trends in Early Cretaceous clay mineralogy in NW Europe. *Zitteliana* 10: 349–357.

Sloan, L. C. and Barron, E. J. (1990) "Equable" climate during Earth history. *Geology* 18: 489–492.

Smiley, C. J. (1967) Paleoclimatic interpretations of some Mesozoic floral sequences. *Am. Assoc. Pet. Geol. Bull.* 51: 849–863.

Smith, A. G. and Briden, J. C. (1977) *Mesozoic and Cenozoic Paleocontinental Maps.* Cambridge: Cambridge University Press. 63 pp.

Smith, A. G., Hurley, A. M. and Briden, J. C. (1981) *Phanerozoic Paleocontinental World Maps.* New York: Cambridge University Press. 102 pp.

Spicer, R. A. (1987) The significance of the Cretaceous flora of Northern Alaska for the reconstruction of the climate of the Cretaceous. *Geol. Jb.* 96: 265–291.

Spicer, R. A. and Parrish, J. T. (1986) Paleobotanical evidence for cool north polar climates in the middle Cretaceous (Albian–Cenomanian) time. *Geology* 14: 703–706.

Spicer, R. A. and Parrish, J. T. (1990) Late Cretaceous–early Tertiary palaeoclimates of the northern high latitudes: a quantitative view. *J. Geol. Soc. Lond.* 147: 329–341.

Spjeldnaes, N. (1961) Ordovician climatic zones. *Nord. Geol. Tidsskr.* 41: 45–77.

Spjeldnaes, N. (1981) Lower Paleozoic Paleoclimatology. In *Lower Paleozoic of the Middle East, Eastern and Southern Africa, and Antarctica*, ed. C. H. Holland, pp. 199–256. Chichester: John Wiley & Sons.

Sprinkle, J. (1981) Diversity and evolutionary patterns of Cambrian echinoderms. *US Geol. Surv. Open-File Report.* 81–0743: 219–221.

Sribrava, V., Bowen, D. Q. and Richmond, G. M. (1986) Quaternary glaciations in the northern hemisphere. *Quat. Sci. Rev.* 5: 513.

Srivastava, S. K. (1981) Evolution of Upper Cretaceous phytogeoprovinces and their pollen. *Rev. Palaeobot. Palynol.* 35: 155–173.

Stanley, S. M. (1984) Marine extinctions: a dominant role for temperature. In *Extinctions*, ed. M. H. Nitecki, pp. 69–117. Chicago: University of Chicago Press.

Stanley, S. M. (1985) Climatic cooling and Plio-Pleistocene mass extinction of molluscs around the margins of the Atlantic. *S. Afr. J. Sci.* 81: 266.

Stauffer, B., Lochbronner, E., Oeschger, H. and Schwander, J. (1988) Methane concentration in the glacial atmosphere was only half that of the pre–industrial Holocene. *Nature* 332: 812–814.

Stauffer, B., Mofer, M., Oescherger, H., Schwander, J. and Siegenthaler, V. (1984) Atmospheric CO_2 concentration during the last glaciation. *Ann. Glaciol.* 5: 160–164.

Stein, R. (1985) Late Neogene changes of paleoclimate and paleoproductivity off northwest Africa (DSDP Site 397). *Palaeogeogr., Palaeoclimatol., Palaeoecol.* 49: 47–59.

Stein, R. (1986) Late Neogene evolution of paleoclimate and paleoceanic circulation in the northern and southern hemispheres: a comparison. *Geol. Rundsch.* 75: 125–138.

Stein, R. and Bleil, U. (1986) Deep-water circulation in the Northeast Atlantic and climatic changes during the Late Neogene (DSDP Site 141). *Mar. Geol.* 70: 191–209.

Stein, R., Sarnthein, M. and Suendermann, J. (1986) Late Neogene submarine events along the northeast Atlantic continental margin. In *North Atlantic Paleoceanography*, ed. C. P. Summerhayes and N. J. Shackleton, pp. 103–118. Geological Society Special Publication.

Steiner, J. and Grillmair, E. (1973) Possible galactic causes for periodic and episodic glaciations. *Geol. Soc. Am. Bull.* 84: 1003–1018.

Stevens, G. R. (1971) Relationship of isotopic temperatures and faunal realms to Jurassic and Cretaceous palaeogeography, particularly of the Southwest Pacific. *J. Roy. Soc. N.Z.* 1: 145–158.

Stevens, G. R. and Clayton, R. N. (1971) Oxygen isotope studies on Jurassic and Cretaceous belemnites from New Zealand and their biogeographic significance. *N.Z. J. Geol. Geophys.* 14: 829–879.

Stott, L. D. and Kennett, J. P. (1989) New constraints on early Tertiary palaeoproductivity from carbon isotopes in foraminifera. *Nature* 342: 526–529.

Street-Perrot, F. A. and Harrison, S. P. (1985) Lake-levels and climate reconstruction. In *Paleoclimate Analysis and Modeling*, ed. A. D. Hecht, pp. 291–340. New York: John Wiley and Sons.

Street-Perrott, F. A. and Roberts, N. (1983) Fluctuations in closed-basin lakes as an indicator of past atmospheric circulation patterns. In *Variations in the Global Water Budget*, ed. F. A. Street-Perrott, M. Beran and R. A. S. Radcliffe, pp. 331 345. Dordrecht. Reidel.

Summerhayes, C. P. (1981) Organic facies of Middle Cretaceous black shales in the deep North Atlantic. *Am. Ass. Petrol. Geol. Bull.* 65: 2364–2380.

Summerhayes, C. P. (1987) Organic-rich Cretaceous sediments from the North Atlantic. In *Marine Petroleum Source Rocks*, ed. J. Brooks and A. J. Fleet, pp. 301–316. London: Geological Society Special Publication.

Sumner, D. Y., Kirschvink, J. L. and Runnegar, B. N. (1987) Soft–sediment paleomagnetic tests of Late Precambrian glaciogenic sediments (abs.). *Eos, Trans. Amer. Geophys. Union*. 68: 1251.

Sundquist, E. T. (1987) The global carbon cycles and Quaternary paleoclimates. *Episodes* 10: 7–10.

Tantiwanit, W., Raksakauwong, L. and Mantajit, N. (1983) The upper Palaeozoic pebbly rocks in southern Thailand. *Geol. Socs. Thailand and Malaysia*. Workshop on Stratigraphic Correlation. pp. 96–104.

Thickpenny, A. and Leggett, J. K. (1987) Stratigraphic distribution and palaeo-oceanographic significance of European Palaeozoic organic-rich sediments. In *Marine Petroleum Source Rocks* ed. J. Brooks and A. J. Fleet, pp. 231–247. Geological Society Special Publication, 26.

Thiede, J., Dean, W. E. and Claypool, G. E. (1982) Oxygen deficient paleoenvironments in the mid-Cretaceous tropical and subtropical Pacific Ocean. In *Nature and Origin of Cretaceous Carbon-Rich Facies*, ed. S. O. Schlanger and M. B. Cita, pp. 79–100. London: Academic Press.

Thomsen, E. (1989) Seasonal variability in the production of Lower Cretaceous calcareous nannoplankton. *Geology* 17: 715–717.

Thunell, R. C. (1979) Pliocene–Pleistocene paleotemperatures and paleosalinity history of the Mediterranean Sea, results from DSDP Sites 125 and 132. *Marine Micropalaeo*. 4: 173.

Thunell, R. C. and Belyea, P. (1982) Neogene planktonic foraminiferal biogeography of the Atlantic Ocean. *Micropalaeontology* 28: 381–398.

Thunell, R. C., Williams, D. F. and Howell, M. (1987) Atlantic-Mediterranean water exchange during the late Neogene. *Paleoceanography* 2: 661–678.

Tissot, B., Deroo, G. and Herbin, J. P. (1979) Organic matter in Cretaceous sediments of the North Atlantic: Contribution to sedimentology and paleogeography. In *Deep Drilling Results in the Atlantic Ocean: Continental margin and paleoenvironment.*, ed. M. Talwani, W. W. Hay and W. B. F. Ryan, pp. 362–374. Washington, D.C.: American Geophysical Union.

Tiwari, R. K. (1987) Higher order eccentricity cycles of the middle and late Miocene climatic variations. *Nature* 327: 219–221.

Toggweiler, J. R. (1990) Bombs and ocean carbon cycles. *Nature* 347: 122–123.

Toggweiler, J. R. and Sarmiento, J. L. (1985) Glacial to interglacial changes in atmospheric carbon dioxide: The critical role of ocean surface water in high latitudes. In *The Carbon Cycle and Atmospheric CO_2: Natural Variations Archean to Present*, ed. E. T. Sundquist and W. S. Broecker, pp. 163–184. Washington, D.C.: American Geophysical Union.

Truswell, E. M. (1978) Palynology of the Permo-Carboniferous in Tasmania; an interim report. *Geol. Surv. Tasmania Bull*. 56: 39.

Truswell, E. M. and Drewry, D. J. (1984) Distribution and provenance of recycled palynomorphs in surficial sediments of the Ross Sea. *Marine Geol*. 59: 187–214.

Tschudy, R. H., Pillmore, C. L., Orth, C. J., Gilmore, J. S. and Knight, J. D. (1984)

Disruption of the terrestrial plant ecosystem at the Cretaceous–Tertiary boundary, Western Interior. *Science* 225: 1030–1032.

Tucker, M. E. and Benton, M. J. (1982) Triassic environments, climate and reptile evolution. *Palaeogeogr., Palaeoclimatol., Palaeoecol.* 40: 361–379.

Tungsheng, L., Shouxin, Z. and Jiaomao, H. (1986) Stratigraphy and paleoenvironmental changes in the loess of central China. *Quat. Sci. Rev.* 5: 489–496.

Turnau, R. and Ledbetter, M. T. (1989) Deep circulation changes in the South Atlantic Ocean: response to initiation of Northern Hemisphere glaciation. *Paleoceanography* 4: 565–583.

Ustritsky, V. I. and Yavshits, G. P. (1971) Middle Carboniferous glaciomarine sediments of the northeastern USSR. *Lithology* 159–161.

Vail, P. R. and Hardenbol, J. (1979) Sea-level changes during the Tertiary. *Oceanus* 22: 71–9.

Vail, P. R., Mitchum, R. M., Jr. and Thompson, S., III. (1977) Seismic stratigraphy and global changes of sea level, 4. Global cycles of relative changes of sea level. *Amer. Assoc. Pet. Geol. Mem.* 26: 83–97.

Vakhrameev, V. A. (1964) Jurassic and Early Cretaceous floras of Eurasia and the paleofloristic provinces of this period. *Tr. Geol. Inst., Moscow.* 102: 1–263. (in Russian)

Vakhrameev, V. A. (1978) The climates of the northern hemisphere in the Cretaceous in the light of palaeobotanical data. *Palaeo. J.* (1978) 2: 143–154.

Vakhrameev, V. A. (1981) Pollen *Classopollis,* indicator of Jurassic and Cretaceous climates. *The Palaeobotanist* 28: 301–307.

Van Campo, E., Duplessy, J.-C. and Rossignol-Stick, M. (1982) Climatic conditions deduced from a 150–kyr oxygen isotope-pollen record from the Arabian Sea. *Nature* 296: 56–59.

Van der Voo, R. (1988) Paleozoic paleogeography of North America, Gondwana and intervening displaced terranes: Comparisons of paleomagnetism with paleoclimatology and biogeographical patterns. *Geol. Soc. Amer. Bull.* 100: 311–324.

Van Houten, F. B. (1985) Oolitic ironstones and contrasting Ordovician and Jurassic paleogeography. *Geology* 13: 722–724.

Van Houten, F. B. and Bhattacharya, D. P. (1982) Phanerozoic oolitic ironstones – geologic record and facies model. *Ann. Rev. Earth. Planet Sci.* 10: 445–451.

Van Zinderen Bakker, E. M. and Mercer, J. H. (1986) Major late climatic events and palaeoenvironmental changes in Africa viewed in a world wide context. *Palaeogeogr., Palaeoclimatol., Palaeoecol.* 56: 217–235.

Veevers, J. J. (1984) *Phanerozoic Earth History of Australia.* Oxford: Clarendon Press. 418 pp.

Veevers, J. J. (1990) Tectonic-climatic supercycle in the billion-year plate-tectonic eon: Permian Pangean icehouse alternates with Cretaceous dispersed-continents greenhouse. *Sed. Geol.* 68: 1–16.

Veevers, J. J. and Powell, C. M. (1987) Late Paleozoic glacial episodes in Gondwanaland reflected in transgressive-regressive depositional sequences in Euramerica. *Geol. Soc. Amer. Bull.* 98: 475–487.

Veevers, J. J. and Ettriem, S. L. (1988) Reconstruction of Antarctica and Australia at breakup (95 ± 5 Ma) and before rifting (160 Ma). *Aust. J. Earth Sci.* 35:

355–362.

Veizer, J. (1985) Carbonate and ancient oceans; isotopic and chemical record on time scales of 10^7–10^9 years. In *The Carbon Cycle and Atmospheric CO_2: Natural Variations Archean to Present*, ed. E. T. Sundquist and W. S. Broecker, pp. 595–601. American Geophysical Union, Geophysical Monograph, 32.

Veizer, J., Fritz, P. and Jones, B. (1986) Geochemistry of brachiopods: Oxygen and carbon isotopic records of Paleozoic oceans. *Geochim. Cosmochim. Acta.* 50: 1679–1696.

Velichko, A. A. (1983) Latitudinal asymmetry of environmental factors during the glacial periods in the Northern Hemisphere. *Polar Geogr. Geol.* 7: 138–158.

Vergnaud Grazzini, C. (1980) Cenozoic paleotemperatures at Site 398, eastern North Atlantic; diagenetic effects on carbon and oxygen isotope signal. *Init. Repts. DSDP* 47: 507–511. Washington (Govt. Printing Office).

Vergnaud Grazzini, C. and Letolle, C. P. R. (1978) Paleoenvironment of the North-East Atlantic during the Cenozoic: oxygen and carbon isotope analysis at DSDP sites 398, 400A and 401. *Oceanologica Acta* 1: 381–390.

Vergnaud Grazzini, C., Grably, M., Pujol, C. and Duprat, J. (1983) Oxygen isotope stratigraphy and paleoclimatology of southwestern Atlantic Quaternary sediments (Rio Grande Rise) at DSDP Site 517. *Init. Repts. DSDP* 72: 871–881. Washington (Govt. Printing Office).

Vincent, E. and Berger, W. H. (1985) Carbon dioxide and polar cooling in the Miocene: The Monterey Hypothesis. In *The Carbon Cycle and Atmospheric CO_2: Natural Variations Archean to Present*, ed. E. T. Sundquist and W. S. Broecker, pp. 455–468. Washington, D.C.: American Geophysical Union.

Vogt, P. R. (1979) Global magmatic episodes: new evidence and implications for the steady-state mid–ocean ridge. *Geology* 7: 93–98.

Vogt, P. R. (1989) Volcanogenic upwelling of anoxic, nutrient-rich water: A possible factor in carbonate-bank/reef demise and benthic faunal extinctions. *Geol. Soc. Amer. Bull.* 101: 1225–1245.

Volk, T. (1987) Feedbacks between weathering and atmospheric CO_2 over the last 100 million years. *Amer. J. Sci.* 287: 763–779.

Volk, T. and Hoffert, M. I. (1985) Oceanic carbon pumps: analysis of relative strengths and efficiencies in ocean-driven atmospheric CO_2 changes. In *The Carbon Cycle and Atmospheric CO_2: Natural Variations Archean to Present*, ed. E. T. Sundquist and W. S. Broecker, pp. 99–110. Washington, D.C.: American Geophysical Union, Geophysical Monograph, 32.

Vozovik, Y. I. (1985) Paleoglaciology of the northern part of the Pacific Ocean during the late Pleistocene. *Polar Geogr. Geol.* 9: 196–202.

Walter, B. (1989) Au Valanginien supérieur, une crise de la faune de bryozoaires: indication d'un important refroidissement dans le Jura. *Palaeogeogr., Palaeoclimatol., Palaeoecol.* 74: 255–263.

Wanless, H. R. and Shepard, F. P. (1936) Sea level and climatic changes related to Paleozoic cycles. *Geol. Soc. Am. Bull.* 47: 1177–1206.

Waples, D. W. (1983) A reappraisal of anoxic and organic richness with emphasis on the Cretaceous North Atlantic. *Am. Ass. Petrol. Geol. Bull.* 67: 963–978.

Warnke, D. A. (1982) Pre-middle Pliocene sediments of glacial and preglacial origin

in the Norwegian-Greenland Seas: results of DSDP Leg 38. *Earth Evol. Sci.* 2: 69–78.

Waterhouse, J. B. and Bonham-Carter, G. F. (1975) Global distribution and character of Permian biomes based on brachiopod assemblages. *Can. J. Earth Sci.* 12: 1085–1146.

Watts, R. G. and Hayder, M. E. (1984) A possible explanation of differences between pre-and post-Jaramillo ice sheet growth. In *Milankovitch and Climate*, ed. A. L. Berger, J. Imbrie, J. Hays, G. Kukla and B. Saltzman, pp. 599–604. Dordrecht: Reidel.

Webb, P.-N. and Harwood, D. M. (1987) The terrestrial flora of the Sirius Formation: its significance in interpreting late Cenozoic glacial history. *Antarctic J. U.S.* 22: 7–11.

Webb, P. N., Harwood, D. M., McKelvey, B. C., Mercer, J. H. and Stott, L. D. (1984) Cenozoic marine sedimentation and ice-volume variation on the East Antarctic craton. *Geology* 12: 287–291.

Webby, B. D. (1984) *Ordovician reefs and climate: a review.* Paleontological Contributions from the University of Oslo. pp. 89–100.

Weertman, J. (1976) Milankovitch solar radiation variations and ice age sheet sizes. *Nature* 261: 17–20.

Wei, K. Y. and Kennett, J. P. (1986) Taxonomic evolution of Neogene planktonic foraminifera and paleoceanographic relations. *Paleoceanography* 1: 67–84.

Weijermars, R. (1988) Neogene tectonics in the Western Mediterranean may have caused the Messinian Salinity Crisis and an associated glacial event. *Tectonophysics* 148: 211–219.

Weissert, H. (1989) C–isotope stratigraphy, a monitor of paleoenvironmental change: a case study from the Cretaceous. *Surveys in Geophysics* 10: 1–61.

Weissert, H. and Channell, J. E. T. (1989) Tethyan carbonate carbon isotope stratigraphy across the Jurassic–Cretaceous boundary: an indicator of decelerated global carbon cycle. *Paleoceanography* 4: 483–494.

Weissert, H., McKenzie, J. A. and Hochuli, P. (1979) Cyclic anoxic events in the Early Cretaceous Tethys Ocean. *Geology* 7: 147–151.

Weissert, H., McKenzie, J. A., Wright, R. C., Clark, M., Oberhansli, H. and Casey, M. (1984) Paleoclimatic record of the Pliocene at DSDP Sites 519, 521, 522 and 523 (Central South Atlantic). *Init. Repts. DSDP* 73: 701–715. Washington (Govt. Printing Office).

Weissert, H. and Oberhansli, H. (1985) Pliocene oceanography and climate: an isotope record from the southwestern Angola Basin. In *South Atlantic Paleoceanography*, ed. K. J. Hsu and H. Weissert, pp. 79–97. Cambridge: Cambridge University Press.

Wenk, T. and Siegenthaler, U. (1985) The high-latitude ocean as a control of atmospheric CO_2. In *The Carbon Cycle and Atmospheric CO_2: Natural Variations Archean to Present*, ed. E. T. Sundquist and W. S. Broecker, pp. 185–194. Washington, D.C.: American Geophysical Union.

Wignall, P. B. and Ruffell, A. H. (1990) The influence of a sudden climatic change on marine deposition in the Kimmeridgian of northwest Europe. *J. Geol. Soc. Lond.* 147: 365–371.

Wilde, P. and Berry, W. B. N. (1984) Destabilization of the oceanic density structure

and its significance to marine "extinction" events. *Palaeogeogr. Palaeoclimatol. Palaeoecol.* 48: 143–162.

Wilde, P. and Berry, W.B.N. (1988) Sulfur isotope anomaly associated with the Frasnian/Famennian extinction, Medicine Lake, Alberta, Canada: Comment. *Geology* 16: 86.

Wilkinson, B. H. and Given, R. K. (1986) Secular variation in abiotic marine carbonates: constraints on Phanerozoic atmospheric carbon dioxide contents and oceanic Mg/Ca ratios. *J. Geol.* 94: 321–333.

Wilkinson, B. H., Owen, R. M. and Carroll, A. R. (1985) Submarine hydrothermal weathering, global eustasy and carbonate polymorphism in Phanerozoic marine oolites. *J. Sed. Petrol.* 55: 171–183.

Williams, D. F., Healy-Williams, N., Thunell, R. C. and Leventer, A. (1983) Detailed stable isotope and carbonate records from the upper Maastrichtian-lower Paleocene of Hole 516F (Leg 72) including the Cretaceous/Tertiary boundary. *Init. Repts. DSDP* 72: 921–930. Washington (Govt. Printing Office).

Williams, D. F., Hodell, D.A. and Vergnaud Grazzini, C. (1985) Synthesis of late Cretaceous and Quaternary stable isotope records of the South Atlantic based on Leg 72 DSDP core material. In *South Atlantic Paleoceanography*, ed. K. J. Hsu and H. J. Weissert, pp. 205–242. Cambridge: Cambridge University Press.

Williams, D. F., Thunell, R. C., Tappa, E., Rio, D. and Rafii, I. (1988) Chronology of the Pleistocene oxygen isotope record: 0–1.88 m.y. BP. *Palaeoegogr., Palaeoclimatol., Palaeoecol.* 64: 221–240.

Williams, G. E. (1975) Possible relation between periodic glaciation and the flexure of the galaxy. *Earth Planet. Sci. Letters.* 26: 361–369.

Wilson, A. T. (1969) The climatic effects of large-scale surges of ice sheets. *Can. J. Earth Sci.* 6: 911–918.

Wilson, A. T. (1973) The great antiquity of some Antarctic landforms–Evidence for an Eocene temperate glaciation in the McMurdo region. In *Palaeoecology of Africa, the Surrounding Islands and Antarctica*, ed. E. M. van Zinderen Bakker Snr, pp. 25–35.

Wolbach, W. S., Gilmour, I., Anders, E., Orth, C. J. and Brooks, R. R. (1988) Global fire at the Cretaceous-Tertiary boundary. *Nature* 334: 665–669.

Wolfe, J. A. (1980) Tertiary climates and floristic relationships at high latitudes in the Northern Hemisphere. *Palaeogeogr., Palaeoclimatol., Palaeoecol.* 30: 313–323.

Wolfe, J. A. (1985) Distribution of major vegetational types during the Tertiary. In *The Carbon Cycle and Atmospheric CO₂: Natural Variations Archean to Present*, ed. E. T. Sundquist and W. S. Broecker, pp. 357–375. Washington, D.C.: American Geophysical Union.

Wolfe, J. A. and Poore, R. Z. (1982) Tertiary marine and non-marine trends. In *Climate in Earth History*, ed. W. H. Berger and J. C. Crowell, pp. 154–158. Washington, D.C.: National Academy Press.

Wolfe, J. A. and Upchurch, G. R. (1986) Vegetation, climatic and floral changes across the Cretaceous-Tertiary boundary. *Nature* 324: 148–152.

Wolfe, J. A. and Upchurch, G. R. (1987) North American non-marine climates and vegetation during the Late Cretaceous. *Palaeogeogr., Palaeoclimatol.,*

Palaeoecol. 61: 33–77.

Woodruff, F. and Savin, S. M. (1989) Miocene deepwater oceanography. *Paleoceanography* 4: 87–140.

Woodruff, F., Savin, S. M. and Douglas, R. G. (1981) A Miocene stable isotopic record: A detailed deep Pacific Ocean study and its paleoclimatic implications. *Science* 212: 665–668.

Worsley, T. R., Nance, R. D. and Moody, J. B. (1986) Tectonic cycles and the history of the Earth's biogeochemical and paleoceanographic record. *Paleoceanography* 1: 233–263.

Wright, J. D. and Thunell, R., C, (1988) Neogene planktonic foraminifera biogeography and paleoceanography of the Indian Ocean. *Micropalaeontology* 34: 193–216.

Wright, V. P. (1980) Climate fluctuation in the Lower Carboniferous. *Naturwischenschaften*. 67: 252–253.

Yasamanov, N. A. (1980) Temperatures of Devonian, Carboniferous and Permian seas in Transcaucasia and the Ural region. *Int. Geophys. Rev.* 23: 1099–1103.

Young, G. M. (1981) Paleozoic tillite of the northern Arabian Peninsula. In *Earth's pre-Pleistocene glacial record*, ed. M. J. Hambrey and W. B. Harland, pp. 338–339. Cambridge: Cambridge University Press.

Young, M. A. and Bradley, R. S. (1984) Insolation gradients and the paleoclimatic record. In *Milankovitch and Climate*, ed. A. L. Berger, J. Imbrie, J. Hays, G. Kukla and B. Saltzman, pp. 707–713. Dordrecht: D. Reidel.

Zachos, J. C. and Arthur, M. A. (1986) Paleoceanography of the Cretaceous/Tertiary boundary event: Inferences from stable isotopic and other data. *Paleoceanography* 1: 5–26.

Zachos, J. C., Arthur, M. A. and Dean, W. E. (1989) Geochemical evidence for suppresion of pelagic marine productivity at the Cretaceous/Tertiary boundary. *Nature* 337: 61–68.

Zastawaniak, E., Wrona, R.A.G. and Birkenmajer, K. (1985) Plant remains from the top part of the Point Hennequin Group (Upper Oligocene), King George Island (South Shetland Islands, Antarctica). *Studia Geologica Polonica* 81: 143–164.

Zharkov, M. A. (1981) *History of Paleozoic Salt Accumulation*. Berlin: Springer. 308 pp.

Zharkov, M. A. (1984) Evolution of halogenic sedimentation through geological history. 27th International Geological Congress. 4: 161–174. Moscow.

Ziegler, A. M., Bambach, R. K., Parrish, F., Barrett, S. F., Gierlowski, E. H., Parker, W. C., Raymond, A. and Sepkowski, J. J. (1981a) Paleozoic biogeography and climatology. In *Paleobotany, Paleoecology and Evolution*, ed. K. J. Niklaus, pp. 231–266. New York: Praeger.

Ziegler, A. M., Barrett, S. F. and Scotese, C. R. (1981b) Paleoclimate, sedimentation and continental accretion. *Phil. Trans. Roy. Soc. London* A301: 253–264.

Ziegler, A. M., Hansen K. S., Johnson, M. E., Kelly, M. E., Scotese, C. R. and Van der Voo, R. (1977) Silurian continental distributions, paleogeography, climatology and biogeography. *Tectonophysics* 40: 13–51.

Ziegler, A. M., Hulver, M. L., Lottes, A. L. and Schmachtenberg, W. F. (1984) Uniformitarianism and palaeoclimates: inferences from the distribution of

carbonate rocks. In *Fossils and Climate*, ed. P. J. Brenchley, pp. 3–25. New York: John Wiley and Sons.

Ziegler, A. M., McKerrow, W. S., Johnson, M. E. and and Bambach, R. K. (1979) Paleozoic paleogeography. *Ann. Rev. Earth Planet. Sci.* 7: 473–502.

Zubakov, V. A. and Borzhenkova, I. I. (1983) *Paleoclimates of Late Cenozoic.* Leningrad: Gidrometeoizdat. 214 pp.

Index

aeolian sediments 48, 170–1
albedo 1, 167
Alpine orogeny 108
Antarctic Bottom Water (AABW) 117,
 121, 135, 171, 178
Antarctic glaciation (Quaternary) 163–4
Aquipollenites province 93
Arcadian orogeny (Devonian) 35
archaeocyathid reefs (Cambrian) 10
Austral Realm 69

Baltica 12, 15
bauxites 10–11, 18–19, 22, 35, 49, 60,
 112
black shales 12, 62, 70, 72–4, 86, 90–2
bioherms 32, 49
Boreal Realm 61, 69–72
Brunhes epoch 170
Bering Sea 160–3
biological pump 151

cadmium 178
calcretes (*see also* soils) 10, 60, 95
Caledonian orogeny 25
Cambrian 7–14
 faunal diversity 12–13
 archaeocyathid reefs 10
 glacials 7–9
 laterites 10
 isotopes 10
 bauxites 10–11
 sea levels 12–13
 evaporites 9–11
 carbonates 9–10
 black shales 12, 26
 oolitic ironstones 13
Camp Century ice core 165
Cancaniri Formation 16

carbon cycle 1–2, 150–2, 193–208,
 214–18
carbon dioxide 1–2, 87, 90–1, 99, 114,
 149–55, 175, 193–5
carbon isotopes 14, 26, 36, 54–5, 63–4,
 72–4, 90–2, 105–7, 127–30, 149–50,
 153, 199–200
carbonates 9–12, 18, 21–5, 28, 33, 43,
 49, 58, 68, 152
carbonate compensation depth (CCD)
 73, 130, 213
Carboniferous 33–5
 carbonates 30–3
 evaporites 30–1
 faunas 33
 floras 50–1
 glacials 33–35, 37–45, 53
 soils 49
Churochnaga Tillite 8
circum-Antarctic current 103, 108–9,
 113
Classopollis 70, 77–8, 97
clay minerals 43, 49–54, 57–8, 79–80
CLIMAP 138, 154, 164–5, 166–9, 172
coal 32–3, 43, 46–53, 59–60, 67, 78, 84,
 97
Cool Modes
 cyclicity 190
 definition 4–5
 development of 191–5
 early Carboniferous to late Permian
 37–55
 early Eocene to late Miocene 99–114
 early Miocene to late Miocene
 99–114
 late Miocene to Holocene 115–88
 late Ordovician to early Silurian
 15–26

271

Cool Modes (*cont.*)
 mid Jurassic to early Cretaceous
 65–82
Cordilleran ice sheet 124–5
Cretaceous 65–82, 83–98
 carbon 72–4
 carbonates 68
 clay minerals 79–80
 evaporites 68
 faunas 72
 floras 77–79, 92–8
 ice rafting 74–7
 isotopes 68–9
 sea levels 84–7
 seasonality 80–2
Cretaceous–Tertiary boundary 89, 92,
 208
cyclicity 12, 70, 77, 82, 108–113, 115,
 189–191, 211
cyclothems 45–7, 52

deglaciation 179–84
Devonian 27–36, 112
 carbonates 29–33
 evaporites 30, 33
 faunas 33
 glacials 34–5, 39
 palaeogeography 28–9
 redbeds 32
 volcanism 35
Dicriodium 60–61
Drake Passage 102–3, 113
Dwyka Tillite 37

Ecca Formation 62
Eocene 99–114
evaporites 6, 9–10, 14, 18, 20–1, 24–5,
 28–33, 43, 47–53, 57, 58–9, 62, 67,
 78–9, 84, 97–8
extinctions 36, 89, 91, 97, 108,
 130–2

faunal diversity 12–13, 23, 34, 50–1,
 70–2, 107, 130–4
faunal provinces 67–72
floras 47, 49–51, 60–1, 77–9, 92–8,
 101–3, 109–113, 137
foraminifera 104, 107–8, 130

glacial deposits 7–9, 11, 15–20, 26–7,
 33–5, 37–53, 57, 62, 64, 76, 101–5,

 113–14, 115–17, 120, 122–7,
 150–164
glendonites 76–7, 123
Glossopteris 47
Gondwana 15, 19, 21, 31, 40–2, 43, 66–7
Gulf Stream 68, 84, 108, 167

Hercynian orogeny 54
Hiatuses 100, 109
high-latitude land 192
Holocene warming 187–8
Huronian 3
hyaloclastite 102

Iapo Formation 16
Ibeleat Group, Mauritania 7
ice cores 165, 174–5, 177
ice decay (Quaternary) 169
ice-rafting 39–43, 65, 74–7, 88–9, 99,
 101–3, 123
ice-sheet growth 167–9
impacts (extra-terrestrial) 36, 89, 93,
 108, 113

Jurassic 56–64, 65–82
 coals 59–60
 evaporites 59
 ice-rafting 74–7
 isotopes 56–7, 68–9
 palaeontology 60–1
 reefs 58
 sea level 61
 soil 60

kaolinite 22, 80
Krakow glaciation 102

lakes 172–3
last glacial maximum (LGM) 148, 157,
 166–71, 179
last interglacial 164–6
late Miocene events 120–1
laterites 35, 49, 60, 96–98, 198
Laurentia 10, 15, 18, 19, 24–5
Laurentide ice sheet 158–160
loess 126, 156, 173–4
Louquan Formation, China 7
Lower Parmeener Supergroup 41
luminosity 1–2

Malvinokaffric Realm 34

manganese 22–23, 49
Mediterranean Basin 104–5, 114–16
Messinian evaporites 21, 120–1
methane 177
Milankovitch 45–7, 52, 59, 126–7, 138,
 140–9, 195
Miocene 99–114, 120–1
models (computer) 67–8, 79, 80–2, 97,
 101, 167–70
Modes (*see under* Cool and Warm)
Monterey Hypothesis 106, 204–5

Neogene 117–22
Neogene ice sheets 122–7
Normapolles province 93
North Atlantic Deep Water (NADW)
 108, 117, 134, 146, 178, 182–3, 186
Northern American glaciation
 (Quaternary) 158–60
Norwegian–Greenland Sea 123
Nothofagidites 93
Nothofagus 103, 110

ocean anoxic events (OAEs) 62, 73,
 90–1, 205–6
oligotaxic oceans 211
onset of Warm Modes 195–201
Ooliths 21, 31
Oolitic ironstones 13, 22–3, 25, 49, 55,
 63
orbital parameters 1–2, 77, 126–7,
 140–9, 151, 171, 185, 208–12
Ordovician
 carbonates 10, 21–2
 evaporites 20–1
 faunal diversity 12–13
 glacials 11, 15–26
 oolitic ironstones 23, 25
 phosphorites 13, 26
 sea level 13
organic-rich shales *see also* black
 shales) 13, 26, 36, 70, 88, 90–2
oxygen isotopes 10, 21–2, 24, 27–8,
 44–5, 56–7, 64, 68–70, 76–7, 87–9,
 99–103, 115–20, 138–9

Pangaea 35, 50, 58–9, 62–4, 66–8, 80
Panthalassa 66
Parana Basin 16
Permian
 carbonates 48–9, 57–8

coals 57–60
evaporites 47–51, 58–9
faunas 50–1
floras 50–1
glacials 37–43, 53–5
isotopes 56–7
phosphorites 55, 63
soils 49
Permian–Triassic boundary 202–3
phosphorites 13–26, 63
Pleistocene 155, 172–7
Pliocene 115–18, 121–38, 130–2
Polar Front 132–5, 145, 182
polytaxic oceans 211
pre-glacial cooling 192–3
Precambrian 3–8
 glacials 7–9
 palaeogeography 8
 tillites 3
 psychrosphere 107–9, 113

Quaternary 138–55
 CO_2 variations and climate 149–55
 ocean circulation 177–9
 late Quaternary climates 155–89
 glaciation 155–64
Queen Maud glaciation 136

redbeds 32, 47–8
reefs 21–4, 58
Ross glaciation 132

sea levels 12–13, 25–6, 35–6, 54–5,
 61–3, 69–71, 84–5, 116, 129,
 100–5
seasonality 72, 76–7, 80–2, 97
Silurian
 carbonates 21–2, 31–2
 evaporites 20–1, 30–3
 glacials 18, 15–26, 30
 graptolites 34
 isotopes 27–8
 oolitic ironstones 23
 phosphorites 26
 reefs 22, 32
Sirius Formation 136
soils 49, 60, 95–6, 112, 171
Straits of Gibraltar 120–1
Strangelove Ocean 207–8, 215
Subtropical Convergence 134, 145,
 167

Taconic orogeny 25
tectonic forcing 212–14
terminations 181–2
Tertiary 83–98, 99–114
 coals 97–8
 evaporites 97–8
 floras 92–8, 100
 glacials 101–4
 isotopes 90–2, 101–5
 sea level 84–7, 100
Tethyan Realm 61, 69–72
Tethys 58, 66, 81, 97
Tibetan Plateau 101
Tiddiline Tillite, Morocco 7
time scale 5, 69
Triassic 56–64
 carbonates 57–8
 coals 57–60
 evaporites 57, 58–9
 isotopes 56–7
 soils 60

upwelling 13–14, 66–7, 98, 154, 167,
 205–6

vertebrates 57–61
Vila Maria Formation 16
volcanism 13, 25, 35, 54, 62–3, 86–7,
 197, 200–1
Vostok ice core 152–4, 174–6

Warm Modes
 cyclicity 190
 definition 4–5
 early Cambrian to late Ordovician
 4–14, 198–9
 late Cretaceous to early Tertiary
 83–98, 196–7
 late Permian to middle Jurassic
 56–64, 196–7
 late Silurian to early Carboniferous
 27–36, 197–8
 onset of Warm Modes 195–201
warm saline bottom waters (WSBW) 22,
 69, 74, 104, 215

Yakataga Formation, Alaska 123–4
Younger Dryas 181–7